Signals and Communication Technology

This series is devoted to fundamentals and applications of modern methods of signal processing and cutting-edge communication technologies. The main topics are information and signal theory, acoustical signal processing, image processing and multimedia systems, mobile and wireless communications, and computer and communication networks. Volumes in the series address researchers in academia and industrial R&D departments. The series is application-oriented. The level of presentation of each individual volume, however, depends on the subject and can range from practical to scientific.

"Signals and Communication Technology" is indexed by Scopus.

More information about this series at http://www.springer.com/series/4748

Arnold McKinley

The Analytical Foundations of Loop Antennas and Nano-Scaled Rings

Springer

Arnold McKinley
University College London
London, UK

ISSN 1860-4862 ISSN 1860-4870 (electronic)
Signals and Communication Technology
ISBN 978-981-13-5891-3 ISBN 978-981-13-5893-7 (eBook)
https://doi.org/10.1007/978-981-13-5893-7

Library of Congress Control Number: 2019931514

This Springer imprint is published by the registered company Springer Nature Singapore Pte Ltd.
The registered company address is: 152 Beach Road, #21-01/04 Gateway East, Singapore 189721, Singapore

This work is dedicated
To my mother who suggested that I join the radio club in high school
so that my father would have something to talk with me about;
and
To my father, who talked with me about
Closed Circuit Resonant Loop Antennas (CCRLs).

Fig. 1 Two Closed Circuit Resonant Loop Antennas from 1963 developed and built by my father. The early loops had one capacitor and used the sub-wavelength resonance. Later ones used six capacitors, where all tuned the loop and modified the loop's radiation pattern. All operated in the 20 m amateur band

Preface

The first attempts at an analytical theory of loop antennas appeared in the literature late in the 19th century, and improved in the early 20th. By the 1960s the theory had advanced well enough to characterise the behaviour of thin-wire, perfectly conducting, closed loops in the radio frequency (RF) and micro-wave (MW) regions. By 1965, even loops multiply loaded with impedances around the periphery, were understood. Much later, in the 1990s, new formulations of the Green's function allowed the calculation of radiation patterns for closed loops.

The work was impressive because it was difficult to derive, but it was also difficult to use. Indeed in 1999, when the first split-ring was conceived, opening meta-materials to the research world, the theory was never used, nor could it be, because the theory had not yet been expanded to include gaps or real lossy materials. Simple inductive-capacitive, *LC* models were used instead for split-rings, with the gap acting like a capacitor; these ultimately showed very strong resonances in the sub-wavelength region where the circumference of the loop is small compared to the driving wavelength. Physicists moved forward with meta-materials into photonics and meta-surfaces using the language of physics, rather than the language of circuit and antenna theory. As they did so, engineers became increasingly interested in communications in the high GigaHertz (GHz) and TeraHertz (THz) regions, and meta-surfaces have recently caught their eye.

The first work extending the analytical theory of loops to lossy materials in the high GHz, THz, low-infrared (IR) and optical regions appeared in 2013. It used a developed theory of surface impedance, incorporating a critical point model of the index of refraction to handle inter-band transitions in the optical region. This allowed several authors in 2017 and 2018 to push the theory of multiply loaded loops and radiation patterns into the optical region. The theory of coupled rings and of thick-wire rings are currently on the cutting edge.

This volume covers the analytical theory of closed and multiply loaded, lossy metal loops and rings from the RF through optical regions. It does so from an engineering antenna theory perspective. It is the first time that all of the historical material has appeared in one place, and it includes new material that only appeared in the last decade.

The author hopes that its presence in the literature opens the way to new applications and new devices that enhance communications, computing, astronomical and bio-physical observations and perhaps even suggests many new fields of endeavour in physics and engineering.

London, UK
October 2018

Arnold McKinley

Acknowledgements

Sometimes scientific discovery requires the insights of those who do not see the world through the same lenses as those in the established scientific community. Modern scientists develop new theory using a complicated, multi-step, process of induction, deduction, and verification that relies on the testing of nature. But scientific criteria are sometimes difficult to understand and many people who are not exposed to a scientific education find the methods somehow mysterious. They often employ simpler, less rigorous and less successful methods for acquiring new knowledge.

Yet, there are some particularly gifted people, who, having taught themselves what they know in their own peculiar way, avoid the normal educational process, but employ the scientific method rigorously and rather naturally. One of those persons was my father. He never finished university, but taught himself electrical engineering and followed in the footsteps of his own inventive father. He could build anything with his lathe and machines and using them built three small businesses in the 1950s. Before the transistor became ubiquitous, he invented the smallest passive element delay lines of the day and the unique machines required to wind the rather novel two-dimensional inductors they used.

In the early 1960s, I left home for a distant boarding school, where I was allowed to phone home only once a week. Not liking the restriction, my father built two amateur radio systems and gave one to the school's radio club. He then designed from scratch, over the course of two experiment filled years, a naturally resonant, single capacitor loop antenna with a very high quality factor, and posted one on his house and one on the roof of the school. He discovered in those years, and exploited, what we now know to be the sub-wavelength resonance of loop antennas. I talked often with him on that ham radio set, and the students at the school talked with people all over the world. It was a marvellous communication system he had invented and it is too bad he never published. He called the antenna a "Closed Circuit Resonant Loop" (CCRL).

I must also acknowledge my doctoral supervisors from the Australian National University, Prof. Kylie Catchpole, Dr. Thomas White, and Dr. Angelika Basch. They taught me many things and I do hope loops or rings may one day be useful on solar cells. I also wish to acknowledge the friends I made among the other Ph.D. students.

I also want to thank Prof. Douglas Werner, Dr. Jogender Nagar, and Mr. Ryan Chaky from Pennsylvania State University, USA, and Dr. Mario Pantoja from the University of Granada, Spain for their encouragement, sharing of material, and useful comments on versions of the manuscript.

Thank you also to Dr. George Fikioris from the School of Electrical and Computer Engineering, National Technical University of Athens, Greece, and to Dr. Panagiotis Papakanellos from the Department of Aeronautical Sciences, Hellenic Air Force Academy, Dekelia Air Force Base, Greece for their useful comments and corrections to Chap. 3 in the manuscript a version or two from print.

Of course, any errors in attribution or in interpretation of actual history are my own.

Contents

List of Figures

Part I
Preliminaries

Abstract A history of the loop antenna covering the period from the late 1800s to the present day precedes a chapter on the mathematical foundations used in this book. The history covers the experimental and the analytical (theoretical) history ending with nano-scaled rings used in meta-materials. Foundations covers the geometrical coordinate systems used in analysing the closed toroidal loop, useful expressions of Maxwell's equations, a detailed look at propagation parameters from the radio frequency (RF) through to the optical regimes, vector and scalar potentials with their respective equations particularly for a Green's perspective, the governing equations for the thick and thin wire loop, and three methods of coupling to a loop.

Chapter 1
General Introduction

1.1 A Motivation for the Study of Loops and Rings as Radiating Structures

The analytical equations describing the characteristic behaviour of circular loop antennas are considered solved problems by many people. This is not entirely true. The equations applied to thin-wire loops were indeed solved in the 1950s and 1960s, but even today the equations for thick wire loops have not been solved and remain on the cutting edge. It was only a few years ago that the equations for thin-wire lossy metal loops were solved. This has now taken metal loops down into the nano-scaled region, but dielectric loops are not yet solved either, and remain a hindrance in the advance of meta-material theory.

Another hindrance is that the classical derivation of the behaviour of loops comes from energising the loop at a given point on the periphery; that is by treating it as a radiator. It is, of course true that loop antennas as receivers are illuminated by plane wave, but that perspective was not important until meta-materials, especially meta-materials in the THz and optical regions. This is because meta-materials were thought to obtain their behaviour from the magnetic field penetrating the loop in a particular orientation of the plane wave. The subtle differences in behaviour have not been studied in much detail, a lack that is covered in this volume in Sects. 2.6.3 and 9.2.

The nano-scaled thick ring has now become an item of some interest; it is a nano-structure that has led to some unexpected and rather curious behaviour. The question is whether there are more unknown and unexpected behaviours waiting to be found. Interestingly, the analytical theory we have about loops has been found deficient in answering these questions of nano-rings, and a thorough going examination of the dielectric ring in the THz and optical regions is now timely, and in fact, overdue.

A. McKinley, *The Analytical Foundations of Loop Antennas and Nano-Scaled Rings*, Signals and Communication Technology,
https://doi.org/10.1007/978-981-13-5893-7_1

1.2 The Experimental History of Loops

Much of the early knowledge of loop antennas was based more on experimentation and experience than on analytical theory. It is true that theory, early on, indicated that the small loop is the "dual" of the small linear dipole, in the sense that the fields radiated by the small loop have the same mathematical form as those radiated by a small linear dipole, except that the magnetic and electric fields are interchanged [1, pp. 153 and 237]. This is because it is easy to guess the current distribution in a small loop (it must be a uniform distribution). As a result, small loops were, and often still are, thought of as magnetic field sensors [2, pp. 7–42]. Consequently loops in the period before the 1960's were used for reception, direction finding, and signal location. One does not find small loops used for transmission in the 1920s to the 1950s, the days of AM radio [3–6]. Indeed, broadcasters used half wavelength dipoles on very tall towers, some 180 m long at 833 kHz [4, 6, 7]. Why did they not use large loops?

Just as with small loops, the current distributions on half-wave and full-wave loops, in polygonal and circular configurations, can be easily guessed, and their radiation patterns calculated at their apparent resonant frequencies. Indeed, their lengths can be adjusted and matched to the source until standing waves appear, to ensure the expected resonance. Yet these large loops were not used, partly because of their areal size, partly because of a difficulty of coupling and impedance matching the loop to the transmitter [8], but mostly because of their radiation patterns.

Small dipoles and small loops (those with circumference under 0.2λ) share the same omni-directional radiation pattern in the azimuthal direction (at $\theta = \pi/2$, when the loop lies in the xy plane), but these patterns diverge as the dipole and loop grow larger with respect to the operating wavelength. While the dipole pattern remains roughly omni-directional, and therefore an optimal pattern for tall towers, a large loop (where circumference $\approx\lambda$) mutates toward a thick lobe facing normal to the surface of the plane of the loop (that is, toward $\theta = 0$ and $\theta = \pi$). So if the plane of the loop were placed horizontally to the earth, it would radiate uselessly straight up and straight down; and if placed at right angles to the earth, two of the four directions would have reduced transmission. The omni-directional pattern of the dipole would be preferred.

It wasn't until FM radio began to challenge the dominance of AM, first in the very high frequency (VHF) band and later in the 1950s in the ultra-high frequency (UHF) band for television that loops became an alternative to dipoles. By that time the importance of shaping radiation patterns by design significantly affected the choice of antenna configuration, and the loop was one among many that were developed and measured. Some of the others, for example, were stacked dipole arrays, the Vee, the Bow-tie, and the Rhombic [9]. The Quad, a rectangular, square or circular loop in Yagi–Uda format with reflectors and a driving loop, was also studied for UHF purposes in some detail [10]. The loop for the Quad was designed to have no gaps in the periphery in order to eliminate regions of voltage, since it was used in its first rendition high in the Andes of Ecuador where gaps suffered coronal discharges and affected transmission.

It is difficult to find anything more than basic theory in the literature for these UHF loops. The earliest useful loop appears to have been a square design by Andrew Alford [11] in 1940. The square is broken into four sections, each as long as a quarter wavelength, and energised at four equidistant points in order to ensure a uniform current distribution around the loop. The goal was to produce an omni-directional pattern with low radiation in the vertical direction.

In 1942, M. W. Scheldorf at General Electric developed a stacked set of circular loops [12], quite by trial and error it seems, starting with two stacked square Alford loops with each arm of length 0.25λ. This evolved into a nested pair of squares, twisted 90° with respect to each other, with one arm on each square missing, as shown in Fig. 1.1. The resulting pattern was omni-directional as needed. Finally the squares evolved into the two stacked circular loops, looking much like a folded dipole shaped into circular form, where the two circles are connected using flat plates at the 180° point opposite the driving gap on the outside circle. The practical example given had diameter 33 in., a circumference about 0.16λ for the operating frequency of 46 MHz.

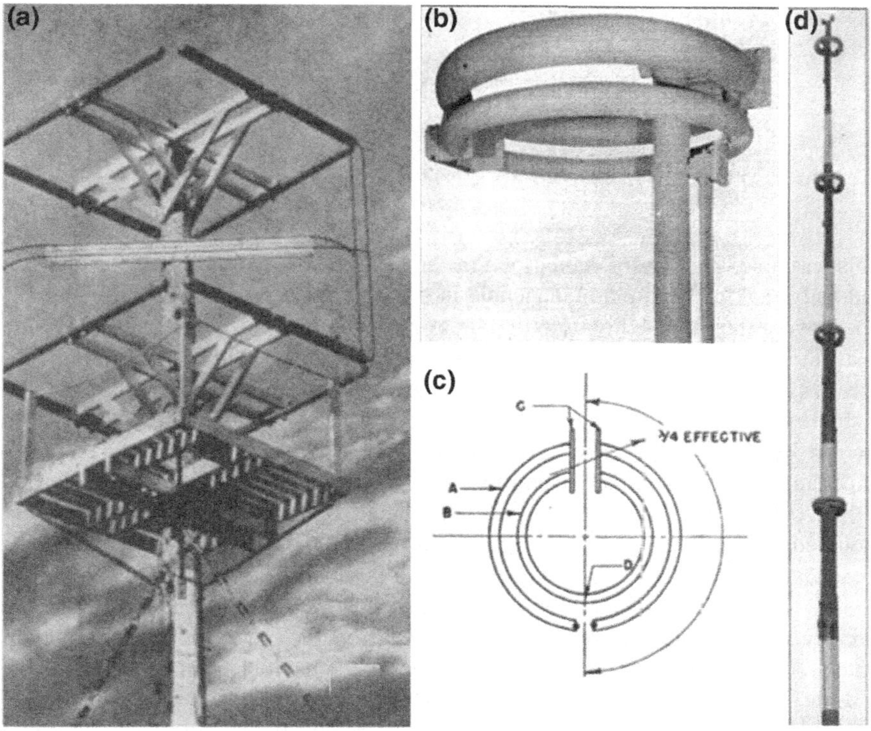

Fig. 1.1 Evolution of the Alford Loop in the 1940s [11]. **a** Sheldorf's extension of the Alford Loop. The Alford Loop is the top square. **b** The eventual loop with **c** its schematic, showing it as a split-ring; probably the first split-ring design. **d** is the implementation as a set of loops on a pole. Courtesy November 1942 QST; copyright ARRL

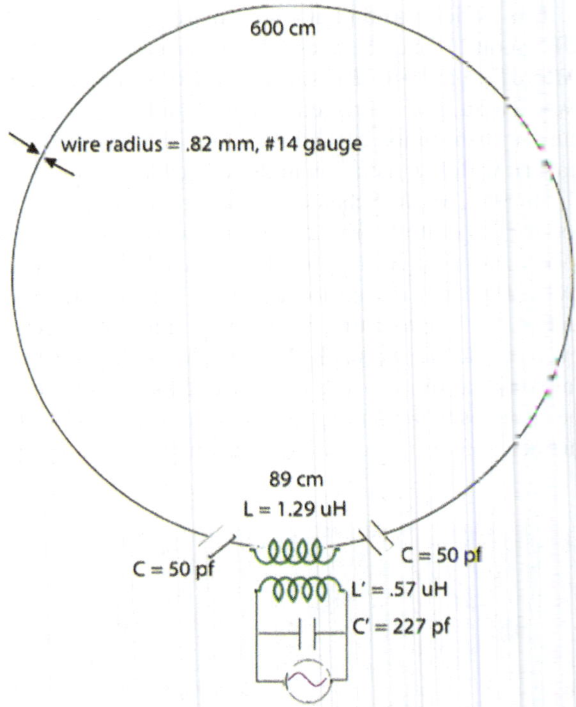

Fig. 1.2 Single capacitor loop antenna design by McKinley from 1963. Redrawn from plans [17]

This was an extraordinary decrease in the size of the loop from the full wavelength, suddenly making the loop manageable in terms of areal size for communications. These were then stacked on a long pole to provide an omni-directional field. Note that the final design was, indeed, a split-ring at MHz frequencies, operating sub-wavelength, probably the first useful split-ring that appears in the literature.

In 1943, George Levy summarised the current state of loop antennas used on aircraft by the military in the Proceedings of the IRE [13]. The preferred "low-impedance" loop carried anywhere from 4 to 20 turns, the high-impedance carried 20 to 70 turns and both were significantly smaller than a wavelength. The loop, mounted on the fuselage outside the aircraft, was kept close to the receiver, so that a capacitor in the receiver could resonate with the loop. The theory applied simple notions of LC resonance and the article spent most of its effort on simplifying the circuit analysis of the loop, receiver and transmission line and on the radiation pattern of the loop rather than on intricate development of behaviour from first principles.

Loop antennas do not appear at all in the rather lengthy *Radio Antenna Engineering* reference by Edmund A. Laport in 1952 [9]. That may be due to the fact that Radio Corporation of America (RCA) strongly supported AM, rather than FM [14]. But even in an historical review of antennas that appeared in the Proceedings of the IRE in 1962, there is little mention of loops antennas except in casual reference [15]. The

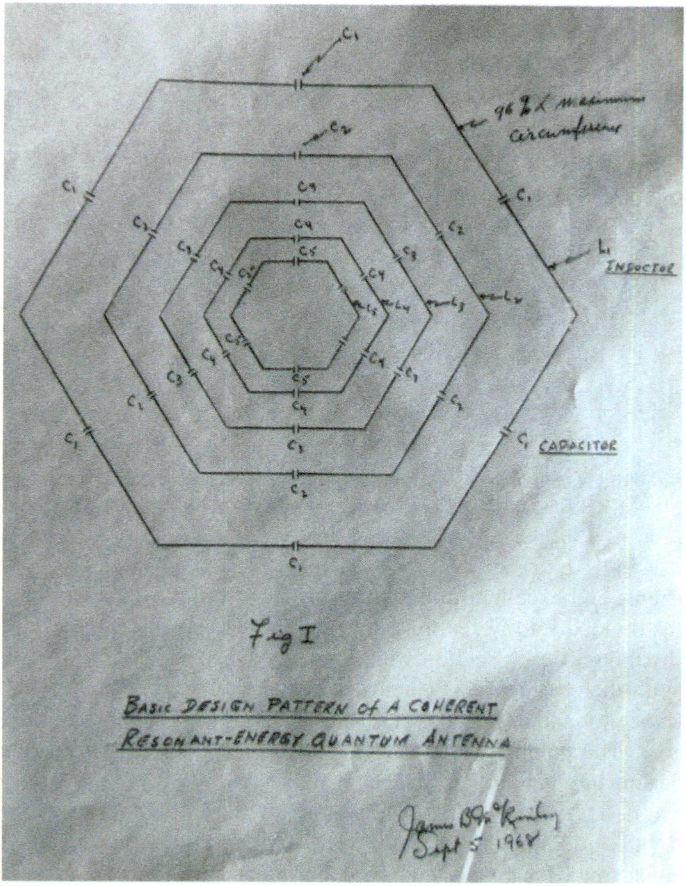

Fig. 1.3 Nested loop antenna design by McKinley from 1968. Reference [17]

1960 edition of the American Radio Relay League's (ARRL) antenna reference book for amateur radio operators describes various useful notes about small, half and full wavelength circular loops in the section under Antenna Fundamentals, but only one loop antenna appears elsewhere in the book, fashioned somewhat like the Levy loop, noted above for aircraft, but many fewer turns [16].

In 1962, James B. McKinley began building and experimenting with closed loop antennas in the 20 m amateur radio band. These evolved into single capacitor circular loops and then multiple capacitor, polygonal loop antennas. He designed each to carry a high Q resonance and he learned experimentally how to match these loops for a 1:1 standing wave ratio on the transmission line. The circumference of his loops were substantially smaller than the operating wavelength, on the order of 0.4λ (see Fig. 1.2). The principle idea was simply that the inductance of the loop resonated with the capacitor and under certain conditions would attain a very high

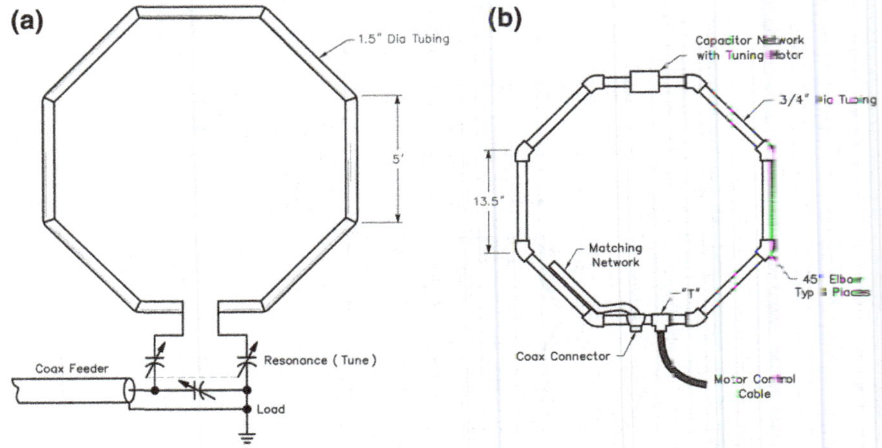

Fig. 1.4 A Simplified Diagram of the Army Loop **a** from 1968 and of the Hart Loop **b** from 1986 described in the ARRL Antenna Book [21]. Courtesy ARRL Antenna Book, 18th Edition, 199[illegible]

Q, high currents and excellent radiation. It is clear from his calculations that when the loop was divided into N sections, by spacing the capacitors evenly around the periphery, he thought each section went into resonance separately; the sectional L resonated with the sectional C. When the loop was energised, he would short an individual capacitor and draw long arcs, showing a very high voltage gradient across the capacitors. The values of the total L and C needed for loop resonance were determined using the standard resonance formula, $f = 1/\sqrt{L_T C_T}$, and the free space impedance $\sqrt{L_T/C_T} = 377$ ohms. The values of the individual capacitors where then given by $C = C_T N$ and $L = L_T/N$. The design eventually evolved various use antennas, consisting of many nested hexagons and arrays of hexagons (see Fig. 1.3).

In an electronics article [18] published in 1967, Kenneth Patterson described the development of a hexagonal loop at the US military's Aberdeen Proving Grounds for use in Vietnam (see Fig. 1.4). The loop resonated with a set of two adjustable capacitors and then impedance matched to the transmission line using an additional capacitor. The "Army Loop", as it came to be called in amateur radio venues achieved a certain notoriety after it appeared in a March 1968 article in QST [19], due to a very strong resonance it exhibited for a small areal size. The circumference was 0.16λ for a design frequency of 3.98 MHz. The article is thought by subsequent amateur authors to be the first of the single capacitor loops that eventually came to be known as a very small, efficient, high-Q antenna for distance communications. That loop and variations on that design began appearing in the ARRL's Antenna Book in about 1988 [20] and has appeared in every edition since then.

The ARRL reference also describes a variant on the Army Loop with design data for 38 loops [22, 23, pp. 5–11]. All of them exhibit a ratio of circumference to incident or driving wavelength ($2\pi b/\lambda$) between 0.06 and 0.40. This is true of the Schelkunoff loop of 1942 (above) and all of the McKinley loops of 1962+ (also above) as well.

In the later literature this region is called the "sub-wavelength" region, because the circumference is smaller than 0.50λ. All exhibit high quality Qs, sometimes as high as several thousand. The high Q was for amateurs, either a blessing, because of the consequent amplification of the signal, or a curse, because the bandwidth is small, requiring an antenna retune and impedance rematch when looking for new contacts.

These sub-wavelength resonances are described in detail in Sect. 7.5.

1.3 The Analytical History of Loops

Large circular closed loops, those with circumferences on the order of the incident or driving wavelength, present complications that are not present in analytical studies of small loops. This is because the current distribution is difficult to guess.

Generally, in small loop analyses, the wire is assumed to be very thin compared with wavelength and the current distribution is assumed to be the same at every point around the periphery of the loop, assumptions which give a uniform magnetic field and, for a closed circular loop, an omni-directional radiation pattern. However, a large closed circular loop can handle many resonant modes and trying to guess which of these is dominant under different circumstances is not straightforward. The simplistic guess is the condition $2\pi b = n\lambda$, where b is the radius of the loop and n is an integer. This is the "Bohr" condition for loops, which comes from Quantum Mechanics [24]. But such a loop can handle many resonant modes and Pocklington [25] in 1897, considering a closed loop illuminated by a plane wave, assumed for the current an infinite series of resonance modes in a Fourier Series. Later solutions for the current based on his approach do show that very thin loops indeed resonate under the Bohr conditions, but a thick wire is not likely to do so for two reasons: (1) toroidal, and less likely radial, currents are more likely to occur in thicker wires and (2) even in metals at low frequencies, which exhibit skin-effect, surface filaments of current on the inside circumference of the loop (at $r = b - a$) travel a shorter distance than those on the outside surface ($r = b + a$), where b is the average radius of the loop (through the middle of the wire) and a is the radius of the wire. Most charge follows a path on the inner radius, thus blue-shifting the resonances slightly. Nevertheless, these assumptions taken together as the "thin-wire" approximation, show a blue-shift in loops as thin as a loop to wire radius of 64.

Hallen [26] extended Pocklington's method using a driven, rather than illuminated, closed loop, finding Fourier solutions for both the current and the impedance, but the coefficients showed a singularity, making the series only "quasi-convergent" and thus limiting the results to small loops. In the mid-century, Storer [27, 28] attempted to solve the convergence problem, but was successful only up to a point; he truncated the series at mode 4 and estimated the rest of the series with a closed function. This method leads to some inaccuracies [29]. Wu [30] discovered the weaknesses in Storer's method and solved the convergence problem completely, allowing for an infinite sum in the Fourier series. These results have been noted in all of the major antenna reference books and in the literature since then, for example [1, 2, 31–35].

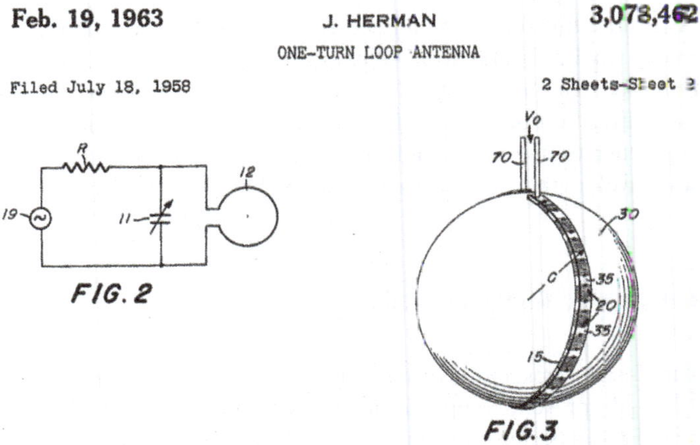

Fig. 1.5 A portion of the cover page to Herman's patent

In 1963 a US patent was granted on papers filed in 1958 by Julius Herman [36] (see Fig. 1.5). It contains a summary of $\vec{E}$ and $\vec{H}$ fields for the closed loop derived under the assumption of an infinite Fourier series representation for the current. The loop is embedded in material media with loss characteristics. Herman was able also to solve the Fourier coefficients and formulate a theory of input impedance using a delta function voltage source, the same approach used by Storer and Wu. Herman also added lumped and distributed impedances around the periphery of the loop. The point of the patent was to devise means of controlling the radiation patterns using these distributed impedances. Unfortunately, Herman never published except in this patent. Since he did not cite any of the above authors, the work apparently is entirely his own.

In 1964, Iizuka [31] extended the Storer theory to include a loading of multiple impedances and driving sources spread around the periphery of the loop. In particular, if only one driving source is used, his work provides the input impedance of the multiply loaded loop. The results are presented in matrix form and the impedances, in particular, form a diagonal matrix. He focused his examples specifically on negative resistive loads. Unfortunately, he used the Storer truncation method on the series and his printed results are therefore a bit inaccurate.

In 1964, Harrington [37] developed a general solution for loaded wire antennas as scatterers using generalised impedances in matrix representation and in 1965 with Ryerson [38] reproduced the Storer/Wu theory and then extended it to the scattering of multiply loaded loops using his new method, noting Iizuka's previous work. However, instead of a driven loop, they used a plane wave and were therefore able to derive the radar cross sections. Then, in a 1968 paper written with Mautz [39] Harrington identified the Fourier coefficients of the infinite series as impedances and as a result, recognised that the input impedance of the loop was the same as that of an infinite network of *parallel* impedances. One of their figures shows the real and imaginary parts of the DC mode impedance and the first three frequency mode impedances. The DC mode is inductive over the region $0 < b/\lambda < 0.40$ (circumference <2.5), while the remaining modes are in some regions capacitive and in other regions inductive. With these results, Harrington and Mautz produced gain patterns for transmitting closed loops and loops loaded with a single capacitor. They also produced forward and backward bi-static scattering plots for the closed and capacitively loaded loop. Harrington and Mautz do not appear to have recognised the sub-wavelength resonance noticed by the amateur community.

In 1966, Iizuka [40] extended the Storer/Wu model to two loops separated in space along the z-axis with planes perpendicular to the z coordinate. Both loops were driven at $\phi = 0$. He solves the two coupled integral equations by splitting the voltage generators into symmetric and anti-symmetric parts and reduces the two integrals to one for solution. This result yields the self and mutual impedances of the two loops. In 1967, Thiele [41] noticed the sub-wavelength resonance when studying loops with a small finite gap. His work was experimental, focusing on the loop as a scatterer and relying on Harrington's 1966 paper.

This theory was not extended again until the second decade of the current century when McKinley et al. [42, 43] recognised that the model impedances identified by Harrington in 1964 could be construed as series resonant RLC circuits. This allowed them to calculate the resistance, inductance and capacitance of each individual mode and the total R, L and C values of the closed loop over the region $0 < 2\pi b/\lambda < 2.5$. They then extended the theory to metallic materials, using the index of refraction as the basis for characterisation. This allowed computations beyond 100 GHz into the optical region. Finally, following Iizuka's method, McKinley extended their results to multiply-loaded loops at low frequency and to multiply-loaded nano-rings in the optical region [44].

Several authors have developed analytical expressions for the near and far field radiation patterns of loops. Rao [32] in 1968 used the Storer theory to calculate the radiation patterns of the far-field, which are good to 2.5λ. Martin [45] in 1960 developed a direct integration procedure for far-zone vector potentials leading to closed form solutions for most current distributions on the loop. In 1996, Werner [46] finally developed an exact integration procedure for the near-zone vector potentials. The last is covered in this volume. In 1997, Li et al. [47] published an alternative vector analysis employing the dyadic Green's functions.

1.4 The Recent History of Nano-Scaled Rings

A surprising development in loops occurred at the turn of the 21st century in the micro-wave (MW) region where metals still conduct well. Veselago [48] in 1968 had suggested that negative index of refraction materials could physically exist and, if constructed, would show novel optical effects. Such a material would be "left-handed", in the sense that the vectors $\vec{E}$, $\vec{H}$, and $\vec{k}$ would follow a left-handed rule of orientation rather than a "right-handed" rule. This result would create strange Doppler

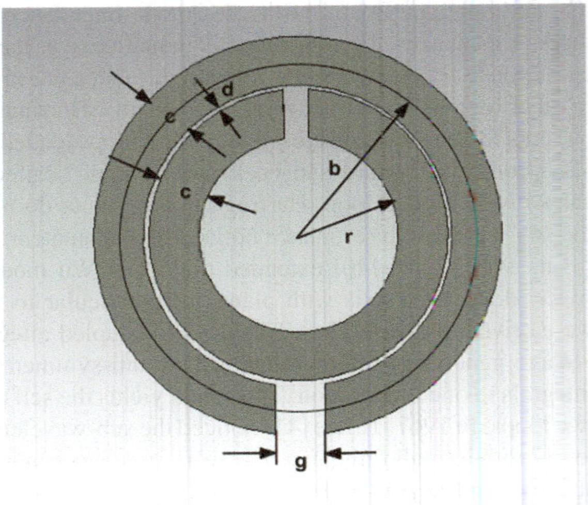

Fig. 1.6 A redrawing of the split rings designed by Pendry for the 3D structure [49]. $a = 10.0$ mm; $c = 1.0$ mm; $d = 0.1$ mm; $l = 2$ mm; $r = 2$ mm. Consequently, the middle radius of the outside ring is $b = r + d + 1.5c = 3.6$ mm

effects, Vasilov-Cerenkov effects, and negative refraction behaviour at boundaries with right-handed materials.

Three seminal papers form the core of efforts to design and build such negative index materials: the works by Pendry et al. [49], Smith et al. [50], and Shelby et al. [51]. After examining and calculating the permeability of various configurations of metallic cylinders and rolled metallic sheets, Pendry's group decided on the flat disk, nested split-ring configuration shown in Fig. 1.6, as the fundamental building block of a 3D structure, which they calculated would show a region of negative permeability at 13.5 GHz. In 2000 they suggested it could be used to create a perfect lens [52].

The Smith group included a small cylinder, just above the ring, to add a negative permittivity to the structure and thus create the required negative index of refraction. The Shelby group built the 3D structure shown in Fig. 1.7 using nested square loops, and showed that it did indeed yield negative index phenomena (see Fig. 1.8). This effect came to be the identifying mark of "meta-materials", human-made electro-magnetic structures for which there is no counter-part in nature.

The outer loop in the Pendry configuration has a circumference of 22.6 mm and a calculated expected resonance was 13.5 GHz, about 22 mm. It resonates therefore at the first mode $n = 1$. Indeed, neither of their two rings resonates in the sub-wavelength region. On the other hand, the nested rings studied by Smith's group does generate a sub-wavelength resonance, at 4.845 GHz where the circumference of the outer ring is 0.25λ; they note this in the text. The Shelby group makes a point of saying that the unit cell dimension of the configuration of square loops that actually demonstrates the negative index effect is a factor 6 less than the resonant wavelength, which puts it on the edge of the sub-wavelength region. Consequently, it may be supposed that the reason for the behaviour of early meta-materials is this very high Q resonance, due to the resonant effect of the inductance of the loop and of the

Fig. 1.7 A Shelby like meta-material structure using flat disk square loops. From Public Domain, NASA; Wikipedia, "Metamaterial", accessed Oct 2018

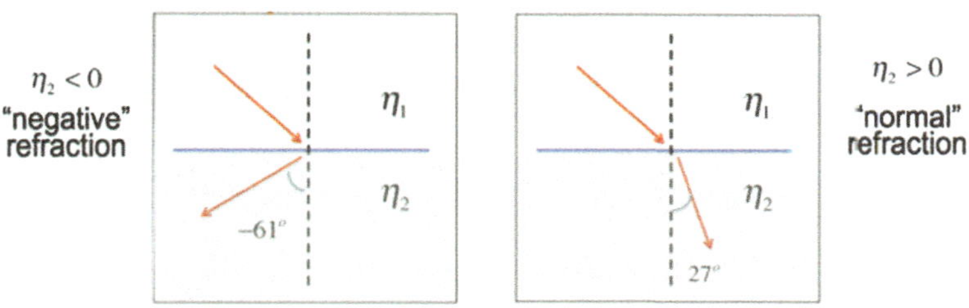

Fig. 1.8 Effects of the negative Index of refraction on incident waves. The waves bend away from the normal at a negative angle, instead of toward the normal at a positive angle. From [51]

gap capacitance. Indeed, all of these authors identified the cause of their resonances as an LC pair created by the inductance of loop and the capacitance of the gap, together with the coupling capacitance between the loops.

In 2004, Shamonin et al. [53] developed a model that used an infinite number of lumped LC circuits in a circular transmission line configuration, where a lumped capacitance was taken for the gap. This allowed them to establish transmission line differential equations relating the voltage and current around the ring. They were able to show how the first, second and third harmonic resonances varied with gap widths. The RLC values per unit length are inputs to the theory.

A few years later, Zhou and Chui [54] claimed more accurate results, finding that Shamonin had left out important capacitances of the system. They used instead a "Quasi-static Approximation (QSA)", which, they claimed, takes into account the inductive and capacitive effects of the system completely and discovers the RLC

elements as a result of calculation. QSA is an assumption specifically meant for structures that are small compared with the incident wavelength. The assumption makes their analysis inherently weaker than that used by Storer and Wu when applied to rings that are large compared to wavelength. Hence their method is analytically strong for small $2\pi b/\lambda$ and weaker for, say, $2\pi b > 0.6\lambda$.

A large number of papers have appeared since then, applying LC circuit models to circular and square rings of varying sizes, attempting to understand the origin of the resonance [55–60]. In 2017, McKinley [44] identified and characterised the sub-wavelength resonance of the single capacitor loop at low frequencies. The rigorous extension of capacitance to gaps, as in the single gap nano-ring at high-frequencies up to the optical region, has not been covered yet in the literature but appears here in Sect. 9.3. These gaps behave capacitively in nano-rings, and follow marginally the flat-plate capacitor model that many workers were using at the time.

Nano-rings have been found useful for high-definition imaging [61], concentrating energy using tiny Fresnel lenses [62], tuning antennas in the THz region [59], controlling loop resonances with dielectric materials [63], shaping the radiation beam pattern [64], emitting single photons [65], bio-sensing [66], capturing light more effectively in solar cells [67], super-scattering [68], and super-directivity [69]. Might not many more uses for these rings be found once we probe ring behaviour more fully? It is only reasonable to expect that a detailed analytical model, describing all magnetic and harmonic resonances of the ring, of various thicknesses, with and without gaps, applicable at all frequencies, will shed light on new applications.

References

1. C.A. Balanis, *Antenna Theory, Analysis and Design*, 4th edn. (Wiley, New York, 2016)
2. L.W. Rispin, D.C. Chang, *Antenna Theory, in Antenna Handbook*, vol. II, 2nd edn., ed. by L.Y. T, S.W. Lee (Van Nostrand Reinhold, New York, 1993), pp. 1–51
3. C. Burrows, Proc. IRE **50**(5), 682 (1962). https://doi.org/10.1109/JRPROC.1962.288097
4. P. Carter, H. Beverage, Proc. IRE **50**(5), 679 (1962). https://doi.org/10.1109/JRPROC.1962.288096
5. L. Chu, Proc. IRE **50**(5), 685 (1962). https://doi.org/10.1109/JRPROC.1962.288098
6. J. Ramsay, IEEE Antennas Propag. Soc. Newslett. **23**(6), 7 (1981). https://doi.org/10.1109/MAP.1981.27575
7. C.H. Sterling, in *Encyclopedia of Radio 3-Volume Set*, ed. by C.H. Sterling (Taylor and Francis, Abingdon, 2004), pp. 137–140
8. J.H. Dunlavy, Us patent 3,588,905: Wide range tunable transmitting loop antenna (1971). http://www.google.co.uk/patents/US3588905
9. E.A. Laport, *Radio Antenna Engineering* (McGraw-Hill Book Company, New York, 1952)
10. S.D.C. Orr, I. William, *All About Cubical Quad Antennas* (Radio Publications, 1959)
11. A. Alford, A.G. Kandoian, Electr. Eng. **59**(12), 843 (1940). https://doi.org/10.1109/EE.1940.6435249
12. A. not specified, QST **XXVI**(11), 19 (1942). http://www.rfcafe.com/references/qst/a-circular-antenna-for-uhf-nov-1942-qst.htm
13. G. Levy, Proc. IRE **31**(2), 56 (1943). https://doi.org/10.1109/JRPROC.1943.234346
14. L. Lessing, *Man of High Fidelity: Edwin Howard Armstrong* (Lippincott, New York City, 1956)

15. I.P.G. on Antennas, Propagation (eds.). *Proceedings of the IRE: Antennas and Propagation*, Section 2 (IRE, 1962)
16. A.L. Budlong (ed.), *The A.R.R.L. Antenna Book*, 9th edn. (The American Radio Relay League, Newington, 1960)
17. J.B. McKinley II, The design of a single capacitor ccrl (1964). Personal papers
18. K. Patterson, Electronics (1967)
19. L.G. McCoy, QST **LII**(3), 17 (1968)
20. J. Kraus (ed.), *The A.R.R.L. Antenna Book*, 15th edn. (The American Radio Relay League, Newington, 1988)
21. A.R.R. League (ed.), *ARRL Antenna Handbook*, cd rom edition edn. (American Radio Relay League). https://www.qrz.ru/schemes/contribute/arrl/chap5.pdf
22. R.T. Hart, QST pp. 33–36 (1986)
23. R.T. Hart, *The ARRL Antenna Book* (1994), p. 11
24. D. Bloom, D.W. Bloom, Phys. Teach. **41**(5), 292 (2003). https://doi.org/10.1119/1.1571265
25. H.C. Pocklington, Proc. Cambridge Phil. Soc. **9**, 324 (1897)
26. E. Hallen, Nova Actae Regiae Soc.Sci. Upsaliensis **Ser. IV, 11**(No.4), 1 (1938)
27. J.E. Storer, Impedance of the thin wire loop. Technical Report 212, Cruft Laboratory, Harvard University (1955)
28. J.E. Storer, Trans. AIEE **75**, 606 (1956)
29. R. King, J. Harrison, C., D. Tingley, IEEE Trans. Antennas Propag. **12**(4), 434 (1964). https://doi.org/10.1109/TAP.1964.1138244
30. T.T. Wu, J. Math. Phys. **3**(6), 1301 (1962)
31. K. Iizuka, IEEE Trans, Antennas Propag. **13**(1), 7 (1965)
32. B.A. Rao, IEEE Trans. Antennas Propag. (1968)
33. M. Kanda, IEEE Trans. Electromagn. Compat. **EMC-26**(3), 102 (1984)
34. L.W. Li, C.P. Lim, M.S. Leong, in *Antennas and Propagation Society International Symposium, 1999. IEEE*, vol. 4 (1999), vol. 4, pp. 2542–2545 vol.4. https://doi.org/10.1109/APS.1999.789327
35. L. RongLin, N.A. Bushyager, J. Laskar, M.M. Tentzeris, IEEE Trans, Antennas Propag. **53**(12), 3920 (2005)
36. J. Herman, *One-Turn Loop Antenna, No. 3078462* (US. Patent Office, 5353 Chillum Place, Washington DC, USA, 1963)
37. R.F. Harrington, Proc. Inst. Electr. Eng. **111**(4), 617 (1964). https://doi.org/10.1049/piee.1964.0111
38. R.F. Harrington, J.L. Ryerson, Radio Sci. **1**, 347 (1966)
39. R. Harrington, J. Mautz, Proc. Inst. Electr. Eng. **115**(1), 68 (1968) https://doi.org/10.1049/piee.1968.0011
40. K. Iizuka, R. King, J. Harrison, C., IEEE Trans. Antennas Propag. **14**(4), 440 (1966). https://doi.org/10.1109/TAP.1966.1138711
41. G.A. Thiele, *Radar cross section of open circular loops* (Technical report, DTIC Document, 1967)
42. A.F. McKinley, T.P. White, I.S. Maksymov, K.R. Catchpole, J. Appl. Phys. **112**(9), 094911 (2012). https://doi.org/10.1063/1.4764104, http://link.aip.org/link/?JAP/112/094911/1
43. A.F. McKinley, T.P. White, K.R. Catchpole, J. Appl. Phys. **114**(4), 044317 (2013). https://doi.org/10.1063/1.4816619, http://scitation.aip.org/content/aip/journal/jap/114/4/10.1063/1.4816619
44. A.F. McKinley, IEEE Trans, Antennas Propag. **65**(5), 2276 (2017)
45. E.J. Martin Jr., IRE Trans. Antennas Propag. **AP-8**(1), 105 (1960)
46. D. Werner, IEEE Trans, Antennas Propag. **44**(2), 157 (1996). https://doi.org/10.1109/8.481642
47. L. Le-Wei, L. Mook-Seng, K. Pang-Shyan, Y. Tat-Soon, IEEE Trans, Antennas Propag. **45**(12), 1741 (1997)
48. V.G. Veselago, Sov. Phys. Uspekhi **10**, 4 (1968)
49. J.B. Pendry, A.J. Holden, D.J. Robbins, W.J. Stewart, I.E.E.E. Trans, Microw. Theory Tech. **47**, 11 (1999)

50. D.R. Smith, W.J. Padilla, D.C. Vier, S.C. Nemat-Nasser, S. Schultz, Phys. Rev. Lett. **84** 4184 (2000). https://doi.org/10.1103/PhysRevLett.84.4184
51. R.A. Shelby, D.R. Smith, S. Schultz, Science **292**(5514), 77 (2001). https://doi.org/10.1126/science.1058847, http://www.sciencemag.org/content/292/5514/77.abstract
52. J.B. Pendry, Phys. Rev. Lett. **85**(18), 3966 (2000)
53. M. Shamonin, E. Shamonina, V. Kalinin, L. Solymar, J. Appl. Phys. **95**(7), 3778 (2004). https://doi.org/10.1063/1.1652251
54. L. Zhou, S.T. Chui, Phys. Rev. B **74**, 035419 (2006). https://doi.org/10.1103/PhysRevB.74.035419
55. A. Radkovskaya, M. Shamonin, C.J. Stevens, G. Faulkner, D.J. Edwards, E. Shamonina, L. Solymar, Microw. Optical Technol. Lett. **46**(5), 473 (2005). https://doi.org/10.1002/mop.21021
56. H. Mosallaei, K. Sarabandi, I.E.E.E. Trans, Antennas Propag. **55**(1), 45 (2007). https://doi.org/10.1109/TAP.2006.886566
57. K. Aydin, E. Ozbay, J. Appl. Phys. **101**, 024911 (2007)
58. O. Sydoruk, E. Tatartschuk, E. Shamonina, L. Solymar, J. Appl. Phys. **105**, 014903 (2009)
59. D.R. Chowdhury, R. Singh, M. Reiten, J. Zhou, A.J. Taylor, J.F. O'Hara, Opt. Express **19**(11), 10679 (2011). https://doi.org/10.1364/OE.19.010679, http://www.opticsexpress.org/abstract.cfm?URI=oe-19-11-10679
60. T.R. Zhan, S.T. Chui, Journal of Applied Physics **115**(14), 144901 (2014). https://doi.org/10.1063/1.4870863, http://scitation.aip.org/content/aip/journal/jap/115/14/10.1063/1.4870363
61. A.L. Koh, D.W. McComb, S.A. Maier, H. Low, J.K.W. Yang, J. Vac. Sci. Technol. B: Microelectr. Nanometer Struct. **28**(6), C6O45 (2010). https://doi.org/10.1116/1.3501351
62. B. Memarzadeh, H. Mosallaei, Opt. Lett. **36**, 2569 (2011)
63. S.Y. Chiam, R. Singh, W. Zhang, A.A. Bettio, Appl. Phys. Lett. **97**, 191906 (2010)
64. A. Ahmadi, H. Mosallaei, Opt. Lett. **35**, 3706 (2010)
65. I.S. Maksymov, A.E. Miroshnichenko, Y.S. Kivshar, Opt. Express **20**(8), 8929 (2012). https://doi.org/10.1364/OE.20.008929, http://www.opticsexpress.org/abstract.cfm?URI=oe-20-8-8929
66. E.M. Larsson, J. Alegret, M. Käll, D.S. Sutherland, Nano Lett. **7**(5), 1256 (2007). https://doi.org/10.1021/nl0701612
67. S. Mokkapati, F.J. Beck, R. de Waele, A. Polman, K.R. Catchpole, J. Phys. D: Appl. Phys. **44**(18), 185101 (2011). http://stacks.iop.org/0022-3727/44/i=18/a=185101
68. M. Fan, G.F. Andradec, A.G. Brolo, Analytica Chimica Acta **693**, 7 (2011)
69. M.F. Pantoja, J. Nagar, B. Lu, D.H. Werner, ACS Photon. **4**(3), 509 (2017). https://doi.org/10.1021/acsphotonics.6b00486

Chapter 2
Foundations

2.1 The Geometry of the Closed Toroidal Ring

The loop or ring is a torus, and therefore the proper geometry for the modelling of internal and surface currents are toroidal coordinates. However, these coordinates are more difficult than needed in this volume, and a more useful system can be constructed from the standard three orthonormal systems which are usually used to model the external field patterns created by these internal currents: the spherical, cylindrical and cartesian systems. These will be used throughout this book, sometimes independently and sometimes combined, depending on the needs and the circumstances.

2.1.1 *Measurements in Different Coordinate Systems*

Invariance is a physical principle that allows an observer using one coordinate system to compare measurements with another observer who uses a different coordinate system. It means the measurements of an object and its physical behaviour will be different for these observers, but the physical properties of the object itself cannot vary just because one observes from a different coordinate system. Therefore, we should be able to transform the measurements of one observer into that of another and arrive at the same result.

Scalar properties are the simplest, since they consist of a pure number, like the impedance of free space, 377 ohms, or the irrational constant π. Scalars are the same in every coordinate system.

Vectors are not so simple. A vector represents the distance between two points in a coordinate system. They are usually expressed by component values related to directional axes established in the system, but the axes in one system may be curvilinear with respect to the axes in another. This makes the transformation a bit more difficult to calculate. The easiest coordinate system is the Cartesian system and that is why most transformations are related to that system.

A. McKinley, *The Analytical Foundations of Loop Antennas and Nano-Scaled Rings*, Signals and Communication Technology,
https://doi.org/10.1007/978-981-13-5893-7_2

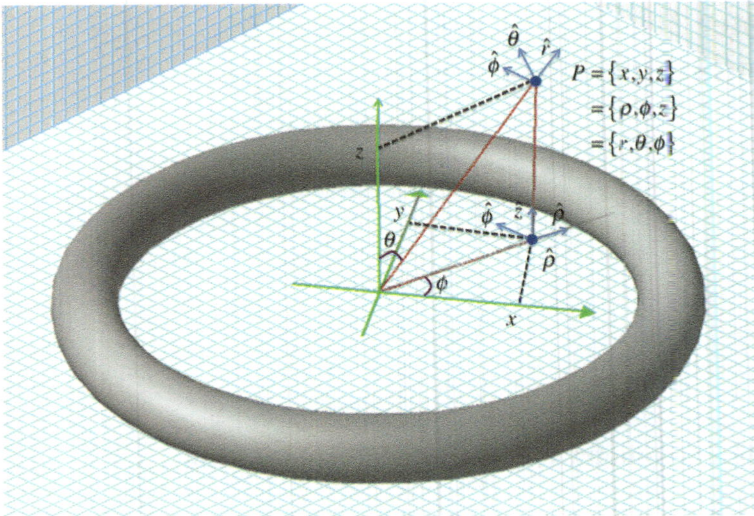

Fig. 2.1 Coordinate systems. The three standard orthonormal coordinate systems

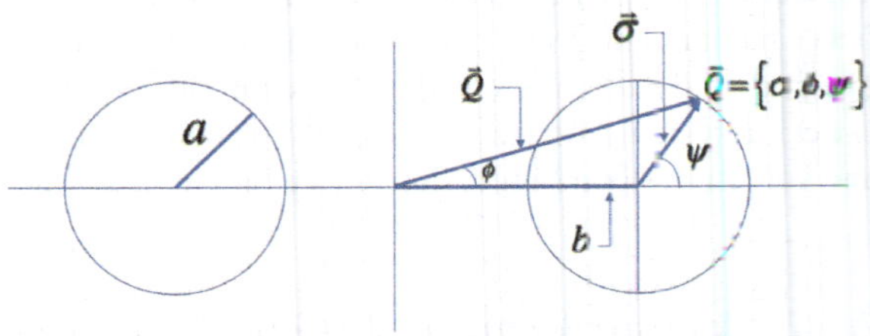

Fig. 2.2 A "wire" coordinate system, an alternative orthonormal system

The three, standard, well-known systems, namely, $\{x, y, z\}$, $\{\rho, \phi, z\}$ and $\{r, \theta, \phi\}$ with their origins, $\{0, 0, 0\}$ at the centre of the loop, are shown in Fig. 2.1. Coordinate ρ is directed along the xy plane, z along the axis perpendicular to the plane. ϕ marks the azimuthal, angular position counterclockwise around the ring with $\phi = 0$ indicated along the x-axis. θ marks the angular position from the zenith above the origin away from the z-axis, with $\theta = 0$ indicated along the z-axis.

The radius of the torus is given by the constant value $\rho = b$, a distance marked from the point of origin to the centre of the loop wire, no matter how thick. The radius of the loop wire is a. The unit vectors are $\hat{x}$, $\hat{y}$, and $\hat{z}$ for the Cartesian system, $\hat{\rho}$, $\hat{\phi}$, and $\hat{z}$ for the cylindrical system, and $\hat{r}$, $\hat{\theta}$, and $\hat{\phi}$ for the spherical system. At times, an assumption will be made that the torus can be represented by a filament of circumference $2\pi b$; that is, by a circle, where the radius of the wire, a, is zero.

The "wire" system, which we shall use throughout the volume, has its origin at the centre of the wire, as in Fig. 2.2. The orthonormal coordinates are $\{\sigma, \phi, \psi\}$. σ marks the distance along the cross-sectional area of the wire, oriented from the centre of the wire outward towards the wire's surface. ϕ is the same as in the ordinary systems given above, and ψ marks the angular position from the centre of the wire counterclockwise, away from the xy plane of the ring, with $\psi = 0$ indicated along the ρ axis of the ring.

The transformation from the Cartesian coordinate system to a curvilinear system, $\{u, v, w\}$, is given by the fundamentals of the Calculus. Given

$$x = x(u, v, w); \quad y = y(u, v, w); \quad z = z(u, v, w) \tag{2.1}$$

Then

$$\begin{aligned}
\Delta x &= \frac{\partial x}{\partial u}\Delta u + \frac{\partial x}{\partial v}\Delta v + \frac{\partial x}{\partial w}\Delta w \\
\Delta y &= \frac{\partial y}{\partial u}\Delta u + \frac{\partial y}{\partial v}\Delta v + \frac{\partial y}{\partial w}\Delta w \\
\Delta z &= \frac{\partial z}{\partial u}\Delta u + \frac{\partial z}{\partial v}\Delta v + \frac{\partial z}{\partial w}\Delta w
\end{aligned} \tag{2.2}$$

where the derivatives are taken at any given point in the system. This can be written more succinctly as:

$$\begin{pmatrix} \Delta x \\ \Delta y \\ \Delta z \end{pmatrix} = \begin{pmatrix} \frac{\partial x}{\partial u} & \frac{\partial x}{\partial v} & \frac{\partial x}{\partial w} \\ \frac{\partial y}{\partial u} & \frac{\partial y}{\partial v} & \frac{\partial y}{\partial w} \\ \frac{\partial z}{\partial u} & \frac{\partial z}{\partial v} & \frac{\partial z}{\partial w} \end{pmatrix} \begin{pmatrix} \Delta u \\ \Delta v \\ \Delta w \end{pmatrix} \equiv \mathbf{J} \begin{pmatrix} \Delta u \\ \Delta v \\ \Delta w \end{pmatrix} \tag{2.3}$$

The inverse transform is

$$\begin{pmatrix} \Delta u \\ \Delta v \\ \Delta w \end{pmatrix} = \begin{pmatrix} \frac{\partial u}{\partial x} & \frac{\partial u}{\partial y} & \frac{\partial u}{\partial z} \\ \frac{\partial v}{\partial x} & \frac{\partial v}{\partial y} & \frac{\partial v}{\partial z} \\ \frac{\partial w}{\partial x} & \frac{\partial w}{\partial y} & \frac{\partial w}{\partial z} \end{pmatrix} \begin{pmatrix} \Delta x \\ \Delta y \\ \Delta z \end{pmatrix} \equiv \mathbf{J}^{-1} \begin{pmatrix} \Delta x \\ \Delta y \\ \Delta z \end{pmatrix} \tag{2.4}$$

where $\mathbf{J}$ is called the Jacobian matrix and $\mathbf{J}^{-1}$ is the inverse Jacobian matrix. Note that $\mathbf{J}\mathbf{J}^{-1} = \mathbf{I}$, the Identity matrix.

Take now a vector of infinitesimal length, for example, $d\vec{R} = \vec{P}_2 - \vec{P}_1$, where $\vec{P}_1 = \{u_0, v_0, w_0\}$ and $\vec{P}_2 = \{u_o + du, v_0 + dv, w_0 + dw\}$. Then,

$$\begin{aligned}
d\vec{R} = \frac{\partial \vec{R}}{\partial x}dx + \frac{\partial \vec{R}}{\partial y}dy + \frac{\partial \vec{R}}{\partial z}dz &= \frac{\partial \vec{R}}{\partial u}\frac{\partial u}{\partial x}dx + \frac{\partial \vec{R}}{\partial v}\frac{\partial v}{\partial y}dy + \frac{\partial \vec{R}}{\partial w}\frac{\partial w}{\partial z}dz \\
&= \begin{pmatrix} \frac{\partial u}{\partial x}dx & \frac{\partial v}{\partial y}dy & \frac{\partial w}{\partial z}dz \end{pmatrix} \begin{pmatrix} \partial \vec{R}/\partial u \\ \partial \vec{R}/\partial v \\ \partial \vec{R}/\partial w \end{pmatrix}
\end{aligned} \tag{2.5}$$

This may also be written

$$d\vec{R} = \frac{\partial \vec{R}}{\partial u}du + \frac{\partial \vec{R}}{\partial v}dv + \frac{\partial \vec{R}}{\partial w}dw = \frac{\partial \vec{R}}{\partial x}\frac{\partial x}{\partial u}du + \frac{\partial \vec{R}}{\partial y}\frac{\partial y}{\partial v}dv + \frac{\partial \vec{R}}{\partial z}\frac{\partial z}{\partial w}dw$$
$$= \left(\frac{\partial x}{\partial u}du \quad \frac{\partial y}{\partial v}dv \quad \frac{\partial z}{\partial w}dw\right)\begin{pmatrix}\partial \vec{R}/\partial x \\ \partial \vec{R}/\partial y \\ \partial \vec{R}/\partial z\end{pmatrix} \quad (2.6)$$

The partials $\partial \vec{R}/\partial x$, $\partial \vec{R}/\partial y$ and $\partial \vec{R}/\partial z$ indicate the change in distance along the (x, y, and z) axes covered by the vector $\vec{R}$ and therefore these are called the "natural bases" of the Cartesian system. Similarly

$$\tilde{u} = \partial \vec{R}/\partial u \quad \tilde{v} = \partial \vec{R}/\partial v \quad \tilde{w} = \partial \vec{R}/\partial w \quad (2.7)$$

are called the "natural bases" for the curvilinear system.

The literature often does not use the natural bases, opting instead for the "unit bases" in which the natural bases are normalized by their length. The unit bases in the Cartesian system are the same as the natural bases, but this is not the case in curvilinear systems. The natural bases need to be normalized by their lengths, often called a "scaling" factor. Generally,

$$\hat{u} \equiv \frac{1}{h_u}\frac{\partial \vec{R}}{\partial u} = \frac{1}{h_u}\tilde{u} \qquad \hat{v} \equiv \frac{1}{h_v}\frac{\partial \vec{R}}{\partial v} = \frac{1}{h_v}\tilde{v} \qquad \hat{w} \equiv \frac{1}{h_w}\frac{\partial \vec{R}}{\partial w} = \frac{1}{h_w}\tilde{w} \quad (2.8)$$

where

$$h_u = \left|\frac{\partial \vec{R}}{\partial u}\right| \quad h_v = \left|\frac{\partial \vec{R}}{\partial v}\right| \quad h_w = \left|\frac{\partial \vec{R}}{\partial w}\right| \quad (2.9)$$

Expanding, the curvilinear unit vectors are:

$$\hat{u} = \frac{1}{h_u}\tilde{u} = \frac{1}{h_u}\frac{\partial R}{\partial u} = \frac{1}{h_u}\left(\frac{\partial x}{\partial u}\hat{x} + \frac{\partial y}{\partial u}\hat{y} + \frac{\partial z}{\partial u}\hat{z}\right)$$
$$\hat{v} = \frac{1}{h_v}\tilde{v} = \frac{1}{h_v}\frac{\partial R}{\partial v} = \frac{1}{h_v}\left(\frac{\partial x}{\partial v}\hat{x} + \frac{\partial y}{\partial v}\hat{y} + \frac{\partial z}{\partial v}\hat{z}\right)$$
$$\hat{w} = \frac{1}{h_w}\tilde{w} = \frac{1}{h_w}\frac{\partial R}{\partial w} = \frac{1}{h_w}\left(\frac{\partial x}{\partial w}\hat{x} + \frac{\partial y}{\partial w}\hat{y} + \frac{\partial z}{\partial w}\hat{z}\right) \quad (2.10)$$

or more succinctly as:

$$\begin{pmatrix} \hat{u} \\ \hat{v} \\ \hat{w} \end{pmatrix} = \begin{pmatrix} \frac{1}{h_u}\partial x/\partial u & \frac{1}{h_u}\partial y/\partial u & \frac{1}{h_u}\partial z/\partial u \\ \frac{1}{h_v}\partial x/\partial v & \frac{1}{h_v}\partial y/\partial v & \frac{1}{h_v}\partial z/\partial v \\ \frac{1}{h_w}\partial x/\partial w & \frac{1}{h_w}\partial y/\partial w & \frac{1}{h_w}\partial z/\partial w \end{pmatrix} \begin{pmatrix} \hat{x} \\ \hat{y} \\ \hat{z} \end{pmatrix} \equiv \mathbf{Q} \begin{pmatrix} \hat{x} \\ \hat{y} \\ \hat{z} \end{pmatrix} \tag{2.11}$$

and

$$\begin{pmatrix} \hat{x} \\ \hat{y} \\ \hat{z} \end{pmatrix} = \begin{pmatrix} h_u \partial u/\partial x & h_v \partial v/\partial x & h_w \partial w/\partial x \\ h_u \partial u/\partial y & h_v \partial v/\partial y & h_w \partial w/\partial y \\ h_u \partial u/\partial z & h_v \partial v/\partial z & h_w \partial w/\partial z \end{pmatrix} \begin{pmatrix} \hat{u} \\ \hat{v} \\ \hat{w} \end{pmatrix} = \mathbf{Q}^{-1} \begin{pmatrix} \hat{u} \\ \hat{v} \\ \hat{w} \end{pmatrix} \tag{2.12}$$

$\mathbf{Q}$ is called the "coordinate transformation matrix". The last step occurs because orthogonal systems have the property that the inverse of their coordinate transform matrix, $\mathbf{Q}^{-1}$, equals the transpose. This also turns out to be the case for $\mathbf{J}$. It is related to the Jacobian through the bases. Comparing with the definition for $\mathbf{J}$, Eq. (2.3),

$$\begin{pmatrix} \hat{x} \\ \hat{y} \\ \hat{z} \end{pmatrix} = \mathbf{Q}^{-1} \begin{pmatrix} \hat{u} \\ \hat{v} \\ \hat{w} \end{pmatrix} = \mathbf{J} \begin{pmatrix} \tilde{u} \\ \tilde{v} \\ \tilde{w} \end{pmatrix} \tag{2.13}$$

Since vector components are distances between two points in the coordinate system, they transform in the same way as Eq. (2.3). Given a vector $\vec{A} = (A_x, A_y, A_z) = (\tilde{A}_u, \tilde{A}_v, \tilde{A}_w)$, these components transform as

$$\begin{pmatrix} A_x \\ A_y \\ A_z \end{pmatrix} = \mathbf{J} \begin{pmatrix} \tilde{A}_u \\ \tilde{A}_v \\ \tilde{A}_w \end{pmatrix} \tag{2.14}$$

and

$$\begin{pmatrix} A_x & A_y & A_z \end{pmatrix} = \begin{pmatrix} \tilde{A}_u & \tilde{A}_v & \tilde{A}_w \end{pmatrix} \mathbf{J}^T \tag{2.15}$$

We then see that a full vector transformation between coordinate systems may be expressed using either the natural bases or the unit bases.

$$\vec{A} = \begin{pmatrix} A_x & A_y & A_z \end{pmatrix} \begin{pmatrix} \hat{x} \\ \hat{y} \\ \hat{z} \end{pmatrix} = \begin{pmatrix} \tilde{A}_u & \tilde{A}_v & \tilde{A}_w \end{pmatrix} \mathbf{J}^T \mathbf{J} \begin{pmatrix} \tilde{u} \\ \tilde{v} \\ \tilde{w} \end{pmatrix} = \begin{pmatrix} \tilde{A}_u & \tilde{A}_v & \tilde{A}_w \end{pmatrix} \begin{pmatrix} \tilde{u} \\ \tilde{v} \\ \tilde{w} \end{pmatrix} \tag{2.16}$$

Here an A with the tilde indicates a component in the natural bases, while the A without the tilde indicates a component in the unit bases. If we realize that

$$\left(\tilde{A}_u\ \tilde{A}_v\ \tilde{A}_w\right)\begin{pmatrix}\tilde{u}\\ \tilde{v}\\ \tilde{w}\end{pmatrix} = \left(h_u\tilde{A}_u\ h_v\tilde{A}_v\ h_w\tilde{A}_w\right)\begin{pmatrix}\hat{u}\\ \hat{v}\\ \hat{w}\end{pmatrix} \tag{2.17}$$

then, we may also define

$$\left(A_u\ A_v\ A_w\right) \equiv \left(h_u\tilde{A}_u\ h_v\tilde{A}_v\ h_w\tilde{A}_w\right) \tag{2.18}$$

and have

$$\vec{A} = \left(A_u\ A_v\ A_w\right)\begin{pmatrix}\hat{u}\\ \hat{v}\\ \hat{w}\end{pmatrix} \tag{2.19}$$

We will be using the unit bases notation for vectors throughout the rest of the volume.

The gradient of a scalar and the divergence of a vector in the curvilinear coordinate system are

$$\begin{aligned}\vec{\nabla}\Phi &= \frac{1}{h_u}\frac{\partial\Phi}{\partial u}\hat{u} + \frac{1}{h_v}\frac{\partial\Phi}{\partial v}\hat{v} + \frac{1}{h_w}\frac{\partial\Phi}{\partial w}\hat{w}\\ \vec{\nabla}\cdot\vec{J} &= \frac{1}{h_u h_v h_w}\left[\frac{\partial(h_v h_w J_u)}{\partial u} + \frac{\partial(h_u h_w J_v)}{\partial v} + \frac{\partial(h_u h_v J_w)}{\partial w}\right]\end{aligned} \tag{2.20}$$

where $\vec{J} = J_u\hat{u} + J_v\hat{v} + J_w\hat{w}$. The Laplace is

$$\vec{\nabla}^2\Phi = \frac{1}{h_u h_v h_w}\left[\frac{\partial}{\partial u}\left(\frac{h_v h_w}{h_u}\frac{\partial\Phi}{\partial u}\right) + \frac{\partial}{\partial v}\left(\frac{h_u h_w}{h_v}\frac{\partial\Phi}{\partial v}\right) + \frac{\partial}{\partial w}\left(\frac{h_u h_v}{h_w}\frac{\partial\Phi}{\partial w}\right)\right] \tag{2.21}$$

And finally, the curl is

$$\vec{\nabla}\times\vec{A} = \frac{1}{h_u h_v h_w}\begin{vmatrix} h_u\hat{u} & h_v\hat{v} & h_w\hat{w}\\ \partial/\partial u & \partial/\partial v & \partial/\partial w\\ h_u A_u & h_v A_v & h_w A_w\end{vmatrix} \tag{2.22}$$

The infinitesimal unit volume, dV, used in integrals, must include the absolute value of the determinant of the Jacobian. Using Eq. (2.2) and comparing with the determinant of **J**:

$$dV = dxdydz = |\mathrm{det}(\mathbf{J})|dudvdw = h_u h_v h_w dudvdw \tag{2.23}$$

2.1.2 Coordinate Systems In Detail

Transformations between the coordinate systems used in this volume are as follows:
Cartesian $\{x, y, z\}$ and Cylindrical $\{\rho, \phi, z\}$ Coordinates:

$$\begin{gathered} x = \rho\cos\phi \quad y = \rho\sin\phi \quad z = z \\ \rho = \sqrt{x^2 + y^2} \quad \tan\phi = y/x \quad z = z \end{gathered} \tag{2.24}$$

The metrics are:

$$h_\rho = 1 \quad h_\phi = \rho \quad h_z = 1$$

yielding the Jacobian and the transformation matrix

$$\begin{gathered} \mathbf{J} = \begin{pmatrix} \cos\phi & -\rho\sin\phi & 0 \\ \sin\phi & \rho\cos\phi & 0 \\ 0 & 0 & 1 \end{pmatrix} \quad \text{and} \quad \mathbf{J}^{-1} = \mathbf{J}^T = \begin{pmatrix} \cos\phi & \sin\phi & 0 \\ -\sin\phi/\rho & \cos\phi/\rho & 0 \\ 0 & 0 & 1 \end{pmatrix} \\ \mathbf{Q} = \begin{pmatrix} \cos\phi & \sin\phi & 0 \\ -\sin\phi & \cos\phi & 0 \\ 0 & 0 & 1 \end{pmatrix} \quad \text{and} \quad \mathbf{Q}^{-1} = \mathbf{Q}^T = \begin{pmatrix} \cos\phi & -\sin\phi & 0 \\ \sin\phi & \cos\phi & 0 \\ 0 & 0 & 1 \end{pmatrix} \end{gathered} \tag{2.25}$$

Cartesian $\{x, y, z\}$ and Spherical $\{r, \theta, \phi\}$ Coordinates:

$$\begin{gathered} x = r\sin\theta\cos\phi \quad y = r\sin\theta\sin\phi \quad z = r\cos\theta \\ r = \sqrt{x^2 + y^2 + z^2} \quad \tan\theta = \sqrt{x^2 + y^2}/z \quad \tan\phi = y/x \end{gathered} \tag{2.26}$$

The metrics are:

$$h_r = 1 \quad h_\theta = r \quad h_\phi = r\sin\theta$$

yielding the Jacobian and the transformation matrix

$$\mathbf{J} = \begin{pmatrix} \sin\theta\cos\phi & r\cos\theta\cos\phi & -r\sin\theta\sin\phi \\ \sin\theta\sin\phi & r\cos\theta\sin\phi & r\sin\theta\cos\phi \\ \cos\theta & -r\sin\theta & 0 \end{pmatrix} \quad \text{and}$$

$$\mathbf{J}^{-1} = \mathbf{J}^{T} = \begin{pmatrix} \sin\theta\cos\phi & \sin\theta\sin\phi & \cos\theta \\ \cos\theta\cos\phi/r & \cos\theta\sin\phi/r & -\sin\theta/r \\ -\sin\phi/r\sin\theta & \cos\phi/r\sin\theta & 0 \end{pmatrix} \tag{2.27}$$

$$\mathbf{Q}^{-1} = \mathbf{Q}^{T} = \begin{pmatrix} \sin\theta\cos\phi & \cos\theta\cos\phi & -\sin\phi \\ \sin\theta\sin\phi & \cos\theta\sin\phi & \cos\phi \\ \cos\theta & -\sin\theta & 0 \end{pmatrix} \quad \text{and}$$

$$\mathbf{Q} = \begin{pmatrix} \sin\theta\cos\phi & \sin\theta\sin\phi & \cos\theta \\ \cos\theta\cos\phi & \cos\theta\sin\phi & -\sin\theta \\ -\sin\phi & \cos\phi & 0 \end{pmatrix} \tag{2.28}$$

Cartesian $\{x, y, z\}$ and Wire $\{\sigma, \phi, \psi\}$ Coordinates:

$$x = (b + \sigma\cos\psi)\cos\phi \quad y = (b + \sigma\cos\psi)\sin\phi \quad z = \sigma\sin\psi$$
$$\sigma = \sqrt{(\sqrt{x^2 + y^2} - b)^2 + z^2} \quad \tan\phi = y/x \quad \tan\psi = z/(\sqrt{x^2 + y^2} - b) \tag{2.29}$$

The metrics are:

$$h_\sigma = 1 \quad h_\phi = b + \sigma\cos\psi \quad h_\psi = \sigma$$

yielding the Jacobian and the transformation matrix

$$\mathbf{J} = \begin{pmatrix} \cos\phi\cos\psi & -(b + \sigma\cos\psi)\sin\phi & -\sigma\cos\phi\sin\psi \\ \sin\phi\cos\psi & (b + \sigma\cos\psi)\cos\phi & -\sigma\sin\phi\sin\psi \\ \sin\psi & 0 & \sigma\cos\psi \end{pmatrix} \quad \text{and}$$

$$\mathbf{J}^{-1} = \mathbf{J}^{T} = \begin{pmatrix} \cos\phi\cos\psi & \sin\phi\cos\psi & \sin\psi \\ -\sin\phi/(b + \sigma\cos\psi) & \cos\phi/(b + \sigma\cos\psi) & 0 \\ -\cos\phi\sin\psi/\sigma & -\sin\phi\sin\psi/\sigma & \cos\psi/\sigma \end{pmatrix} \tag{2.30}$$

$$\mathbf{Q}^{-1} = \mathbf{Q}^{T} = \begin{pmatrix} \cos\phi\cos\psi & -\sin\phi & -\cos\phi\sin\psi \\ \sin\phi\cos\psi & \cos\phi & -\sin\phi\sin\psi \\ \sin\psi & 0 & \cos\psi \end{pmatrix} \quad \text{and}$$

$$\mathbf{Q} = \begin{pmatrix} \cos\phi\cos\psi & \sin\phi\cos\psi & \sin\psi \\ -\sin\phi & \cos\phi & 0 \\ -\cos\phi\sin\psi & -\sin\phi\sin\psi & \cos\psi \end{pmatrix} \tag{2.31}$$

2.1.3 A Measure of Thickness of the Toroidal Ring: Ω

The thickness measure, $\Omega \equiv 2\ln(2\pi b/a)$ expresses the aspect ratio of the toroid, b/a, in a simple way. The natural log turns a very large aspect ratio into easy to remember numbers; the larger the number, the thinner the toroid. Loops with $\Omega \geq 10$ may be considered "thin". Table 2.1 gives equivalent aspect ratios for various thickness measures (Fig. 2.3).

Table 2.1 Equivalent aspect ratios for various values of Ω with same loop radius, b

Ω	12	11	10	9	8	7	6
b/a	64.2	38.9	23.6	14.3	8.7	5.3	3.2

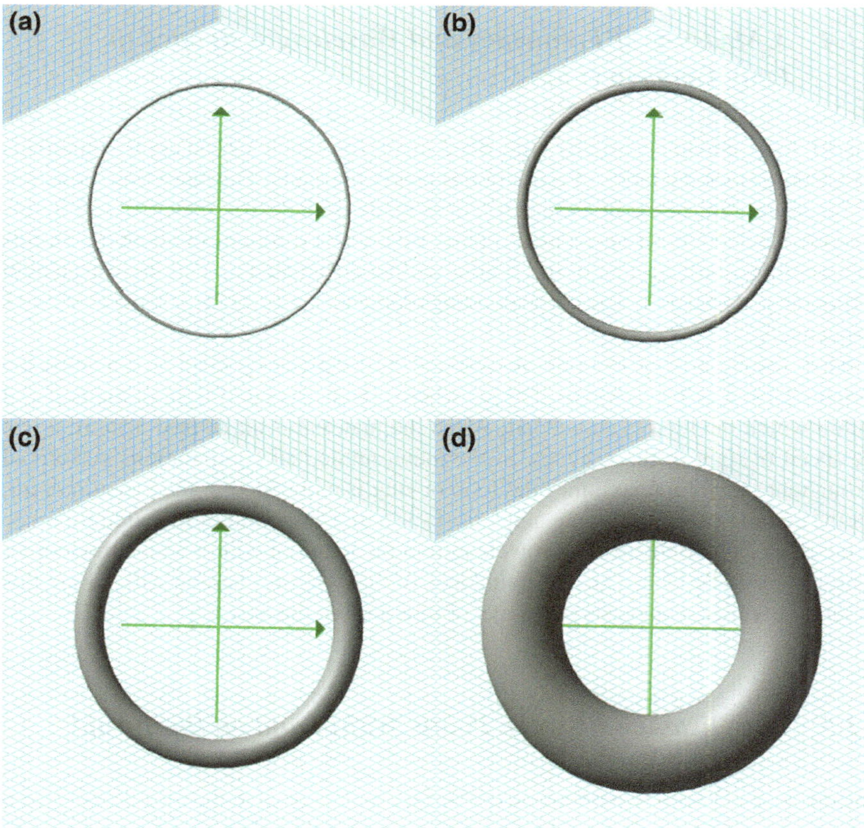

Fig. 2.3 Rings with the same radii, but different aspect ratios. Specifically, **a** $\Omega = 12$, **b** $\Omega = 10$, **c** $\Omega = 8$, and **d** $\Omega = 6$

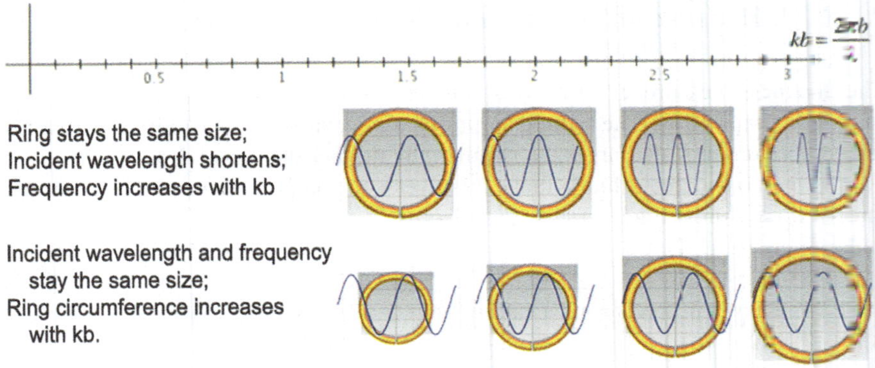

Fig. 2.4 Two interpretations of the variable k_b

2.1.4 A Measure of Frequency and Wavelength Related to Loop Geometry: k_b

The RF loop antenna is usually embedded in air, for which a good approximation is the vacuum since the index of refraction is nearly the same. The nano-ring however, is sometimes embedded in a dielectric or is made of dielectric material, and in this case the wavelength of an incident plane wave has a different momentum. In this volume, then, the wavelength in vacuum will be designated λ_0 and the wavelength within the material as λ. The corresponding momenta are $k_0 \equiv 2\pi/\lambda_0$ and $k = 2\pi/\lambda$ respectively.

Most results arising from an analysis of the loop can be displayed against the variable $k_b \equiv k \times b = \omega b/v = 2\pi b/\lambda$, where v is the speed of light of the material within which the loop is embedded. When the material is air or the vacuum, $v = c$ the speed of light in the vacuum. In the main, numerical computations performed throughout the volume assumes the loop or ring is embedded in the vacuum or air, but the equations are kept more general.[1] k_b is a unit-less ratio of the circumference of the ring, $2\pi b$, to the driving wavelength in the material within which the loop or nano-ring is embedded. The horizontal axis of many figures hereon will feature this variable. Figure 2.4 suggests two ways of interpreting the axis.

2.1.5 The Distance Between Points Within and on the Toroid

The distance between two points on the surface and internal to the toroid arises from time to time in the analyses. The two points are usually taken as a "source" point with vector $\vec{S}$ from the origin and "observation" point, $\vec{O}$ from the origin with the

[1] King et al. have an interesting paper on the topic written in 1964 [1].

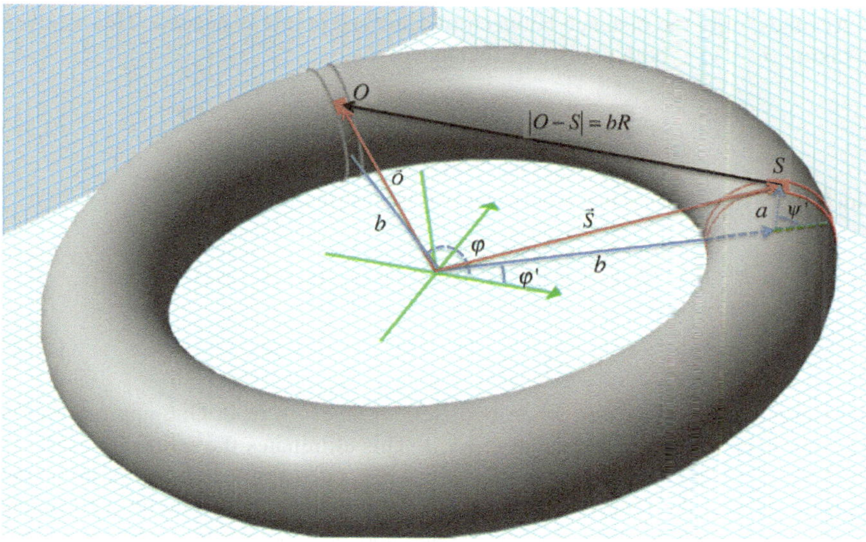

Fig. 2.5 The distance, bR, from source to observation

former coordinates represented by primed coordinates, $\{\sigma', \phi', \psi'\}$ and the latter by unprimed coordinates, $\{\sigma, \phi, \psi\}$. Using this terminology, for example, the exact distance from the source to the observation point, R, in Fig. 2.5 may be determined in the following way.

Using Eq. (2.29), the vector from the centre of the loop to the source point is

$$\vec{S} = [b + \sigma'\cos\psi']\cos\phi'\hat{x} + [b + \sigma'\cos\psi']\sin\phi'\hat{y} + \sigma'\sin\psi'\hat{z} \tag{2.32}$$

and from the centre to the observation point

$$\vec{O} = [b + \sigma\cos(\psi)]\cos(\phi)\hat{x} + [b + \sigma\cos(\psi)]\sin(\phi)\hat{y} + \sigma\cos(\psi)\hat{z} \tag{2.33}$$

After some algebra, the length of the vector $(\vec{O} - \vec{S})$ is found to be

$$|\vec{O} - \vec{S}|^2 \equiv R^2 = 4b\left[b + \sigma'\cos\psi' + \sigma\cos\psi + \frac{1}{b}(\sigma'\cos\psi')(\sigma\cos\psi)\right]\sin^2(\phi - \phi') \\ + \sigma'^2 + \sigma^2 - 2\sigma\sigma'\cos(\psi - \psi') \quad \text{Exact} \tag{2.34}$$

This is exact for all thicknesses of wire. The value of R goes to zero at the center of the wire at the source point. This may introduce either an indeterminacy (a removable singularity) or lead to a true pole, since R appears in the denominator. It is however the standard singularity that appears in all $1/R$ physical problems, as occurs often in E&M. In our case, we attempt to eliminate it.

At times, only the distance between two surface points is needed. This is given by taking $\sigma \to a$ at both points, we then have

$$\sigma'^2 + \sigma^2 - 2\sigma\sigma'\cos(\psi - \psi') \to 4a^2\sin^2\frac{(\psi - \psi')}{2} \tag{2.35}$$

which gives the exact distance between two points on the surface

$$|\vec{O} - \vec{S}|^2 \equiv R^2 = 4b\left[b + a(\cos\psi' + \cos\psi) + \frac{a}{b/a}(\cos\psi'\cos\psi)\right]\sin^2\frac{(\phi - \phi')}{2} + 4a^2\sin^2\frac{(\psi - \psi')}{2} \quad \text{Surface to Surface exact} \tag{2.36}$$

The value of R goes to zero when both sets of sines go to 0. One approximation is to eliminate the terms in second order a. This eliminates the influence of the ψ variables. The maximum error for R^2 is about 5%, occurring when points are on the opposite sides of the loop. This remains the case until the very unimportant points where $\phi \approx \phi'$, then the difference between ψ and ψ' in the second order term plays a small role.

$$R^2 = 4b\left[b + a(\cos\psi' + \cos\psi)\right]\sin^2\frac{(\phi - \phi')}{2} \quad \text{Surface to surface approximation 1} \tag{2.37}$$

Alternatively, another approximation keeps both sets of sines, but can be in substantial error between most points on the ring. When points are on opposite sides of the loop, R^2 can be off by as much as 40%.

$$R^2 = 4b^2\sin^2\frac{(\phi - \phi')}{2} + 4a^2\sin^2\frac{(\psi - \psi')}{2} \quad \text{Surface to surface approximation 2} \tag{2.38}$$

A third approximation, reduces the error to less than 3% in the worst case.

$$R^2 = 4b\left[b + a(\cos\psi' + \cos\psi)\right]\sin^2\frac{(\phi - \phi')}{2} + 4a^2\sin^2\frac{(\psi - \psi')}{2} \quad \text{Surface to surface approximation 3} \tag{2.39}$$

Another approximation, reduces the error to less than 1% between most points along the wire, but approaches a 5% error rather abruptly as the sine goes to zero. This is perhaps the best approximation.

$$R^2 = 4b\left[b + a(\cos\psi' + \cos\psi)\right]\sin^2\frac{(\phi - \phi')}{2} \quad \text{Surface to surface approximation 4} \tag{2.40}$$

The root of this is

$$R = 2b\sqrt{1 + \frac{a}{b}(\cos\psi' + \cos\psi)}\left|\sin\frac{(\phi - \phi')}{2}\right| \tag{2.41}$$

or to very little error, and perhaps more useful

$$R = 2b\left(1 + \frac{a}{2b}(\cos\psi' + \cos\psi)\right)\left|\sin\frac{(\phi - \phi')}{2}\right| \tag{2.42}$$

A final approximation produces an error of 20% in R^2 between most points along the wire than the exact form, but eliminates the dependence of ψ.

$$R^2 = 4b\left[b + a\cos\psi'\right]\sin^2\frac{(\phi - \phi')}{2}$$
$$\text{Surface to surface approximation 5} \tag{2.43}$$

There are also times when the exact R is used under the thin-wire approximation, explained more fully in Sect. 2.5.3. In this case, $\sigma \rightarrow 0$ everywhere and we have only

$$R^2 = 4b^2\sin^2\frac{(\phi - \phi')}{2} \quad \text{Thin-wire approximation A} \tag{2.44}$$

This is singular as $\phi \rightarrow \phi'$; the addition of the actual wire radius, a, a small value in this case can be used to prevent the singularity if desired:

$$R^2 = 4b^2\sin^2\frac{(\phi - \phi')}{2} + a^2 \quad \text{Thin-wire approximation B} \tag{2.45}$$

2.2 Useful Expressions of Maxwell's Equations for Toroidal Rings

Some physical parameters of materials are needed throughout the text. They express several important properties that determine how materials respond to electromagnetic phenomena. Although these properties arise from the chemistry of the material, a mathematical formalism predicts the material's behaviour and therefore describes these properties in terms of functions and constants.

Maxwell's equations are nearly impossible to solve for real materials, because of these characteristics. The electric field, $\vec{E}$, and the magnetic field, $\vec{H}$, represent the average of the effect created by their source, a charge distribution of some sort. Yet the fields have meaning in their own right, distinct from and independent of the charge distributions that created them. Moving charge (currents) radiate energy and momentum, and these are carried away by an electric field and a magnetic field.

When these fields pass through the surface of a material, they affect the charge distributions within the material, causing them to move, and these effects in turn affect the incoming fields themselves. This infinite regress ultimately means that Maxwell's Equations within the material are non-linear.

These non-linear effects are captured in Maxwell's equation by the electric charge displacement vector, $\vec{D}$, and the magnetic induction vector, $\vec{B}$. The displacement vector arises from the movement of charge "bound" to the atoms that embody the molecular structure of the material. Easily polarised atoms constitute materials that respond to external $\vec{E}$ fields by swinging themselves against the applied vector field, pivot-like in place in the 3D molecular structure, leaving the total field inside the material weaker than the applied field. The rate at which this swinging or displacement occurs, $\partial\vec{D}/\partial t$, across the material constitutes an additional current that adds to the "free" charge current, $\vec{J}$, in the material. The free charge arises from electrons so energetic that they cannot remain bound to the atomic structure of the material and hence move from one atom to another. The applied $\vec{E}$ field directly dictates their direction and acceleration. We work in this volume only with the linear effects.

Maxwell's equations under all conditions at all frequencies are given by

$$\vec{\nabla}\times\vec{E} = -\frac{\partial\vec{B}}{\partial t} \qquad \vec{\nabla}\cdot\vec{D} = \rho_v$$
$$\nabla\times\vec{H} = \frac{\partial\vec{D}}{\partial t} + \vec{J} \qquad \vec{\nabla}\cdot\vec{B} = 0 \tag{2.46}$$

Here, ρ_v is a volume density of bound charge. The current and charge densities are related by the charge conservation equation

$$\vec{\nabla}\cdot\vec{J} = -\frac{\partial\rho_v}{\partial t} \tag{2.47}$$

where the surface enclosing the charge volume surrounds a single point and $\vec{\nabla}\cdot\vec{J}$ is the divergence of the current at that point. Power flow is given by Poynting's vector

$$\vec{S} = \vec{E}\times\vec{H} \tag{2.48}$$

Easily remembered units for $\vec{E}$ are "volts/metre"; for $\vec{D}$, "(ampere-seconds)/metre^2"; for H, "amperes/metre"; and for B, "(volt-seconds)/metre^2".

2.2.1 Materials Linear in Maxwell's Equations

Many experiments have shown that the vacuum responds linearly to the imposition of energetic electromagnetic fields. That is, the relationships between $\vec{D}$ and $\vec{E}$ and between $\vec{H}$ and $\vec{B}$ are linear expressions: $\vec{D} = \varepsilon_0\vec{E}$ and $\vec{B} = \mu_0\vec{H}$ respectively, where

$\varepsilon_0 = 8.854 \times 10^{-12}$ farads/metre and $\mu_0 = 4\pi \times 10^{-7}$ henries/metre. These give correct quantitative results for the vacuum. This linearity in general is not preserved inside materials; instead the relationship is much more complicated there. A very detailed and complicated model is given in Jackson [2, Sect. 1.1]. A slightly less complicated model, but a model as exacting for most materials (and perhaps easier to understand), uses two 3×3 tensors, $\boldsymbol{\varepsilon}_{\mathbf{b}}$ and μ, in which the elements are spatial functions dependent on direction within the medium. In other words,

$$\vec{D} = \varepsilon_b \vec{E} \quad \texttt{and} \quad \vec{B} = \mu \vec{H} \tag{2.49}$$

Here, $\boldsymbol{\varepsilon}_{\mathbf{b}}$ is called the "permittivity tensor" of the material due to the bound charge, and μ is called the "permeability tensor". Equations (2.49) are called the "constitutive equations" of the material.

Fortunately, many materials commonly used for loops and rings are not so complicated, so that these two tensors become diagonal and, in fact, carry the same value or function of frequency in all three coordinate directions; that is, these materials are isotropic. In other words,

$$\begin{aligned} \vec{D} &= \varepsilon_b \vec{E} \equiv \varepsilon_{rb}\varepsilon_0 \vec{E} \\ \vec{B} &= \mu \vec{H} \equiv \mu_r \mu_0 \vec{H} \end{aligned} \tag{2.50}$$

The functionality is swept up into the two unit-less functions of frequency $\mu_r(\omega)$ and $\varepsilon_{rb}(\omega)$. These are established so that in the vacuum, $\mu_r = 1.0$ and $\varepsilon_{rb} = 1.0$. With this linearization, Maxwell's Equations become

$$\begin{aligned} &(a) \quad \vec{\nabla} \times \vec{E} = -\frac{\partial \vec{B}}{\partial t} \quad (b) \quad \vec{\nabla} \cdot \vec{E} = \frac{\rho_v}{\varepsilon_b} \\ &(c) \quad \vec{\nabla} \times \vec{B} = \mu\varepsilon_b \frac{\partial \vec{E}}{\partial t} + \mu \vec{J} \quad (d) \quad \vec{\nabla} \cdot \vec{B} = 0 \\ &\texttt{where} \quad \mu = \mu_r \mu_0 \quad \texttt{and} \quad \varepsilon_b = \varepsilon_{rb}\varepsilon_0 \end{aligned} \tag{2.51}$$

The choice of writing the linear Maxwell's equations using the magnetic induction vector, $\vec{B}$, instead of the magnetic field, $\vec{H}$, is due to preference. It comes from a preference in writing the force equation that appears in many textbooks as:

$$\vec{F} = q(\vec{E} + \vec{v} \times \vec{B}) \tag{2.52}$$

Here q is the charge affected by the force and $\vec{v}$ is its instantaneous velocity.

2.2.2 Linear Wave Equations

Substituting Maxwell's Equations, one into another, and rearranging leads to wave equations, one for $\vec{E}$ and one for $\vec{B}$. Taking first the curl of Eq. (2.51) (c), substituting (a), and using a vector identity leads to:

$$\vec{\nabla} \times \vec{\nabla} \times \vec{B} = \mu\varepsilon_b \frac{\partial(\vec{\nabla} \times \vec{E})}{\partial t} + \mu(\vec{\nabla} \times \vec{J})$$
$$\vec{\nabla}(\vec{\nabla} \cdot \vec{B}) - \vec{\nabla}^2 \vec{B} = -\mu\varepsilon_b \frac{\partial^2 \vec{B}}{\partial t^2} + \mu(\vec{\nabla} \times \vec{J}) \tag{2.53}$$

The first term on the left-hand side is zero by Eq. (2.51) (d). Free charge follows the applied $\vec{E}$ field linearly and therefore, the free charge current density, $\vec{J}$, is directly proportional to the electric field. The proportionality has units of conductivity, and $\vec{J}$ is therefore written $\sigma\vec{E}$. This can be written as a 3×3 tensor function of position. When the material is isotropic, the tensor reduces to $\vec{J} = \sigma\vec{E}$, and σ becomes a function of frequency, $\sigma = \sigma(\omega)$. Its DC value is σ_0. Substitute into Eq. (2.53), use Eq. (2.51) (a) again and then rearrange:

$$\vec{\nabla}^2 \vec{B} - \mu\varepsilon_b \frac{\partial^2 \vec{B}}{\partial t^2} + \mu\sigma(\vec{\nabla} \times \vec{E}) = 0$$
$$\left(\vec{\nabla}^2 - \mu\varepsilon_b \frac{\partial^2}{\partial t^2} - \mu\sigma \frac{\partial}{\partial t}\right)\vec{B} = 0 \tag{2.54}$$

Similarly, the electric field has a corresponding equation

$$\left(\vec{\nabla}^2 - \mu\varepsilon_b \frac{\partial^2}{\partial t^2} - \mu\sigma \frac{\partial}{\partial t}\right)\vec{E} = 0 \tag{2.55}$$

The material is isotropic and therefore the solution for $\vec{B}$ is a wave moving in the direction of power flow, $\hat{r}$, of the form

$$|\vec{B}| = B_o e^{j\omega t} e^{\pm j\vec{k}\cdot\vec{r}} \tag{2.56}$$

where k is the propagation constant along the direction $\hat{r}$ (see Sect. 2.3 following). The negative sign on $\vec{k}$ gives waves moving positively along the direction $\vec{r}$ while the positive sign gives waves moving negatively. At times the literature uses one or the other, and it is important to know which is being used in the given context.

Orienting the coordinate system along $\vec{r}$, using positively travelling waves, $|\vec{E}| = B_o e^{j(\omega t - kr)}$, and $\nabla^2 = \partial^2/\partial r^2$ Eq. (2.54) becomes

$$\left[k^2 - \omega^2\mu\varepsilon_b\left(1 - j\frac{\sigma}{\omega\varepsilon_b}\right)\right]B_o = 0 \tag{2.57}$$

Equation (2.57) can only be zero generally if the portion within the brackets is zero, which gives the complex propagation parameter:

$$k = \omega\sqrt{\mu\left(\varepsilon_b - j\frac{\sigma}{\omega}\right)} \equiv \omega\sqrt{\mu\varepsilon} \tag{2.58}$$

Remember that the permeability, permittivity and conductivity are all functions of frequency.

2.3 Propagation Parameters

To keep everything clear, four forms of the permittivity functions may be distinguished and are often used in the literature. The point is to distinguish the free charge from the bound charge, since both influence the permittivity and the conductivity, and both give rise to the fundamental behaviour of the material. That behaviour can be quite different depending upon the ratio of bound to free charge. Using $e^{j(\omega t - kr)}$:

1. ε_0, the permittivity of free space, given earlier as 8.85×10^{-12} farads/metre,
2. ε_b, the permittivity due only to the bound charge in the material. A highly dielectric material has a large value of bound charge and ε_b is positive.
3. $\varepsilon \equiv \varepsilon_b - j\sigma/\omega$, the effective permittivity of the material due to both the bound charge and the conductivity of the free charge, as if all of the effects were due to displacement. This is often used for a "lossy material", meaning a material in which some absorption and heating occurs due to mobile electrons.
4. $\varepsilon_f \equiv -\sigma/\omega$, the effective permittivity of the material for very good conductors where $(\sigma/\omega) \gg \varepsilon_b$. The free charge for a perfectly conducting material (PEC) is very large, leading to high conductivity.

We will be using perfectly conducting materials, dielectrics and lossy materials throughout the volume. Understanding the physical differences is critical to understanding the behaviour.

Notice that $\varepsilon = \varepsilon_b + j\varepsilon_f = (\varepsilon_{rb} + j\varepsilon_{rf})\varepsilon_0$, which shows the complex nature of the relative permittivities. Typically the literature writes this as $\varepsilon = (\varepsilon_1 + j\varepsilon_2)\varepsilon_0$ or $(\varepsilon' + j\varepsilon'')\varepsilon_0$. ε_{rb} and ε_{rf} are values relative to ε_0 and therefore called "relative" permittivities.

These distinctions lead to four variations on the propagation parameter, k.

1. $k_0 = \omega\sqrt{\mu_0\varepsilon_0}$, the propagation parameter of the vacuum;
2. $k_{\varepsilon b} = \omega\sqrt{\mu\varepsilon_b}$, the propagation parameter due to the bound charge only (i.e., in a purely dielectric material);
3. $k = \omega\sqrt{\mu\varepsilon}$, as in Eq. (2.58), the propagation parameter using the effective permittivity of a lossy material;

4. $k_f = \omega\sqrt{j\mu\varepsilon_f} = \sqrt{-j\omega\mu\sigma} = (1-j)\sqrt{\omega\mu\sigma/2}$, the propagation parameter for very good conductors. Given a good conductor, where μ is constant, the propagation depends entirely on the conductivity.

The speed of the wave is given by setting the exponent of the travelling wave to zero, since that specifies the periodicity. Doing so, gives

$$kr = \omega t \quad \text{or} \quad v = \frac{r}{t} = \frac{\omega}{k} \tag{2.59}$$

There are, therefore, four variations:

1. $v_0 = \omega/k_0 = 1/\sqrt{\mu_0\varepsilon_0} \equiv c$, the speed in the vacuum (2.99792×10^8 m/s);
2. $v_b = \omega/k_{\varepsilon b} = 1/\sqrt{\mu\varepsilon_b}$, the speed due to the bound charge only (i.e., in a purely dielectric material);
3. $v = \omega/k = 1/\sqrt{\mu\varepsilon}$, the speed using the effective permittivity of a lossy material;
4. $v_f = \omega/k_f = 1/\sqrt{j\mu\varepsilon_f} = 1/\sqrt{-j\omega\mu\sigma} = (1+j)/\sqrt{2\mu\sigma/\omega}$, the speed in very good conductors. This, of course, goes to 0 at DC, since there is no wave propagation.

Since the index of refraction is defined

$$\eta \equiv c/v = \sqrt{(\mu/\mu_0)(\varepsilon/\varepsilon_0)} \equiv \sqrt{\mu_r\varepsilon_r} \tag{2.60}$$

it is related to the permeability, permittivity, and conductivity characteristics of the material. There are four variations:

1. $\eta_0 = c/v_0 = 1.0$, the index in the vacuum;
2. $\eta_b = c/v_b = \sqrt{(\mu/\mu_0)(\varepsilon_b/\varepsilon_0)} = \sqrt{\mu_r\varepsilon_{rb}}$, the index due to the bound charge only (i.e., in a purely dielectric material);
3. $\eta = c/v = \sqrt{(\mu/\mu_0)(\varepsilon/\varepsilon_0)} = \sqrt{\mu_r\varepsilon_r}$, the index using the effective permittivity of a lossy material;
4. $\eta_c = c/v_f = \sqrt{(\mu/\mu_0)(j\varepsilon_f/\varepsilon_0)} = \sqrt{j\mu_r\varepsilon_{rf}}$, the index for very good conductors.

The most general propagation constant, due to both the bound charge and the free charge, is complex, making the index of refraction generally complex.

$$k = \frac{\omega}{v} = \frac{\omega}{c}\eta = k_0(n - j\kappa) \tag{2.61}$$

Since $k = \omega\sqrt{\mu(\varepsilon_b + j\varepsilon_f)} = k_0\sqrt{\mu_r(\varepsilon_{rb} + j\varepsilon_{rf})}$,

$$n^2 - \kappa^2 = \mu_r\varepsilon_{rb}; \quad 2n\kappa = -\mu_r\varepsilon_{rf} \tag{2.62}$$

leading to

$$n^2 = \frac{\mu_r \varepsilon_{rb}}{2} \pm \frac{\mu_r}{2}\sqrt{\varepsilon_{rb}^2 + \varepsilon_{rf}^2} = \frac{\mu_r \varepsilon_{rb}}{2} \pm \frac{\mu_r}{2}|\varepsilon_r|;$$
$$\kappa^2 = -\frac{\mu_r \varepsilon_{rb}}{2} \pm \frac{\mu_r}{2}\sqrt{\varepsilon_{rb}^2 + \varepsilon_{rf}^2} = -\frac{\mu_r \varepsilon_{rb}}{2} \pm \frac{\mu_r}{2}|\varepsilon_r| \quad (2.63)$$

where the subscript "r" refers to the relative values. For non-magnetic materials, $\mu_r = 1.0$, leading to

$$\eta^2 = \varepsilon_r \quad (2.64)$$

2.3.1 Material Impedance and Absorption

By Maxwell's Equation Eq. (2.51) (a), the electric field must also have the same form as $|\vec{B}|$ in Eq. (2.56) with integration constant E_0. Substituting Eq. (2.56) into Eq. (2.51) (c) and using Eq. (2.49) gives

$$\frac{E_0}{H_o} = \frac{k}{\omega\varepsilon} = \frac{\sqrt{\mu\varepsilon}}{\varepsilon} = \sqrt{\frac{\mu}{\varepsilon}} \equiv Z \quad (2.65)$$

This ratio, Z, characterises a balance between the electric field and the magnetic field within the material; it is commonly called the "impedance" of the material. Notice that this can be written

$$Z = \sqrt{\frac{\mu_r}{\varepsilon_r}}\sqrt{\frac{\mu_0}{\varepsilon_0}} = b_x\sqrt{\frac{\mu_0}{\varepsilon_0}} \equiv \xi \quad (2.66)$$

The parameter b_x refers to the balance between the relative permeability and relative permittivity of the material; in this volume it will be called the "balanced reactance" of the material. Since its constitutent parameters are functions of frequency, so it is itself a function of frequency. When the material medium is the vacuum, $b_x = 1.0$; the impedance then equals the impedance of the vacuum:

$$Z_0 = \sqrt{\frac{\mu_0}{\varepsilon_0}} \equiv \xi_0 = 376.773 \ \mathtt{ohms} \quad (2.67)$$

The propagation parameter, k is generally complex.

$$k = k_0\sqrt{\mu_r(\varepsilon_{rb} + j\varepsilon_{rf})} = k_0(n - j\kappa) = \beta + j\alpha \quad (2.68)$$

A substitution of this into Eq. (2.56), gives

$$B_o e^{j(\omega t \pm kr)} = B_o e^{j(\omega t \pm \beta r)} e^{\mp \alpha r} \quad (2.69)$$

The parameter β gives the propagation constant in the material. $\pm\alpha$ implies one wave increasing in magnitude and one wave decaying in magnitude depending upon the sign used. The former is physically impossible. Therefore $-\alpha$ is used, and implies the decay of the wave within the material, an absorption of the incident energy. It determines the degree to which the field dies out, falling off to $1/e$ in the distance $r = 1/\alpha$. The intensity of the field is related to $|E|^2$ and therefore the literature often uses an absorption coefficient of $K = 2\alpha$.

By Eq. (2.68), the complex components of the index of refraction are directly related to the propagation and the absorption:

$$\beta = k_0 n \quad \texttt{and} \quad \alpha = -k_0\kappa \tag{2.70}$$

The real part of the index is simply called "the index" and is associated with the propagation of the wave. A strong index means a strong propagation in the material. κ is called the "extinction coefficient" since it is related to the absorption. A stronger extinction coefficient means a stronger dampening of signal.

The wave penetrates the material to a depth determined by the inverse of the absorption as noted above. This measure is called the "skin depth".

$$\delta \equiv \frac{1}{\alpha} = -\frac{1}{k_0\kappa} = \frac{-c/\omega}{\mathtt{Im}(\eta)} = \frac{-\lambda}{2\pi\,\mathtt{Im}(\eta)} \tag{2.71}$$

Of more immediate interest is the relationship of the skin depth to the wire size of the ring, a. Using the aspect ratio and the thickness measure described in Sect. 2.1.3 and the measure of frequency, k_b described in Sect. 2.1.4:

$$\frac{\delta}{a} = \frac{-b/a}{k_b\mathtt{Im}(\eta)} = \frac{-e^{\Omega/2}}{2\pi k_b\mathtt{Im}(\eta)} \tag{2.72}$$

The skin depth to wire radius ratio will provide a good understanding of whether currents are moving inside the loop material or not. For thin loops, the aspect ratio is very large, thus extinction must also be quite large to prevent deep current flow within the material. As the ring thickens, the aspect ratio in the numerator becomes smaller, implying a shallower current flow. If $k_b < 1.0$, penetration also increases.

2.3.2 Frequency Dependence of Material Parameters

Materials are usually classified as dielectrics, conductors, or insulators. In dielectrics, charge that is bound to ionised molecules dominates current behaviour, while in conductors, charge that is free of molecular and atomic bonds dominate. In insulators molecules are not ionised and charge is tightly bound to the host lattice, making either displacement or free charge current flow impossible under room temperature conditions. Semi-conductors are dielectrics until exposed to IR and visible photons, which push electrons in the outer valence bands of the molecular crystalline structure into the conduction band. They then become conducting, at least until the electrons fall back to the valence band, radiating energy as they do so.

These differences in materials lead to variations in response over the frequency spectrum from DC to the optical region and this response is measured in different ways in each frequency regime. Indeed, the equipment to do so becomes significantly more complicated as frequency increases. The THz region is the last and most recent to receive concentrated attention.

One simple way to measure this behaviour involves an incident EM wave at an angle, ϕ from the vacuum, upon the surface of a material with complex index of refraction, η [3, Sect. 2.2.2] [4, Sect. 3.2.2] [5, Sect. 2.7.1]. Fresnel's formulas give the ratios of the normal and parallel components of the reflected Electric field.

$$\begin{aligned}\frac{E_{rn}}{E_n} &= \frac{\cos\phi - \sqrt{\eta^2 - \sin^2\phi}}{\cos\phi + \sqrt{\eta^2 - \sin^2\phi}} \equiv r_n \\ \frac{E_{rp}}{E_p} &= \frac{\eta^2\cos\phi - \sqrt{\eta^2 - \sin^2\phi}}{\eta^2\cos\phi + \sqrt{\eta^2 - \sin^2\phi}} \equiv r_p\end{aligned} \tag{2.73}$$

Here, the ratios r_n and r_p are the amplitude reflection coefficients. The intensity reflection coefficients are $R_n = |r_n|^2$ and $R_p = |r_p|^2$ respectively. Using Snell's Law, $\sin\phi = \eta\sin\alpha$, the reflection coefficients become

$$R_n = \frac{\sin^2(\alpha - \phi)}{\sin^2(\alpha + \phi)} \quad \text{and} \quad R_p = \frac{\tan^2(\alpha - \phi)}{\tan^2(\alpha + \phi)} \tag{2.74}$$

In the case of normal incidence, $\phi = 0$, the Fresnel formulas indicate that these coefficients are fully dependent on the index of refraction of the material:

$$R_n = R_p = R = \frac{(n-1)^2 + \kappa^2}{(n+1)^2 + \kappa^2} \tag{2.75}$$

Keeping $\phi = 0$ and rewriting the Fresnel formulas as

$$r_n = r_p = r = \frac{n - j\kappa - 1}{\eta - j\kappa + 1} = |r|e^{j\theta} \tag{2.76}$$

gives the reflection coefficients in polar form. The angle θ is the phase shift of the incident wave on reflection. Rationalizing this gives the tangent of the shift:

$$\tan\theta = \frac{-2\kappa}{n^2 + \kappa^2 - 1} \tag{2.77}$$

Equations (2.75) and (2.77) allow for the direct determination of the complex index using reflectance and transmittance measurements [6]; also see [7, Chap. 4.8].

Low Frequency Results—Dielectrics. In this frequency range, the index of refraction for dielectrics depends upon the response of the bound charge. Typically the permittivity is measured directly using capacitive techniques [8] and the index is calculated from that. The permittivity is real and constant. Since $\varepsilon_{rb} \gg \varepsilon_{r\_}$, it may be represented as ε_{rb} in this regime. Therefore, $\eta = \sqrt{\varepsilon_{rb}}$, $\kappa = 0$, $\alpha = 0$, and $R = (\sqrt{\varepsilon_{rb}} - 1)^2/(\sqrt{\varepsilon_{rb}} + 1)^2$. The reflectance is between 50 and 60% for both silicon and germanium in this region. With the absorption at 0, waves deeply penetrate the material.

Low Frequency Results—The Noble Metals. On the other hand, the low frequency response for metals depends upon the response of the free charge; a measurement of DC conductivity via charge flow is therefore straightforward. A noble metal, such as copper, silver or gold, has high conductivity in the low frequency regime. The DC conductivity, σ_0, is roughly 10^7 or 10^8 Siemens-meter^{-1}. Since the bound charge in perfect metals is zero, the bound charge permittivity $\varepsilon_{rb} = 1.0$ at DC. And since $\mu_r = 1.0$ for the Noble metals, the index and extinction coefficient are about equal according to Eq. (2.63): $n^2 = \kappa^2 \approx \sigma_0/(2\omega\varepsilon_0)$. The free charge permittivity, $\varepsilon_{rf} = -\sigma_0/(\omega\varepsilon_0)$, undefined at DC, is finite, yet very high in magnitude, in the low frequency regime into the 100 GHz region (see Fig. 2.6). The model is too simplistic to capture the proper upper cutoff frequency. It suggests that the free charge permittivity for gold ($\sigma_0 = 45 \times 10^6$) drops to a factor of 100 times more than the bound charge permittivity at $f = \sigma_0/(2\pi 100\varepsilon_0) = 8 \times 10^{15}$ Hz, much higher than the more detailed model given below. The cause of the cutoff is the inertia of the free charge itself, as discussed below.

Mid-frequency Range—Dielectrics. The real, constant nature of the index for dielectrics remains out to the THz region unless it shows a type of resonance prominent with ionic bonding in some compounds. These are called "reststrahlen resonances" (residual radiation resonances). Gallium Arsenide shows a number of resonances at about 8 THz, for example. There can be resonances of other sorts throughout the GHz region as well. These may be modelled using Lorentz resonances, of the kind that is described below when the inter-band optical resonances are discussed. The permittivity of dielectrics may continue to be represented through the mid-range by their low frequency model unless they show these resonances.

Mid-frequency Range—The Noble Metals. The low frequency model in metals begins to falter when the metal's free charge begins to lag the incident wave, causing a phase shift in the resulting current and therefore an imaginary part to the measurement. Defect scattering in the crystalline structure also occurs, leading to energy loss. The very large real values of the permittivity plunges into measurable territory at about 150–250 GHz depending on the metal. Here the real part of the conductivity begins to drop while the imaginary part rises to a maximum, only to drop off again slowly as it passes into the THz range (see Fig. 2.6). Metals are generally represented in this region using the Drude model, a solution to the classical 2nd order damping equation

$$m\ddot{x} + \frac{m}{\tau}\dot{x} = -eE_0 e^{j\omega t} \tag{2.78}$$

where m is the mass of the electron (and in the optical region, the effective mass gleaned from band theory). The parameter τ is called the "damping time" and is due to absorption. The latter is related to a damping frequency $\Gamma_0 = 1/2\tau$. The solutions for the electron's classical position and velocity are

$$x = \frac{eE_0}{m\omega(\omega - j/\tau)} e^{j\omega t} \quad \text{and} \quad v = \dot{x} = \frac{-eE_0\tau}{m(1 + j\omega\tau)} e^{j\omega t} \tag{2.79}$$

The current density, that is, the number of electrons per square metre passing a given perpendicular surface per second, is related to the electron speed and the free charge conductivity by

$$J = -Nev = \frac{Ne^2 E_0 \tau}{m(1 + j\omega\tau)} \equiv \sigma E_0 = \left(\frac{\sigma_0}{1 + j\omega\tau}\right) E_0 \quad \text{where} \quad \sigma_0 = Ne^2\tau/m \tag{2.80}$$

The parameter N is the free electron density, which metals have, of course, at the low frequencies. Dielectrics do not have a free electron density, and this is the one reason that the Drude model is not appropriate for dielectrics. Another is that dielectrics exhibit resonances that metals do not, so the governing equation includes those resonances (see below). The conductivity is simplified by rationalizing Eq. (2.80):

$$\sigma(\omega) = \frac{\sigma_0}{\sqrt{1 - (\omega\tau)^2}} - j\frac{\sigma_0\omega\tau}{\sqrt{1 - (\omega\tau)^2}} = \sigma' + j\sigma'' \tag{2.81}$$

The phase shift in the current is therefore negative, indicating a lag behind the impinging electric field; this is due to the inherent inertia of the electrons in the metal. All Drude like models show this lag in the current.

From Sect. 2.3, the permittivity for metals is related to the free charge, so the loss at the mid-frequency range can be expressed in a finite value for the conductivity; it becomes a function of frequency:

$$\begin{aligned} \varepsilon = \varepsilon_0 - j\frac{\sigma(\omega)}{\omega} &= \varepsilon_0\left(1 - j\frac{\sigma_0}{\omega\varepsilon_0(1 + j\omega\tau)}\right) \\ &= \varepsilon_0\left(1 - \frac{\omega_p^2}{\omega(\omega - j2\Gamma_0)}\right) = \varepsilon_0\varepsilon_r \end{aligned} \tag{2.82}$$

An alternative way of writing the conductivity then is

$$\sigma(\omega) = j\omega\varepsilon_0(\varepsilon_r - 1) = \frac{-j\omega_p^2\varepsilon_0}{\omega - j2\Gamma_0} \tag{2.83}$$

Note that

$$\sigma(0) = \frac{\omega_p^2 \varepsilon_0}{2\Gamma_0} = \sigma_0 \tag{2.34}$$

The parameter $\omega_p^2 = Ne^2/m\varepsilon_0$ is called the "plasma" frequency. It must be in radians for proper computation of the conductivity, but it is usually given in eV, so be careful of the placement of the 2π when converting.

Generally, the material's plasma frequency is calculated from physical characteristics and Γ_0 is then determined by calculation. The characteristic parameters of gold, silver and copper are all about the same, giving a plasma frequency about 9 eV and a damping frequency of about 0.01 eV.

Optical Frequency Response—Dielectrics and the Noble Metals. In the IR and the visible regions, measurements of the index of refraction for metals and dielectrics exhibit resonances called "inter-band transitions", where valence electrons in the outer shells energised by incoming photons jump to the crystal's conduction band. In the dielectrics, electrons come from the outermost valence bands, but in the metals, the outer bands have been stripped of electrons at room temperature, and so the transition electrons must come from deeper inner orbits. As a result, these donated electrons are fewer and less energetic in the metals, leading to a lower value conductivity at these high frequencies than the conductivity for the dielectrics and semi-conductors. Conductivity due to the inter-band transitions in the noble metals is at least two orders of magnitude less than at DC (see Fig. 2.8). For example, gold conductivity suddenly jumps from 0 to 0.2 S-m^{-1} at 2.8 eV due to the inter-band transitions. Silicon, on the other hand jumps from 0 to 1.7 at 3.4 eV and eventually reaches 2.6 S-m^{-1} at 4.2 eV.

Many analytical models exist for these inter-band transitions, which can quite accurately represent the permittivity and which are also consistent with the Kramers–Kronig relations of complex entities. The models generally rely on Lorentz-type resonances, the types of responses appearing in classically damped harmonic equations (compare with Eq. (2.79)):

$$\begin{aligned} m\ddot{x} + \frac{m}{\tau_n}\dot{x} + m\omega_n^2 x &= -eE_0 e^{j\omega t} \\ x &= \frac{-eE_0}{m(\omega_n^2 - \omega^2 + j\omega\Gamma_n)} e^{j\omega t} \end{aligned} \tag{2.35}$$

Here $\tau_n \equiv 1/\Gamma_n$. Since the bound charge remains near zero for the Noble metals in the optical range, $e_{rb} = 1.0$ and the associated permittivity is

$$\varepsilon = \varepsilon_0 - j\frac{\sigma}{\omega} = \varepsilon_0 \left(1 + \frac{\omega_p^2}{(\omega_n^2 - \omega^2 + j\omega\Gamma_n)}\right) \tag{2.36}$$

The form found most useful comes by assuming that this is not a single transition but a bandwidth of transitions. The term then needs to be "broadened". This can be done reasonably by separating the term into two parts and applying partial fractions to obtain

$$\frac{\omega_p^2}{(\omega_n^2 - \omega^2 + j\omega\Gamma_n)} \approx \frac{\omega_p^2}{2\omega_n}\left[\frac{1}{\omega_n - \omega + j\Gamma_n} + \frac{1}{\omega_n + \omega - j\Gamma_n}\right] \tag{2.87}$$

since $\Gamma_n \ll (\omega \text{ and } \omega_n)$ in the region where these transitions occur. This model is called the Drude Critical Point (DCP) model [9–14]. We modify it a bit below for our purposes.

Full Spectrum Response Model. The full frequency spectrum is vast and no single model can be expected to cover the entire range accurately. Customarily, then, the approach is to use the model best suited for the range of interest and these models have been indicated above. The range, however, from about 500 GHz through the low THz and into the IR is just now being measured for many materials because of the paucity of methods and lack of equipment required to do so. For this region then, a composite model is appropriate which ties both the low frequency DC conductivity and/or permittivity together with the high frequency optical end and therefore it attempts to map the mid range effectively. The point of this subsection, then, is to provide a full spectrum model, which should be used only in second instance to the models given above.

To do this, an additional damping frequency in the Drude model is needed. The permittivity model for the full spectrum, whether for metals or for semi-conductors where $\mu_r = 1.0$, then appears as:

$$\eta^2 = \varepsilon_r = 1 - f_0\frac{\omega_p^2}{\omega}\left(\frac{1}{\omega - j2\Gamma_0} + \frac{\alpha}{\omega - j2\beta\Gamma_0}\right) + \sum_{n=1}^{\infty}\frac{f_n\omega_p^2}{2\omega_n}\left[\frac{e^{j\pi/\gamma_n}}{(\omega_n - \omega + j\Gamma_n)^{\alpha_n}} + \frac{e^{-j\pi/\gamma_n}}{(\omega_n + \omega - j\Gamma_n)^{\alpha_n}}\right] \tag{2.88}$$

The Drude term is only for metals; hence the coefficient $f_0 = 1$ for metals and 0 for dielectrics. The plasma frequency, ω_p is set by physical characteristics of the material (see Eq. (2.82), whether metal or dielectric. The Drude damping frequency, Γ_0 is varied to help match DC conductivity on the low spectrum end with the interband Lorentz resonance data on the optical end. The DC conductivity with the additional term is now given by

$$\sigma_0 = \frac{f_0\omega_p^2\varepsilon_0}{2\Gamma_0}\left[1.0 + \frac{\alpha}{\beta}\right] \tag{2.89}$$

In the numerator of the Lorentz resonances, the exponential functions are related to the probabilities of the particular band transitions. Coefficients f_n, the resonance parameters ω_n, the transition parameters γ_n and α_n, and the damping frequency parameters Γ_n, used in this volume are listed in Table 2.2 for several noble metals

Table 2.2 Parameter fits for Eq. (2.88) for Gold, Silver and Copper. The parameters Γ_1 and β for Ag were changed slightly from Table 2.1 in [17] to rid the conductivity of a negative imaginary component in the region of 4 eV. The parameter α_3 for Cu was also modified for a better fit

	ω_p (eV)	f_0	Γ_0 (eV)	f_1	ω_1 (eV)	Γ_1 (eV)	γ_1	a_1	f_2	ω_2 (eV)	Γ_2 (eV)	γ_2	a_2
Au	9.0	0.37	0.005	0.20	2.62	0.60	4.00	1.0	0.35	3.70	1.10	4.00	1.0
Ag	9.0	0.94	0.010	0.14	4.32	0.50*	4.62	1.0	0.45	5.50	1.40	12.30	1.0
Cu	8.4	1.00	0.064	0.23	2.30	0.52	3.53	1.0	0.22	3.14	0.95	2.55	1.[illegible]
	σ_0	α	β	f_3	ω_3	Γ_3	γ_3	a_3					
	$\times 10^6$				(eV)	(eV)							
Au	45	1.540	13.18	0.60	7.00	2.20	4.0	1.0					
Ag	63	0.100	0.433*	0.40	7.53	2.10	4.0	1.0					
Cu	60	0.035	0.005	0.32	4.87	1.10	2.7	1.6*					

Table 2.3 Parameter fits for Eq. (2.88) for Silicon, Germanium and Gallium Arsenide

	ε_{rb}	f_1	ω_1 (eV)	Γ_1 (eV)	γ_1	a_1	f_2	ω_2 (eV)	Γ_2 (eV)	γ_2	a_2
Si	11.90	0.017	3.37	0.12	4.20	1.05	0.063	3.77	0.60	5.2[illegible]	1.27
Ge	16.80	0.014	2.11	0.12	11.3	0.71	0.105	2.66	1.54	1.53	1.15
GaAs	12.20	0.040	2.92	0.19	20.00	0.62	0.053	4.39	0.18	35.0	0.40
	ω_p	f_3	ω_3	Γ_3	γ_3	a_3	f_4	ω_4	Γ_4	γ_4	a_4
	(eV)		(eV)	(eV)				(eV)	(eV)		
Si	32.18	0.079	4.27	0.19	−6.61	0.77	0.032	5.20	0.58	6.49	0.98
Ge	30.65	0.140	4.35	0.12	−1.76	0.36	1.760	2.38	0.05	−2.03	0.01
GaAs	31.48	0.010	1.52	4.62	−2.15	0.12	0.100	4.93	0.193	−1.50	0.428

and in Table 2.3 for important semi-conductors. Since there are many parameters to fit, many sets of numbers can fit the data, and one focuses on a particular fit in a particular range where the model will be used the most. The parameters given in the tables are suggested to fit the metallic optical data measured by [15] and the semi-conductor optical data measured by [16].

2.3.3 *Characteristics of Metals and Dielectrics From RF to Optical Wavelengths*

The low frequency conductivity of gold is shown in Fig. 2.6 as an example of the behaviour of a Noble metal. A roll-off of the real part from its DC value appears at about 250 GHz and drops by a factor of 4 as it approaches the low THz region. The imaginary part indicates a negative phase shift.

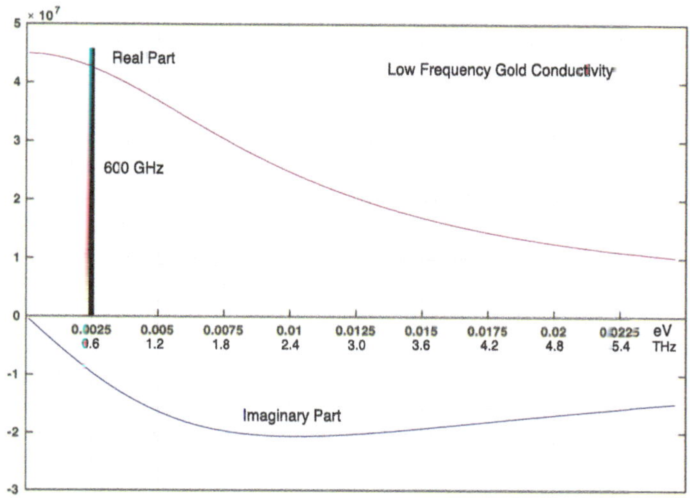

Fig. 2.6 Gold's low frequency conductivity given by the theoretical model. Conductivity at the low frequencies begins to pull away from its constant real value of 45×10^6 at about 0.0001−0.0002 eV (24–48 GHz) to a value four times lower at 0.025 eV (6 THz). Drop off begins about 200 GHz

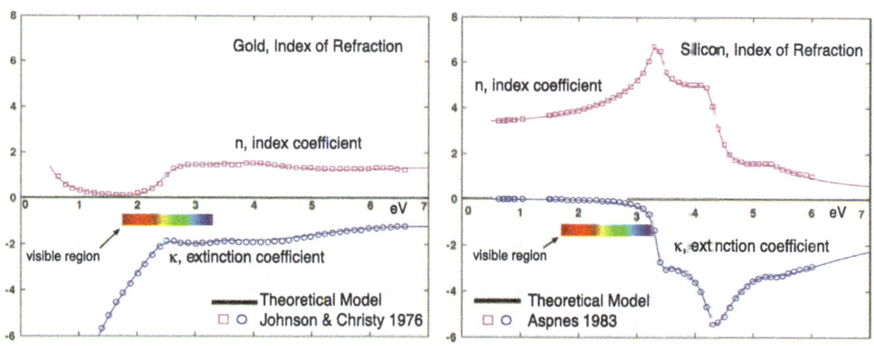

Fig. 2.7 A Comparison of the the Index of Refraction for a Noble Metal and a Semi-conductor in the Optical Region. The two main differences are (1) the larger real part of the index in the semi-conductor throughout, and (2) a negatively very large extinction coefficient in the metal while it is 0 in the semi-conductor until the optical region. Johnson and Christy and Aspnes use positive extinction coefficients

Comparisons of measured data for the index of refraction with results of the model are shown in Fig. 2.7 for gold and for silicon. The index of refraction with corresponding permittivity, and reflectance are shown in the optical region. Conductivity and absorption are shown in Figs. 2.8 and 2.9 for both low frequency and optical regions. Absorption begins its upward climb there as well. This behaviour is due to the precipitous drop in the permittivity. Reflectance remains at 100% through this region and into the optical region until, at the inter-band transitions in the yellow region, it drops to half its value, giving gold its characteristic colour with the

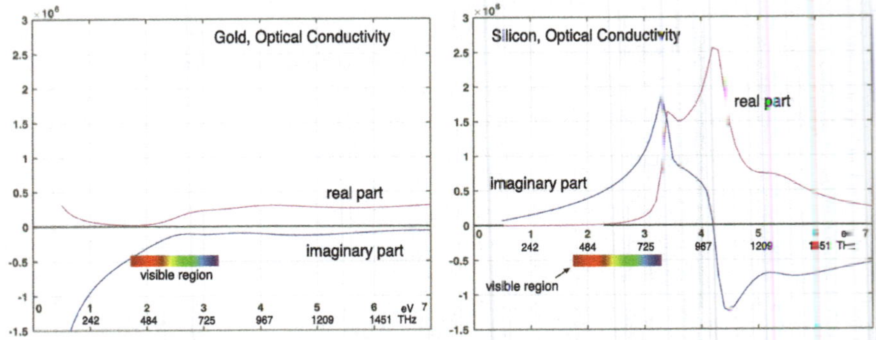

Fig. 2.8 A Comparison of the the conductivity of a noble metal and a semi-conductor in the optical region. The inter-band transitions in gold occur in the yellow region, giving rise to its golden colour. The reason is enhanced conductivity due to the extra electrons in the conduction band. The conductivity of silicon rises sharply in the UV to a value 5 times that of gold in the same region

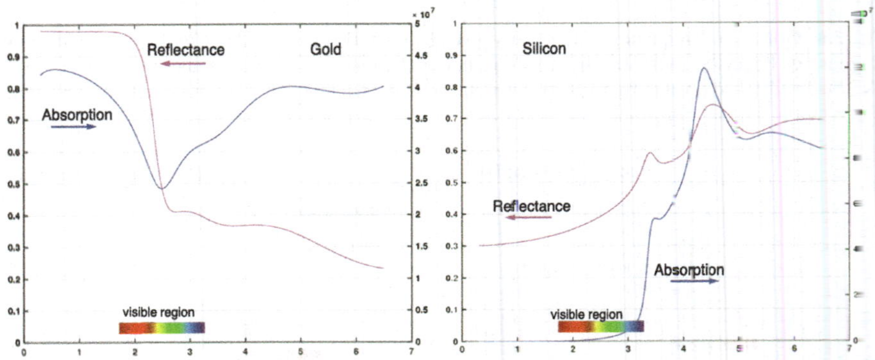

Fig. 2.9 A Comparison of the reflectance and absorption of a Noble Metal and a Semi-conductor in the Optical Region. One sees readily the drop in Reflectance for gold in the yellow region and a sharp increase in Absorption in silicon in the UV

absorption of the blue spectra. Conductivity drops 4 orders of magnitude as it moves toward the optical, and at the transitions picks up some additional conductivity well into the UV.

The skin depth, given by Eq. (2.90) is shown in Fig. 2.10 for several thickness measures at $k_b = 1.0$ for gold and silicon. It may be generally assumed that the field penetrates dielectric wires entirely for even the thickest wires at all frequencies. Not so for metal wires, for which the field does not begin to penetrate significantly until roughly 0.01 EV (2.4 THz) and this for only very thin wires. The field penetrates rapidly however after that for very thin wires. By 24 THz an $\Omega = 12$ wire is thoroughly penetrated. The gradient for thicker wires is less steep, so penetration is not as marked. We can consider an $\Omega = 10$ wire fully penetrated after roughly 48 THz, and an $\Omega = 8$ wire after 145 THz. This is considered the far-infrared region. The optical

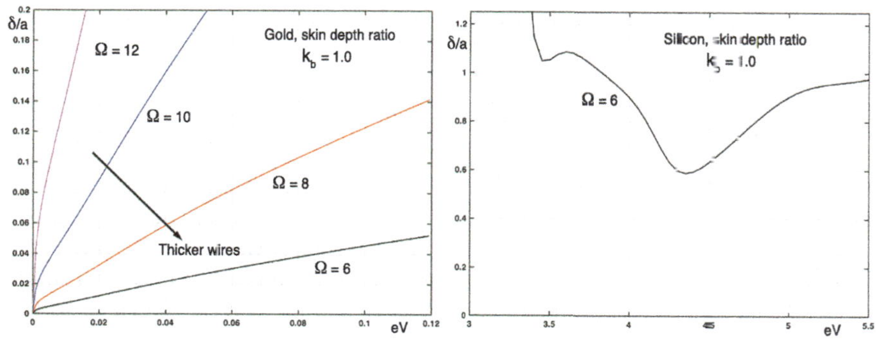

Fig. 2.10 The skin depth ratio for gold and silicon loops of wire radius a. The field penetrates the loop more than 10% beginning at about 0.01 eV (2.4 THz) for very thin wires and at about 0.12 eV (24 THz) for very thick wires. This never occurs for dielectrics; in which case the field penetrates the entire wire for all frequencies, except for the thickest of wires

region begins about 1.5 eV, roughly 360 THz. By the optical region, all wires of whatever material are fully penetrated by impinging $\vec{E}$ fields.

$$\frac{\delta}{a} = \frac{-b/a}{k_b \mathtt{Im}(\eta)} = \frac{-e^{\Omega/2}}{2\pi k_b \mathtt{Im}(\eta)} \tag{2.90}$$

2.4 Vector and Scalar Potentials

The magnetic flux density is always solenoidal; that is, its divergence is always zero: $\nabla \cdot \vec{B} = 0$ (see Maxwell's Equation (2.51) d). Since the divergence of a curl is always 0, $\vec{B}$ can be represented as the curl of a vector potential

$$\vec{B} \equiv \vec{\nabla} \times \vec{A} \tag{2.91}$$

Therefore, from Eq. (2.51) (a)

$$\begin{aligned} \vec{\nabla} \times \vec{E} = -\vec{\nabla} \times \frac{\partial \vec{A}}{\partial t} \quad \mathtt{or} \\ \vec{\nabla} \times \left(\vec{E} + \frac{\partial \vec{A}}{\partial t} \right) = 0 \end{aligned} \tag{2.92}$$

The kernel can be represented by the gradient of a scalar potential, $-\vec{\nabla}\Phi$, because the curl of a gradient is always zero and therefore

$$\vec{E} + \frac{\partial \vec{A}}{\partial t} = -\vec{\nabla}\Phi \quad \text{or}$$

$$\vec{E} = -\vec{\nabla}\Phi - \frac{\partial \vec{A}}{\partial t} \tag{2.93}$$

The negative sign on the gradient is conventional to match the $\vec{E}$ fields of static charge. All of these are vector quantities and therefore have components in three directions. Equation (2.93) is simply a compact form of the two noted Maxwell's Equations. To ensure that the other two equations are included, return to Eq. (2.51) (c), substitute Eqs. (2.91) and (2.93), use the identity $\vec{\nabla} \times \vec{\nabla} \times \vec{A} = \vec{\nabla}(\vec{\nabla} \cdot \vec{A}) - \vec{\nabla}^2 \vec{A}$ and use $e^{j\omega t}$:

$$\begin{aligned}
\vec{\nabla} \times \vec{\nabla} \times \vec{A} &= \mu\varepsilon \frac{\partial}{\partial t}\left(-\vec{\nabla}\Phi - \frac{\partial \vec{A}}{\partial t}\right) + \mu\vec{J} \\
\vec{\nabla}(\vec{\nabla} \cdot \vec{A}) - \vec{\nabla}^2\vec{A} &= -\vec{\nabla}\left[\mu\varepsilon\frac{\partial \Phi}{\partial t}\right] - \mu\varepsilon\frac{\partial^2 \vec{A}}{\partial t^2} + \mu\vec{J} \\
\vec{\nabla}(\vec{\nabla} \cdot \vec{A} + \mu\varepsilon\frac{\partial \Phi}{\partial t}) &= \left(\vec{\nabla}^2 + \omega^2\mu\varepsilon\right)\vec{A} + \mu\vec{J}
\end{aligned} \tag{2.94}$$

Note that the conductivity is buried inside $\vec{J}$ and therefore the permittivity here is the bound charge permittivity, ε, given in the enumerations of Sect. 2.3 (2). There is freedom left on the vector potential, in the sense that the curl, but not the divergence, of $\vec{A}$ has been defined. Jackson [18] describes several different gauges and shows that all of them lead to the same $\vec{E}$ and $\vec{B}$ fields; he also shows how to transform from one to another. If the Lorentz gauge,

$$\vec{\nabla} \cdot \vec{A} \equiv -\mu\varepsilon\frac{\partial \Phi}{\partial t} \tag{2.95}$$

is used, the term on the left hand side will go to zero, leaving the inhomogeneous Helmholtz equation,

$$\left(\vec{\nabla}^2 + k^2\right)\vec{A} = -\mu\vec{J} \tag{2.96}$$

where $k = \omega\sqrt{\mu\varepsilon}$, which we see from the enumerations in Sect. 2.3 is indeed the propagation parameter. Similarly, beginning with Eq. (2.51) (b) and using the identity $\vec{\nabla} \cdot \vec{\nabla}\Phi = \vec{\nabla}^2$ with the Lorentz gauge, obtain

$$\left(\vec{\nabla}^2 + k^2\right)\Phi = -\frac{\rho_v}{\varepsilon} \tag{2.97}$$

Equations (2.93), (2.96), and (2.97) represent a system of equations that encapsulates Maxwell Equations under conditions of linearity in the constitutive equations and the Lorentz gauge. They are presented in the form of a governing equation relating the Electric field with the potentials and two wave equations, one for each potential.

Equations (2.96) and (2.97) have the same form, a form called the Helmholtz Wave Equation. More generally, they have the form

$$Lu(\vec{x}) = -f(\vec{x}) \tag{2.98}$$

where $u(x)$ is the response to the forcing function $f(x)$ under the action of a dynamical system given by the linear differential operator L, in this case $(\vec{\nabla}^2 + k^2)$ and $\vec{x}$ represents a general point in some selected coordinates.

It often happens that differential operators have inverses that are integral operators and so we expect a solution of the form

$$u(\vec{x}) = \int G(x, x') f(\vec{x}') dx' \tag{2.99}$$

where x' is the variable of integration in the same coordinates. The integral can be thought of as a summation of impacts at x created by the forcing function situated at each source point x'. We therefore seek the form of $G(x, x')$. It is found by forming

$$LG(\vec{x}, \vec{x}') = -\delta(\vec{x} - \vec{x}') \tag{2.100}$$

and then integrating. The sifting property of the delta distribution comes into play:

$$\int LG(\vec{x}, \vec{x}') f(\vec{x}') dx' = -\int \delta(\vec{x} - \vec{x}') f(\vec{x}') dx' = -f(\vec{x}) = Lu(\vec{x}) \tag{2.101}$$

Since the operator L is linear and operates with respect to $\vec{x}$, it may be pulled out of the integration over $\vec{x}'$. We then have

$$Lu(\vec{x}) = L \int G(\vec{x}, \vec{x}') f(\vec{x}') dx' \quad \texttt{yielding} \quad u(\vec{x}) = \int G(\vec{x}, \vec{x}') f(\vec{x}') dx' \tag{2.102}$$

The solution to the general Helmholz equation

$$LG(\vec{x}, \vec{x}') = (\vec{\nabla}^2 + k^2) G(\vec{x}, \vec{x}') = -\delta(\vec{x} - \vec{x}') \tag{2.103}$$

is a retarded wave travelling outward from the source point $\vec{x}'$ to the observation point $\vec{x}$. Using $e^{j\omega t}$ (see Eq. (2.56), this solution appears as

$$G(\vec{x}, \vec{x}') = \frac{e^{-jkR}}{4\pi R} \quad \texttt{where} \quad R = |\vec{x} - \vec{x}'| \tag{2.104}$$

Jackson shows all of the details [2, see Sect. 6.4] as do other textbooks. The solution requires that the point not be on a boundary; these waves move away symmetrically, that is spherically. This boundary condition leads to an inconsistency in the use of the solution for electromagnetic fields generated on the surface of wires, since points on the surface of a wire are on a boundary between air and another material, such as metal or a dielectric. The inconsistency is dealt with by distinguishing the regions of applicability: the interior of the wire is a different material than the outside, but waves travel there as well, with the speed, and sometimes the direction, given by the material characteristics. We shall take the solution, Eq. (2.104), as is and assume linear material characteristics, leaving the permeability, μ, and the permittivity, ε, as general parameters in the equations.

The solution for $\boldsymbol{\Phi}$. Applying Eq. (2.102), the solution for Φ is:

$$\Phi = \frac{1}{4\pi\varepsilon}\int_{v'} \rho_{v'} \frac{e^{-jkR}}{R} dv' \tag{2.105}$$

where $R = |\vec{x} - \vec{x}'|$. The integration is over the volume of the source, v', using the differential volume dv'. The parameter k is the appropriate wave propagation function, given in the enumeration of Sect. 2.3. The volume charge density, $\rho_{v'}$ is the density of charge at the source.

The derivation of the Green's function is in general coordinates; one must only be consistent in application. Since the charge density is a scalar function, we may write the function in any coordinate system we wish, as long as we do so in the remainder of the integral as well. The best coordinate basis for the source is the wire system described in Sect. 2.1: $\{\sigma', \phi'$ and $\psi'\}$. The current density, $\vec{J}$ is associated with the charge density through the relationship

$$\vec{\nabla}' \cdot \vec{J}(\sigma', \phi', \psi') = -\frac{\partial \rho_{v'}}{\partial t} = -j\omega\rho_{v'} \tag{2.106}$$

where the prime on ∇' refers to derivatives with respect to the source variables. The current density therefore must also be written in the selected coordinate system. The scalar potential then is

$$\Phi = \frac{j}{4\pi\omega\varepsilon}\int_{V'} (\vec{\nabla}' \cdot \vec{J}(\sigma', \phi', \psi')) \frac{e^{-jkR}}{R} dv' \tag{2.107}$$

The gradient of this is

$$\vec{\nabla}\Phi = \frac{j}{4\pi\omega\varepsilon}\int_{V'} \vec{\nabla}\left((\vec{\nabla}' \cdot \vec{J}(\sigma', \phi', \psi'))\frac{e^{-jkR}}{R}\right) dv' \tag{2.108}$$

where we note that the gradient is not over the source variables.

The solution for $\vec{A}$. Applying Eq. (2.102), the solution for $\vec{A}$ is:

$$\vec{A} = \frac{\mu}{4\pi} \int_{v'} \vec{J}(\sigma', \phi', \psi') \frac{e^{-jkR}}{R} dv' \tag{2.109}$$

The current vector that appears in A is the current at the source in wire coordinates, but it is in Cartesian bases and must be transformed. In other words, we have,

$$\begin{aligned}\vec{J} &= J_x\hat{x} + J_y\hat{y} + J_z\hat{z} \\ &= J_\sigma(\sigma, \phi, \psi)\hat{\sigma} + J_\phi(\sigma, \phi, \psi)\hat{\phi} + J_\psi(\sigma, \phi, \psi)\hat{\psi} \\ &= J_\sigma(\sigma', \phi', \psi')\hat{\sigma} + J_\phi(\sigma', \phi', \psi')\hat{\phi} + J_\psi(\sigma', \phi', \psi')\hat{\psi}\end{aligned} \tag{2.110}$$

The transformation is effected between the source and the observation point by using Eq. (2.31).

$$J_x J_y J_z = \left(J_\sigma \; J_\phi \; J_\psi \right) \mathbf{Q}(\sigma, \phi, \psi) = (J_{\sigma'} \; J_{\phi'} \; J_{\psi'})\mathbf{Q}(\sigma', \phi', \psi') \tag{2.111}$$

and therefore

$$(J_\sigma \; J_\phi \; J_\psi) = (J_{\sigma'} \; J_{\phi'} \; J_{\psi'})\mathbf{Q}(\sigma', \phi', \psi')\mathbf{Q}^T(\sigma, \phi, \psi) \tag{2.112}$$

where

$$\begin{aligned}&\mathbf{Q}(\sigma', \phi', \psi')\mathbf{Q}^T(\sigma, \phi, \psi) \\ &= \begin{pmatrix} \cos\phi'\cos\psi' & \sin\phi'\cos\psi' & \sin\psi' \\ -\sin\phi' & \cos\phi' & 0 \\ -\cos\phi'\sin\psi' & -\sin\phi'\sin\psi' & \cos\psi' \end{pmatrix} \begin{pmatrix} \cos\phi\cos\psi & -\sin\phi & -\cos\phi\sin\psi \\ \sin\phi\cos\psi & \cos\phi & -\sin\phi\sin\psi \\ \sin\psi & 0 & \cos\psi \end{pmatrix}\end{aligned} \tag{2.113}$$

Expanding gives the three components

$$\begin{aligned}J_\sigma &= J_{\sigma'}\left(\cos\psi'\cos\psi\cos(\phi - \phi') + \sin\psi'\sin\psi\right) + J_{\phi'}\cos\psi\sin(\phi - \phi') \\ &\quad - J_{\psi'}\left(\sin\psi'\cos\psi\cos(\phi - \phi') - \cos\psi'\sin\psi\right) \\ J_\phi &= -J_{\sigma'}\cos\psi'\sin(\phi - \phi') + J_{\phi'}\cos(\phi - \phi') + J_{\psi'}\sin\psi'\sin(\phi - \phi') \\ J_\psi &= -J_{\sigma'}\left(\cos\psi'\sin\psi\cos(\phi - \phi') - \sin\psi'\cos\psi\right) - J_{\varphi'}\sin\psi\sin(\phi - \phi') \\ &\quad + J_{\psi'}\left(\sin\psi'\sin\psi\cos(\phi - \phi') + \cos\psi'\cos\psi\right)\end{aligned} \tag{2.114}$$

The derivative with respect to time of the vector potential, assuming $e^{j\omega t}$ and using $k = \omega\sqrt{\mu\varepsilon}$, is

$$\frac{\partial \vec{A}}{\partial t} = \frac{j}{4\pi\omega\varepsilon} \int_{v'} k^2 \vec{J} \frac{e^{-jkR}}{R} dv' \tag{2.115}$$

2.5 The Governing Equation of the Loop Antenna and Nano-Scaled Ring

2.5.1 The General Equation

Solution for $\vec{E}$. Substituting $\partial\vec{A}/\partial t$ and $\vec{\nabla}\Phi$ into Eq. (2.93) gives for the Electric field in the ring $\vec{E}$:

$$\vec{E} = \frac{-j}{4\pi\omega\varepsilon}\int_{V'}\left[\vec{\nabla}\left(\left(\vec{\nabla}'\cdot\vec{J}\right)\frac{e^{-jkR}}{R}\right)+k^2\vec{J}\frac{e^{-jkR}}{R}\right]dv' \tag{2.116}$$

The gradient, expanded in wire coordinates using Eq. (2.20), becomes:

$$\vec{\nabla} = \frac{\partial}{\partial\sigma}\hat{\sigma}+\frac{1}{(b+\sigma\cos\psi)}\frac{\partial}{\partial\phi}\hat{\phi}+\frac{1}{\sigma}\frac{\partial}{\partial\psi}\hat{\psi} \tag{2.117}$$

The volume element is $dv' = \sigma'(b+\sigma'\cos\psi')d\sigma' d\phi' d\psi'$ and $R = R(\sigma', \phi', \psi')$ is the distance magnitude between points of source and observation. The divergence may be derived from Eq. (2.20).

$$\nabla'\cdot J = \frac{1}{\sigma'(b+\sigma'\cos\psi')}\left[\frac{\partial(\sigma'(b+\sigma'\cos\psi')J_{\sigma'})}{\partial\sigma'}+\frac{\partial(\sigma' J_{\phi'})}{\partial\phi'}+\frac{\partial((b+\sigma'\cos\psi')J_{\psi'})}{\partial\psi'}\right] \tag{2.118}$$

Expanding into the three components of $\vec{E}$, we have **the General Governing Equation** for loop antennas and nano-scaled rings:

$$\begin{aligned}
E_\sigma &= \frac{-j}{4\pi\omega\varepsilon}\int_{V'}\left[\frac{\partial}{\partial\sigma}\left(\left(\vec{\nabla}'\cdot\vec{J}\right)\frac{e^{-jkR}}{R}\right)+k^2\vec{J}_\sigma\frac{e^{-jkR}}{R}\right]dv' \\
E_\phi &= \frac{-j}{4\pi\omega\varepsilon}\int_{V'}\left[\frac{1}{(b+\sigma\cos\psi)}\frac{\partial}{\partial\phi}\left(\left(\vec{\nabla}'\cdot\vec{J}\right)\frac{e^{-jkR}}{R}\right)+k^2\vec{J}_\phi\frac{e^{-jkR}}{R}\right]dv' \\
E_\psi &= \frac{-j}{4\pi\omega\varepsilon}\int_{V'}\left[\frac{1}{\sigma}\frac{\partial}{\partial\psi}\left(\left(\vec{\nabla}'\cdot\vec{J}\right)\frac{e^{-jkR}}{R}\right)+k^2\vec{J}_\psi\frac{e^{-jkR}}{R}\right]dv'
\end{aligned} \tag{2.119}$$

The general governing equations for the current on the loop, derived from the linear Maxwell equations, are given in Eq. (2.119).

2.5.2 The Governing Equation for Perfectly Conducting (PEC) Metals

Since we are dealing with perfectly conducting wire, there can be no $\vec{E}$ field internally and therefore the E_σ component can never exist; the electrons move instantaneously to counter the field. The divergence of the current that appears in each of the remaining two field equations for E_ϕ and for E_ψ are deduced from Eq. (2.113). If we assume that the electric field does not penetrate the wire up to 24 THz, as noted in Sect. 2.3.3, then we may take $\sigma, \sigma' \rightarrow a$ and $(b + \sigma \cos\psi), (b + \sigma' \cos\psi') \rightarrow (b + a\cos\psi), (b + a\cos\psi')$. This causes the current surface density ($amperes/m^2$) to reduce to a linear current density ($amperes/m$), since the current is not accumulating along the face of the wire cross-section. This leaves

$$\nabla' \cdot J = \frac{1}{(b + a\cos\psi')} \frac{\partial J_{\phi'}}{\partial \phi'} + \frac{1}{a(b + a\cos\psi')} \frac{\partial (b + a\cos\psi') J_{\psi'}}{\partial \psi'}$$

The functional dependence of R is given in Eq. (2.36). Moreover,

$$\begin{aligned} &dv' \rightarrow ds' = a(b + a\cos\psi') d\phi' d\psi' \quad \text{and} \\ &R = R(a, \phi', \psi') \end{aligned} \tag{2.120}$$

Substituting these, using Eq. (2.66), and using the appropriate components of Eq. (2.114), we have for the field equations.

$$\begin{aligned} E_\sigma &= 0 \\ E_\phi &= \frac{-j}{4\pi\omega\varepsilon} \int_{V'} \left[\frac{1}{(b + a\cos\psi)} \frac{\partial}{\partial \phi} \left(\left(\vec{\nabla}' \cdot \vec{J} \right) \frac{e^{-jkR}}{R} \right) + k^2 \vec{J}_\phi \frac{e^{-jkR}}{R} \right] dv' \\ &= \frac{-j\xi}{k(b + a\cos\psi)} \left[\frac{\partial}{\partial \phi} \int_{S'} \left(\vec{\nabla}' \cdot \vec{J} \right) \frac{e^{-jkR}}{4\pi R} a(b + a\cos\psi') d\phi' d\psi' \right. \\ &\quad \left. + k^2 (b + a\cos\psi) \int_{S'} \vec{J}_\phi \frac{e^{-jkR}}{4\pi R} a(b + a\cos\psi') d\phi' d\psi' \right] \\ &= \frac{-j\xi}{k(b + a\cos\psi)} \left[\frac{\partial}{\partial \phi} \int_{S'} \frac{e^{-jkR}}{4\pi R} \left(a \frac{\partial J_{\phi'}}{\partial \phi'} + \frac{\partial (b + a\cos\psi') J_{\psi'}}{\partial \psi'} \right) d\phi' d\psi' \right. \\ &\quad + k^2 a (b + a\cos\psi) \int_{S'} \frac{e^{-jkR}}{4\pi R} (b + a\cos\psi') \times \\ &\quad \left. \left(J_{\phi'} \cos(\phi - \phi') + J_{\psi'} \sin\psi' \sin(\phi - \phi') \right) d\phi' d\psi' \right] \end{aligned} \tag{2.121}$$

$$\begin{aligned}E_\psi &= \frac{-j}{4\pi\omega\varepsilon}\int_{V'}\left[\frac{1}{a}\frac{\partial}{\partial\psi}\left(\left(\vec{\nabla}'\cdot\vec{J}\right)\frac{e^{-jkR}}{R}\right)+k^2\vec{J}_\psi\frac{e^{-jkR}}{R}\right]dv' \\ &= \frac{-j\xi}{k}\left[\frac{\partial}{\partial\psi}\int_{S'}\frac{e^{-jkR}}{4\pi R}(b+a\cos\psi')\left(\left(\vec{\nabla}'\cdot\vec{J}\right)\right)d\phi' d\psi'\right. \\ &\quad\left.+k^2\int_{S'}\frac{e^{-jkR}}{4\pi R}a(b+a\cos\psi')\vec{J}_\psi\right]d\phi' d\psi' \\ &= \frac{-j\xi}{k}\left[\frac{\partial}{\partial\psi}\int_{S'}\frac{e^{-jkR}}{4\pi R}\left(\frac{\partial J_{\phi'}}{\partial\phi'}+\frac{1}{a}\frac{\partial(b+a\cos\psi')J_{\psi'}}{\partial\psi'}\right)d\phi' d\psi'\right. \\ &\quad+k^2\int_{S'}\frac{e^{-jkR}}{4\pi R}a(b+a\cos\psi')\times\Big(-J_{\phi'}\sin\psi\sin(\phi-\phi') \\ &\quad\left.+J_{\psi'}\left(\sin\psi'\sin\psi\cos(\phi-\phi')+\cos\psi'\cos\psi\right)\Big)d\phi' d\psi'\right] \end{aligned} \tag{2.122}$$

2.5.3 The Governing Equation for Thin-Wire, PEC Loops and Rings

The Thin, Perfectly Conducting Wire Approximation. A physical loop has a definite wire radius, a. The ***thin-wire approximation*** says that the circumference of the wire is quite a bit less than the wavelength of the source frequency: $2\pi a \lll \lambda$, or equivalently, $ka \ll 1$. If we write the ratio of the ring circumference to the source wavelength, $kb = 2\pi b/\lambda$, then the approximation may be stated equivalently $kb \lll b/a$. Since the region of interest for all loops in this text is $kb < 2.5$, the aspect ratio, b/a, should be greater than 25. Table 2.1 indicates that this occurs for $\Omega \geq 10$. The thin-wire approximation, therefore, limits computed results using it to fairly thin loops and rings.

Ultimately this approximation simplifies two problems of analysis: (1) resonances along the wire circumference in the $\hat{\psi}$ direction at some radius within the wire (where $\lambda \approx 2\pi a$); and (2) resonances across the cross-sectional face of the wire in the $\hat{e}$ direction (where $\lambda \approx a$).

As noted in Sect. 2.3.3, it may be generally assumed that any $\vec{E}$ field source penetrates dielectric wires entirely for even the thickest wires at all frequencies, But not so for metals, for which the field does not begin to penetrate 10% of the wire radius until roughly 0.1 EV (24 THz) and this for only thin wires, $\Omega > 10$. Thick metal wires are not generally affected by skin depth until the incident frequency is above the optical region.

Therefore, for metals wires, it may be assumed that only the surface components of the incident $\vec{E}$ field are present in the wire and when the thin-wire approximation is added, the field calculated at the center of the wire is distributed uniformly around the circumference, $2\pi a$.

For dielectric wires, it may be assumed that all three components of the incident $\vec{E}$ field are active and generating current. When the thin-wire approximation is applied, the $\hat{\sigma}$ component will have no gradient variation across the face, certainly not for $kb < 2.5$, and therefore will not generate currents in that direction. Nor will the $\hat{\psi}$ field component have a gradient variation around the circumference, as for metals, and there will therefore be no current components in that direction either.

Under all thin-wire configurations then, the incident $\vec{E}$ field may be considered to have only its $\hat{\phi}$ component active. It allows the assumption of a filamentary current in the loop at the centre of the wire, directed only in the $\hat{\phi}$ direction. In this case, the current is usually given as a variation in ϕ: $I = I_\phi(\phi)$ amperes. The approximation also allows for the current to be given as a surface current distributed uniformly around the circumference of the wire. This looks like $J_\phi = I(\phi)/(2\pi a)$ amperes/metre. One approach is to think of the surface current as consisting of an infinite number of filaments spread evenly around the surface each one remaining the same distance from the centre of the loop [19, see for example, Sect. 8.8]. The filaments on the inside circumference of the loop have length $2\pi(b-a)$, while those on the outside have length $2\pi(b+a)$, lending a slight "uncertainty", shall we say, to the exact de Broglie resonance condition of the loop. As it turns out, thicker loops tend to have their resonances blue-shifted from the thin-wire de Broglie conditions.

The ***perfectly conducting wire approximation*** assumes the wire has infinite conductivity, an approximation which may be lent to noble metals at frequencies below 150–200 GHz, according to Fig. 2.6. The approximation must be dropped for dielectric wires at all frequencies and for noble metals above this cut-off.

The Governing Equation for Thin-wire, Perfectly Conducting Loop Antennas to 150 GHz. The governing equation under the thin-wire approximation appears as a reduction of Eqs. (2.121) and (2.122). These may be reduced if we apply the thin-wire approximation. The terms $(b + a\cos\psi)$, $(b + a\cos\psi') \approx b$, since $b/a > 10$ for wires where $\Omega \leq 9$. There is no current in $\hat{\psi}$, and the current is not a function of ψ' nor a variation in ψ'; that is, $J_{\psi'} = 0$, $\vec{J}|_{src} \neq J(\psi')$, $\partial\vec{J}/\partial\psi' = 0$. The current then flows only in a thin layer on the outer surface of the wire, so that $J_\phi = I_{\phi'}(\phi)/(2\pi a)$.

These approximations reduce the governing equations, Eqs (2.121) and (2.122) to the single component, E_ϕ.

$$
\begin{aligned}
E_\sigma &= E_\psi = 0;\\
E_\phi &= \frac{-j\xi}{4\pi}\left[\frac{1}{kb}\frac{\partial}{\partial\phi}\int_{-\pi}^{\pi}\left(\frac{1}{2\pi}\int_{-\pi}^{\pi}\frac{e^{-jkR}}{R}d\psi'\right)\frac{\partial I_{\phi'}}{\partial\phi'}d\phi'\right.\\
&\quad\left. + kb\int_{-\pi}^{\pi}\left(\frac{1}{2\pi}\int_{-\pi}^{\pi}\frac{e^{-jkR}}{R}d\psi'\right)I_{\phi'}\cos(\phi-\phi')d\phi'\right]\\
&= \frac{-j\xi}{4\pi}\int_{-\pi}^{\pi}\left[\frac{1}{kb}\frac{\partial I_{\phi'}}{\partial\phi'}\frac{\partial}{\partial\phi} + kbI_{\phi'}\cos(\phi-\phi')\right]\left(\frac{1}{2\pi}\int_{-\pi}^{\pi}\frac{e^{-jkR}}{R}d\psi'\right)d\phi'
\end{aligned}
\tag{2.123}
$$

2.5.4 *The Governing Equation for Thick-Wire, PEC Loops and Rings*

We seek now a way to reduce the general equations of current on a loop or ring, Eqs. (2.121) and (2.122), for thick loops and rings, where the thin-wire approximation no longer holds. With a surface integral, the current densities become surface densities. The sum of these two current vectors can yield helical, sometimes called "toroidal", currents along the surface of the wire. The term $(b + a\cos\psi) = a(b/a + \cos\psi)$, which for $b/a \gg 1$ is $\approx b$. However, for $\Omega = 6$, $b/a = 3.2$; $\Omega = 7, b/a = 5.3$; and for $\Omega = 8, b/a = 8.7$. Since these are not much greater than 1, the term must be used as is. We can break out the surface integral and rearrange the above equations term by term:

$$
\begin{aligned}
E_\sigma &= 0 \\
E_\phi &= -\frac{j\xi}{2}\int_{-\pi}^{\pi}\left[\frac{a}{k(b+a\cos\psi)}\frac{\partial J_{\phi'}}{\partial\phi'}\frac{\partial}{\partial\phi}\left(\frac{1}{2\pi}\int_{-\pi}^{\pi}\frac{e^{-jkR}}{R}d\psi'\right)\right]d\phi' \\
&\quad -\frac{j\xi}{2}\int_{-\pi}^{\pi}\left[kaJ_{\phi'}\cos(\phi-\phi')\left(\frac{1}{2\pi}\int_{-\pi}^{\pi}\frac{e^{-jkR}}{R}(b+a\cos\psi')d\psi'\right)\right]d\phi' \\
&\quad -\frac{j\xi}{2}\int_{-\pi}^{\pi}\left[\frac{1}{k(b+a\cos\psi)}\frac{\partial(b+a\cos\psi')J_{\psi'}}{\partial\psi'}\frac{\partial}{\partial\phi}\left(\frac{1}{2\pi}\int_{-\pi}^{\pi}\frac{e^{-jkR}}{R}d\phi'\right)\right]d\psi' \\
&\quad -\frac{j\xi}{2}\int_{-\pi}^{\pi}\left[ka(b+a\cos\psi')J_{\psi'}\sin\psi'\left(\frac{1}{2\pi}\int_{-\pi}^{\pi}\frac{e^{-jkR}}{R}\sin(\phi-\phi')d\phi'\right)\right]d\psi'
\end{aligned}
\tag{2.124}
$$

The ψ component of the E field, given in Eq. (8.2) is expanded by rearrangement to

$$
\begin{aligned}
E_\psi &= \frac{-j\xi}{2}\int_{-\pi}^{\pi}\left[\frac{1}{k}\frac{\partial J_{\phi'}}{\partial\phi'}\frac{\partial}{\partial\psi}\left(\frac{1}{2\pi}\int_{-\pi}^{\pi}\frac{e^{-jkR}}{R}d\psi'\right)\right]d\phi' \\
&\quad +\frac{j\xi}{2}\int_{-\pi}^{\pi}\left[kaJ_{\phi'}\sin\psi\sin(\phi-\phi')\left(\frac{1}{2\pi}\int_{-\pi}^{\pi}\frac{e^{-jkR}}{R}(b+a\cos\psi')d\psi'\right)\right]d\phi' \\
&\quad -\frac{j\xi}{2}\int_{-\pi}^{\pi}\left[\frac{1}{ka}\frac{\partial(b+a\cos\psi')J_{\psi'}}{\partial\psi'}\frac{\partial}{\partial\psi}\left(\frac{1}{2\pi}\int_{-\pi}^{\pi}\frac{e^{-jkR}}{R}d\phi'\right)\right]d\psi' \\
&\quad -\frac{j\xi}{2}\int_{-\pi}^{\pi}\left[ka(b+a\cos\psi')J_{\psi'}\sin\psi'\sin\psi\left(\frac{1}{2\pi}\int_{-\pi}^{\pi}\frac{e^{-jkR}}{R}\cos(\phi-\phi')d\phi'\right)\right]d\psi' \\
&\quad -\frac{j\xi}{2}\int_{-\pi}^{\pi}\left[ka(b+a\cos\psi')J_{\psi'}\cos\psi'\cos\psi\left(\frac{1}{2\pi}\int_{-\pi}^{\pi}\frac{e^{-jkR}}{R}d\phi'\right)\right]d\psi'
\end{aligned}
\tag{2.125}
$$

These are partially solved in Sect. 8.2. These equations for the thick wire ring have not appeared in the literature as of this writing.

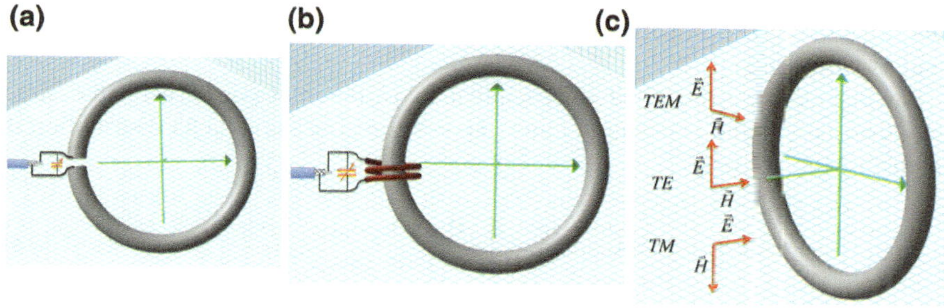

Fig. 2.11 Three methods of Coupling. **a** Direct; **b** Inductive; and **c** Illuminated by three types of transverse waves

2.6 Coupling a Driving Source to the Loop or Ring

Loops are coupled to a source using one of three different methods: (1) Direct coupling, (2) Inductive "frill" coupling, or (3) Illumination by a plane wave. They are illustrated in Fig. 2.11.

They produce a voltage source for the loop and these voltage sources need to be modelled in the mathematics. To do so, we represent the source in its Fourier expansion, focusing only on the ϕ component:

$$E_\phi = \sum_{n=-\infty}^{\infty} E_n e^{jn\phi} \tag{2.126}$$

leading to the Fourier coefficients

$$E_n = \frac{1}{2\pi} \int_{-\pi}^{\pi} E_\phi(\phi) e^{-jn\phi} d\phi \equiv E_n \tag{2.127}$$

2.6.1 *Direct Coupling*

Direct coupling refers to a transmission line tied across a small gap; a capacitor may be applied across the gap to help with impedance matching. Inductive coupling refers to the use of a small coil turned about the loop wire at one point. A capacitor may be placed across the inductor to aid with impedance matching. These two methods can be employed for either transmission or reception at frequencies in the MW range or lower, although it would be physically possible with today's technologies to use direct coupling at THz where the size of the ring is approaching nano-scale sizes. Illuminating the loop with a plane wave is reserved for reception purposes and to determine scattering properties. The method is particularly useful when the loop is a nano-scaled ring.

Direct coupling is modelled analytically in two different ways: (1) using the Dirac delta distribution and (2) using a finite gap voltage. The Dirac delta distribution is defined by the property

$$\delta(x-x_0)=\begin{cases}0 \text{ if } x\neq x_0\\ \infty \text{ if } x=x_0\end{cases} \tag{2.128}$$

with the specification that the area under the distribution be exactly 1; that is,

$$\int_{x_1}^{x_2}\delta(x-x_0)dx=\begin{cases}0 \text{ if } x<x_1 \text{ or } x>x_2\\ 1 \text{ if } x_1<x<x_2\end{cases} \tag{2.129}$$

The relation

$$\int_{x_1}^{x_2} f(x)\delta(x-x_0)dx=f(x_0) \tag{2.130}$$

is true because $\delta(x-x_0)$ is zero everywhere except at $x=x_0$, at which point $f(x)=f(x_0)$ and it is no longer a function of x, coming therefore out of the integral. The specifications Eqs. (2.128) and (2.129) for the Dirac delta distribution gives the right hand side.

Recall that the standard boundary condition on the tangential $\vec{E}$ field at the surface of the PEC loop requires $E_\phi=0$ everywhere, except where the voltage source is located. Therefore [20],

$$\int_{-\pi}^{\pi} E_\delta(\phi)b d\phi=-V_0 \tag{2.131}$$

The model for the field therefore must be

$$E_\delta(\phi)=-\frac{V_0\delta(\phi)}{b} \tag{2.132}$$

such that when integrated, the Dirac delta forces the desired voltage. The source is oscillatory: $V=V_0e^{j\omega t}$. The Fourier coefficients of the distribution, Eq. (2.127), are calculated as

$$E_{\delta n}=-\frac{1}{2\pi}\int_{-\pi}^{\pi}\frac{V_0}{b}\delta(\phi)e^{-jn\phi}d\phi=-\frac{V_0}{2\pi b} \tag{2.133}$$

This delta distribution implies an infinitesimal gap width across which the voltage source is placed, and this further implies an infinite capacitance at the source. Such an argument gives rise to an infinite input susceptance, which has caused problems with the analysis of cylindrical dipole antennas, of which there is a long history in the literature [21–26]. Identical problems arise for the analysis of the loop [27, 28]. See Tastsoglou's serial papers [25, 26] for a history of the gap problem.

Some difficulties extend to the finite gap voltage generator model. Using the gap width = γ in radians, the model appears as

$$E_{\Delta}(\phi) = \begin{cases} -V_0/(b\gamma), & \texttt{if} \ \ |\phi| < \gamma/2 \\ 0, & \texttt{otherwise} \end{cases} \tag{2.134}$$

The Fourier coefficients are

$$E_{\Delta n} = \frac{1}{2\pi}\int_{-\gamma/2}^{\gamma/2} \frac{V_0}{b\gamma} e^{-jn\phi} d\phi = \begin{cases} -V_0 \texttt{sin}(n\gamma/2)/(\pi b n\gamma) & \texttt{if} \ \ n \neq 0 \\ -V_0/(2\pi b), & \texttt{if} \ \ n = 0 \end{cases} \tag{2.135}$$

The voltage due to direct coupling will be oriented in the ϕ direction along the periphery of the ring.

2.6.2 Inductive Coupling

Since inductive coupling consists of an inductive coil wrapped around the loop, it is usually modelled by the "frill" generator [29, 30]. The generator creates a magnetic surface density current given by

$$M(\rho) = \frac{V_0}{\rho \texttt{ln}(a_0/a)} \quad \texttt{for} \ \ a < \rho < a_0 \tag{2.136}$$

where a_0 is the outer radius of the frill, a the radius of the wire and ρ is the radial component in toroidal coordinates. The outside radius is chosen so that the impedance of the feed transmission line, $Z_{in} = \xi_0 \texttt{ln}(a_o/a)$. According to Fikioris [31], the corresponding $\vec{E}$ field is

$$E_f = \frac{jkV_0}{2b\xi_0 \texttt{ln}(b/a)} \left[\frac{e^{-jkR_a}}{R_a} - \frac{e^{-jkR_b}}{R_b} \right] \tag{2.137}$$

where the appropriate variable R uses either a or b. The corresponding Fourier coefficients depend on the choice of R from among those suggested in Sect. 2.1.5. Fikioris [27], using

$$R \approx \sqrt{4b^2 \texttt{sin}^2(\frac{\phi - \phi'}{2}) + a^2} \tag{2.138}$$

developed a set of simple coefficients:

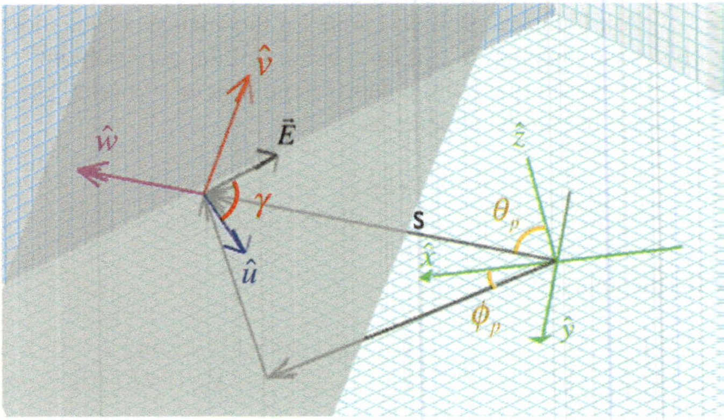

Fig. 2.12 The coordinate system on an illuminating wave plane

$$E_n = E_{fn}(\phi) = \frac{V}{2\pi b\,\mathtt{ln}(a_o/a)}\left[(K_{o,n+1} + K_{o,n-1}) - (K_{n+1} + K_{n-1})\right] \quad (2.139)$$

where K_n are those shown in Eq. (3.57). Here, K_o is equal to K_n but uses a_o rather than a and $x_o = 2b/a_o$ instead of x.

2.6.3 *Illuminated Coupling*

An illuminating plane wave may be applied in Transverse EM (TEM), Transverse E (TE), or Transverse M (TM) mode. Wave propagation in the far-field in real media, such as in air or vacuum, occurs as TEM waves. Such modes appear in a coaxial transmission line as well. TE and TM modes appear only in waveguides, which usually have circular or rectangular openings. An impinging TE or TM mode plane wave can only occur when the loop is near the opening of a waveguide.

Establish a cylindrical coordinate system with origin, O, at the center of the ring; use coordinates as given by Eq. (2.24). Find the plane of an incoming TEM wave, as shown in Fig. 2.12 and establish on the plane an orthonormal coordinate system with unit vectors $\{\hat{u}, \hat{v}$, and $\hat{w}\}$ where $\hat{w}$ defines the normal axis. Define the angle θ_p as the angle between $\hat{w}$ and $\hat{z}$; then $\hat{w} \cdot \hat{z} = \mathtt{cos}\theta_p$. Define the angle ϕ_p as the angle between the projection of $\hat{w}$ on the $\{xy\}$ plane and the x axis, so that the projection of $\hat{w}$ on the $\{xy\}$ plane is $\mathtt{sin}\theta_p\left(\mathtt{cos}\phi_p\hat{x} + \mathtt{sin}\phi_p\hat{y}\right)$. The unit vector in the $\{xy\}$ plane is $\hat{\rho}_p$ and the normal to the plane in standard Cartesian coordinates is $\hat{w}$. Therefore,

$$\hat{\rho}_p = \mathtt{cos}\phi_p\hat{x} + \mathtt{sin}\phi_p\hat{y} \qquad \hat{w} = \mathtt{sin}\theta_p\hat{\rho}_p + \mathtt{cos}\theta_p\hat{z} \quad (2.140)$$

Define, now, the related two coordinate unit vectors, $\hat{u}$ normal to the $\{zw\}$ plane and thus in the $\{xy\}$ plane, and $\hat{v}$ normal to the $\{uw\}$ plane.

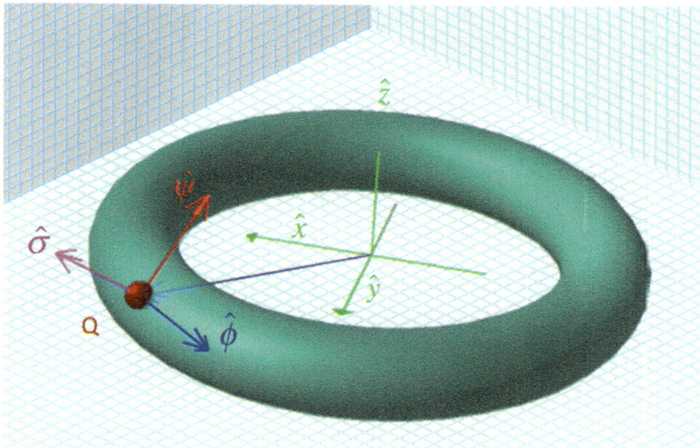

Fig. 2.13 The coordinate system on the thick ring

$$\hat{u} = (\hat{z} \times \hat{w})/\left|\hat{z} \times \hat{w}\right| = -\sin\phi_p\hat{x} + \cos\phi_p\hat{y}$$
$$\hat{v} = (\hat{w} \times \hat{u})/\left|\hat{w} \times \hat{u}\right| = -\cos\theta_p\cos\phi_p\hat{x} - \cos\theta_p\sin\phi_p\hat{y} + \sin\theta_p\hat{z} \quad (2.141)$$

Place a point S on the plane in space such that the normal line of the plane through S intersects the center of the ring. This is the vector $\vec{S} = d_0\hat{w}$ from the origin. The plane is then given by $P_{uv} = u\hat{u} + v\hat{v} + d_0\hat{w}$.

Now, using wire coordinates defined in Sect. 2.1.1 and shown in Fig. 2.2, establish a point $Q = \{\sigma_q\hat{\sigma}, \phi_q\hat{\phi}, \psi_q\hat{\psi}\}$ somewhere on the ring, as shown in Fig. 2.13. Then, $Q = b\hat{\rho} + \sigma_q\hat{\sigma}$. The unit vectors at Q on the ring using Eq. (2.30) are

$$\begin{aligned}
\hat{\sigma} &= \cos\phi_q\cos\psi\hat{x} + \sin\phi_q\cos\psi\hat{y} + \sin\psi\hat{z} \\
\hat{\phi} &= -\sin\phi_q\hat{x} + \cos\phi_q\hat{y} \\
\hat{\psi} &= -\cos\phi_q\sin\psi\hat{x} - \sin\phi_q\sin\psi\hat{y} + \cos\psi\hat{z}
\end{aligned} \quad (2.142)$$

The point on the plane closest to Q is normal to the plane and is given by the vector from the origin to S plus the projection of $\vec{Q}$ onto the unit vectors of the plane, namely $\hat{u}$ and $\hat{v}$ (Fig. 2.14).

$$\vec{C} = \vec{S} + (\vec{Q} \cdot \hat{u})\hat{u} + (\vec{Q} \cdot \hat{v})\hat{v} \quad (2.143)$$

This distance is the difference between the two vectors $\vec{C}$ and $\vec{Q}$.

$$\begin{aligned}
\vec{R} = \vec{Q} - \vec{C} &= (Q \cdot \hat{w})\hat{w} - \vec{S} = ((b\hat{\rho} + \sigma_q\hat{\sigma}) \cdot \hat{w})\hat{w} - d_0\hat{w} \\
&= \left((b + \sigma_q\cos\psi_q)\sin\theta_p\cos(\phi_q - \phi_p) + \sigma_q\sin\psi_q\cos\theta_p - d_0\right)\hat{w} \\
&= \left((x_q\cos\phi_p + y_q\sin\phi_p)\sin\theta_p + z_q\cos\theta_p - d_0)\right)\hat{w}
\end{aligned} \quad (2.144)$$

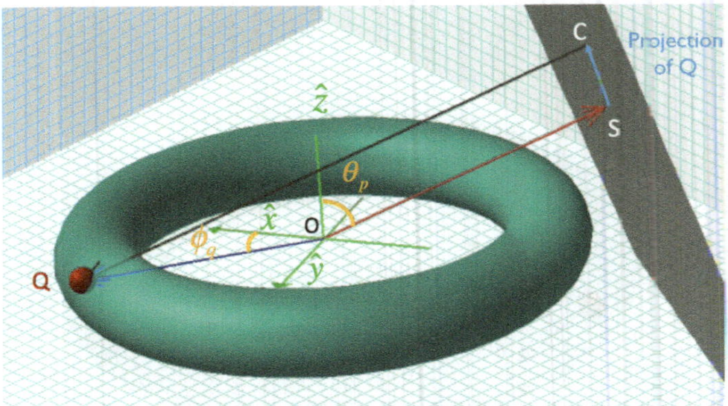

Fig. 2.14 The coordinate system relating the illuminating plane and the thick ring

The wave spreads through space with time along $-\hat{w}$ to the point Q with momentum $\vec{k} = -(2\pi/\lambda)\hat{w}$. This gives a variation in the $\vec{E}$ field with time. Choosing a particular d_0 effectively chooses a particular point in time and gives us the $\vec{E}$ field at points on the ring at that time. Some points on the ring will feel the plane at an earlier time than others, depending entirely on the tilt of the plane wave and the thickness of the ring.

We are now ready to find the $\vec{E}$ field at the point Q on the ring. First establish the $\vec{E}$ field on the plane. Take γ as the angle between the $\hat{u}$ axis and $\vec{E}$. Take E_0 as the amplitude of the field at the plane. Then, using $e^{j\omega t}$ as we have been doing, see Eq. (2.56), the spatial variation is

$$e^{-j\vec{k}\cdot\vec{R}} = e^{jk\left[(b+\sigma_q\cos\psi_q)\sin\theta_p\cos(\phi_q-\phi_p)+\sigma_q\sin\psi_q\cos\theta_p-d_0\right]} \tag{2.145}$$

We have (Fig. 2.15)

$$\vec{E}_Q = E_0\left(\cos\gamma\;\hat{u} + \sin\gamma\;\hat{v}\right)e^{-j\vec{k}\cdot\vec{R}} \tag{2.146}$$

We assume no dampening due to the medium. We want the $\vec{E}$ field at the point Q on the ring when it arrives and it arrives full strength. The plane containing the wave is centered on the ring, but tilted between 0 and π. Transforming to **Cartesian coordinates** gives

$$\begin{aligned}\vec{E}_Q = E_0 e^{-j\vec{k}\cdot\vec{R}}\big[&(-\cos\gamma\sin\phi_p - \sin\gamma\cos\theta_p\cos\phi_p)\;\hat{x}\\ &+(\cos\gamma\cos\phi_p - \sin\gamma\cos\theta_p\sin\phi_p)\;\hat{y}\\ &+\sin\gamma\sin\theta_p\;\hat{z}\big]\end{aligned} \tag{2.147}$$

Transforming to **Cylindrical coordinates** gives the result

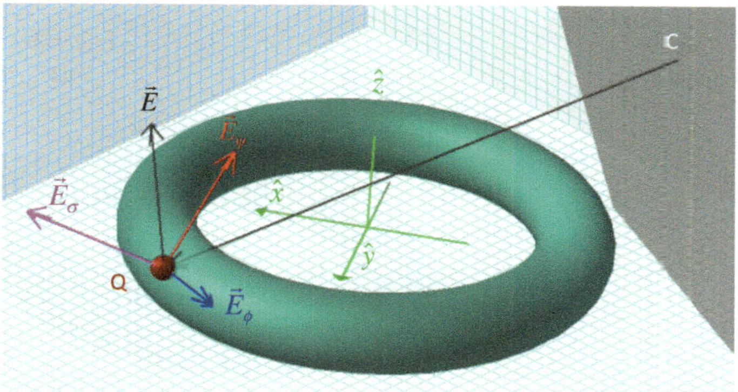

Fig. 2.15 The E field on the thick ring as illuminated by the E field of the plane

$$\vec{E}_Q = E_0 e^{-j\vec{k}\cdot\vec{R}}\big[(\cos\gamma\sin(\phi_q-\phi_p)-\sin\gamma\cos\theta_p\cos(\phi_q-\phi_p))\,\hat{\rho} \\ +(\cos\gamma\cos(\phi_q-\phi_p)+\sin\gamma\cos\theta_p\sin(\phi_q-\phi_p))\,\hat{\phi} \\ +\sin\gamma\sin\theta_p\,\hat{z}\big] \quad (2.148)$$

which cannot be simplified. Further transformation to **Wire coordinates** and using $\alpha = (\phi_q - \phi_p)$ gives

$$\vec{E}_Q = E_0 e^{-j\vec{k}\cdot\vec{R}} \\ \times\big[(\cos\gamma\cos\psi_q\sin\alpha-\sin\gamma\cos\theta_p\cos\psi_q\cos\alpha+\sin\gamma\sin\theta_p\sin\psi_q)\,\hat{\sigma} \\ +(\cos\gamma\cos\alpha+\sin\gamma\cos\theta_p\sin\alpha)\,\hat{\phi} \\ -(\cos\gamma\sin\psi_q\sin\alpha-\sin\gamma\cos\theta_p\sin\psi_q\cos\alpha-\sin\gamma\sin\theta_p\cos\psi_q)\,\hat{\psi}\big] \quad (2.149)$$

Special Orientations. There are three interesting orientations of the plane wave, which arise because of the symmetry of the closed loop. The first of these, $\theta_p = 0$, yields a wave illuminating the loop broadside, so the $\vec{E}$ and $\vec{H}$ are both oriented in the plane of the loop. The second, $\theta_p = \pi/2$, yields a wave illuminating from the side; the orientation of $\vec{E}$ with respect to its $\hat{u}$ axis will yield $\vec{E}$ partly in the plane and partly perpendicular to the plane. Simultaneously the magnetic field, $\vec{H}$, will be partly perpendicular and partly in the plane. In other words, if γ is selected once as 0 and then later as $\pi/2$, one then the other orientation can be studied.

(1) In the first case, $\theta_p = 0$; the wave appears broadside. The effect on the equations is made simpler if at the same time γ and ϕ_p are also set to 0. This puts $\vec{E}$ along the $\hat{u}$ axis on the plane and brings the plane wave in along the $-\hat{z}$ axis. Substituting into Eq. (2.149) yields the intensity when the wave arrives at point Q on the loop (see Fig. 2.16):

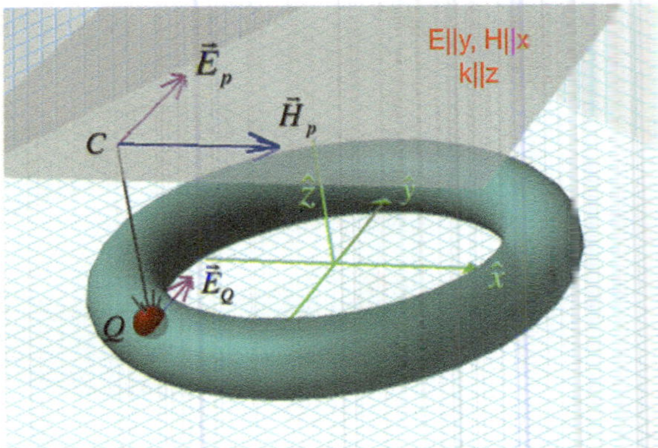

Fig. 2.16 Case 1: Broadside Illumination. Such illumination stimulates a broad resonance around $k_b = 1.0$

$$\begin{aligned}\vec{E}_Q &= E_0 e^{-jkR}\hat{y} \\ &= E_0 e^{-jkR}\left[\mathtt{sin}\phi_q\hat{\rho} + \mathtt{cos}\phi_q\hat{\phi}\right] \\ &= E_0 e^{-jkR}\left[\mathtt{cos}\psi_q\mathtt{sin}\phi_q\hat{\sigma} + \mathtt{cos}\phi_q\hat{\phi} - \mathtt{sin}\psi_q\mathtt{sin}\phi_q\hat{\psi}\right]\end{aligned} \tag{2.150}$$

The field vector points along the $\hat{y}$ axis no matter where Q is on the loop. If the ring is thick, the wave arrives at different z points on the ring at different times.

$$R = \sigma_q \mathtt{sin}\psi_q - d_0 = z_q - d_0 \quad \text{where} \quad k = -\frac{2\pi}{\lambda} \tag{2.151}$$

If the ring is thin, then $R = -d_o$, and the wave arrives equally strong at all points on the ring. In either case, the magnetic field has no affect on the current distribution of the loop, since there is no Faraday action. To induce currents, the field must penetrate the plane of the loop.

The electric field, however, does induce currents and moves the charge along the $\hat{y}$-axis with the curvature forcing the current into a $\mathtt{cos}\phi$ configuration with $\phi = 0$ at the x-axis. As the wave passes through, the current pulsates like a standing wave on the circumference of the loop, as in Fig. 2.17.

Indeed, the ring reradiates like two half-circle dipoles. The circle splits at $y = \pm b$, because at those points the circle has little room for the current to move along the $\hat{y}$ axis. This is a resonance as one might find, for example, with the first fundamental current stimulated by a delta function generator. The simulations shown in Fig. 2.18, indicate a broad, weak resonance ($Q \approx 4$–8) around $k_b = 1.0$. It can be explained as an important consequence of the split at $y = \pm b$ and the ring current "pulsing" at the illuminating frequency.

The $\vec{E}$ field is the same at every point in the plane wave, and every point in the plane wave hits every point in the ring at the same time, broadside. In this configuration

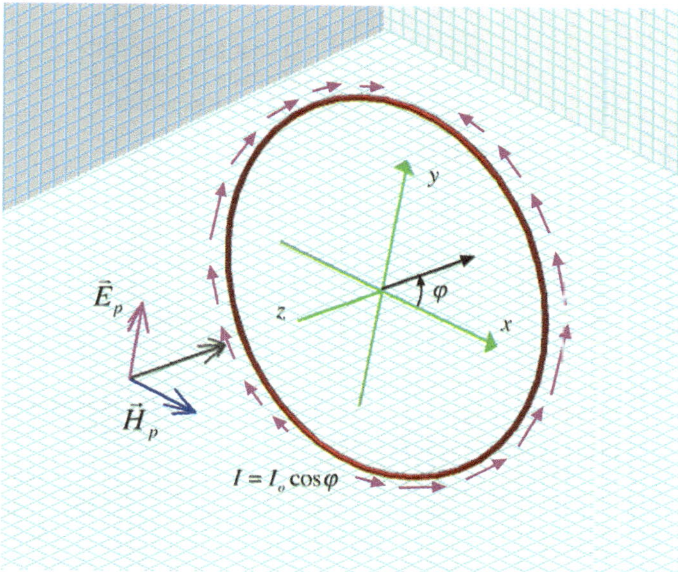

Fig. 2.17 The effects of broadside illumination on an $\Omega = 12$ loop. The $\vec{H}$ field has no effect. The $\vec{E}$ field induces current that flows in a $\cos\phi$ pattern only when the illuminating wave acts at $k_b = 1.0$. As the wave moves through the loop, the current oscillates. There is a fairly good resonance at this k_b but not at any other harmonic, because the field stimulates only the two sides and cannot stimulate a quadrupole or anything more complicated. See Fig. 2.18

then, the only resonance is the pulsed one, near $k_b = 1.0$, which will be slightly red-shifted for thicker rings. Also see [32].

(2) In the second case, $\theta_p = \pi/2$, the wave approaches from the side and there are two cases.

1. Setting $\gamma = 0$ puts the $\vec{E}$ field along $\hat{y}$ in the plane of the loop and the $\vec{H}$ field perpendicular to the plane of the loop along $\hat{z}$. See Fig. 2.19. Due to the symmetry of the closed loop, it doesn't matter what value one chooses for ϕ_p. Electing $\phi_p = 0$ moves the incoming wave plane in the $-\hat{x}$ direction. Then, in free space

$$\vec{E}_Q = E_0 e^{-jkR}\hat{y} = E_0 e^{-jkR}\left[\cos\psi_q \sin\phi_q \hat{\sigma} + \cos\phi_q \hat{\phi} - \sin\psi_q \sin\phi_q \hat{\psi}\right]$$
$$\vec{H}_Q = -\frac{E_0}{\xi_0} e^{-jkR}\hat{z} \tag{2.152}$$

The $\vec{E}$ field appears as in case 1, however, the plane touches different points on the loop at different times unlike case 1. In particular,

$$R = x_q - d_0 = (b + \sigma_q \cos\psi_q)\cos\phi_q - d_0 \tag{2.153}$$

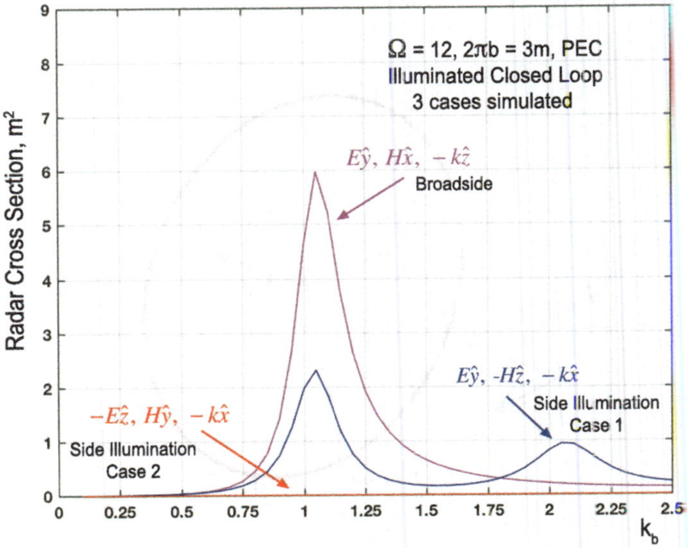

Fig. 2.18 Simulations showing the resonances created by three fundamental illuminations. Note that the broadside illumination has only one resonance at $k_b = 1.0$. The other two are cases in which the wave illuminates from the side, Case 1 with the magnetic field piercing the plane of the loop and Case 2 where it does not

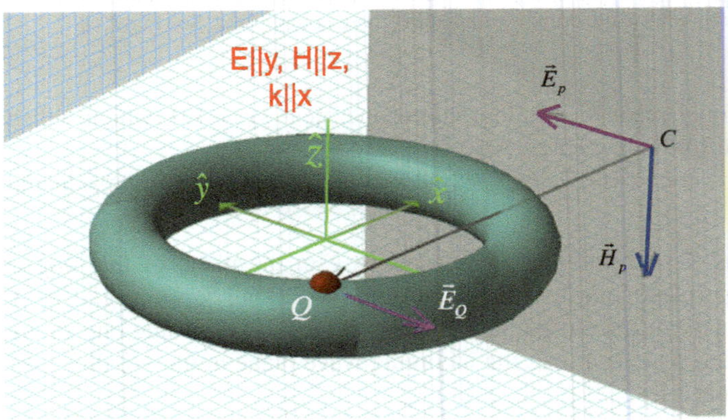

Fig. 2.19 Case 1: Side Illumination with $\gamma = 0$. The Electric field does not stimulate the resonances in this case. However, the standard harmonic resonances occur from currents induced by the magnetic field

The induced current at point Q moves slightly out of phase with the current at a point close to it, and indeed quite out of phase with currents at points on the other side of the loop. The ring experiences a rolling effect in the current, at the incoming wavelength. If the wavelength is long compared to the radius, b of the ring, all points on the ring will pulse at the incoming wavelength. If however, it is similar to the radius, the ring will split into sections where the currents are going in the opposite direction to each other, but, of course, along the $\hat{y}$ axis. Each section acts like its own dipole. If the wavelength is short, there are many such sections of the ring.

The simulations in Fig. 2.18 shows the standard harmonic resonances of the loop as if generated by a delta-function generator. These can only have been stimulated by the magnetic field passing perpendicular to the plane of the loop. Where the electric field generates no useful resonances in this case, the magnetic field does, since it penetrates the loop and generates all of the standard harmonic resonances.

2. Setting $\gamma = \pi/2$ puts the $\vec{E}$ field along $\hat{z}$ perpendicular to the plane of the loop and the $\vec{H}$ field along $\hat{y}$ in the plane of the loop. See Fig. 2.20.

$$\vec{E}_Q = E_0 e^{-jkR}\hat{z} \qquad \vec{H}_Q = \frac{E_0}{\xi} e^{-jkR}\hat{y}$$
$$\text{where} \quad R = x_q - d_0 = (b + \sigma_q \cos\psi_q)\cos\phi_q - d_0 \tag{2.154}$$

The electric field penetrates the plane of the ring and causes no current flow around the ring at all. The magnetic field does not penetrate the ring and therefore stimulates no current flow either. The simulations in Fig. 2.18 verify these assertions. No resonances occur.

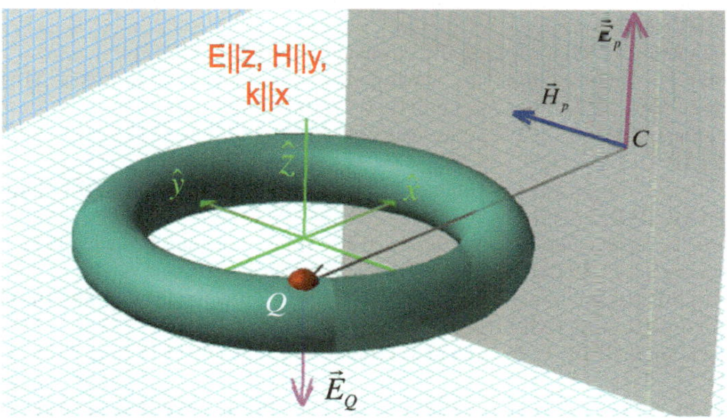

Fig. 2.20 Case 2: Side Illumination with $\gamma = \pi/2$. This case generates no resonances of any sort

References

1. R. King, J. Harrison, C., D. Tingley, IEEE Trans. Antennas Propag. **12**(4), 434 (1964). https://doi.org/10.1109/TAP.1964.1138244
2. J.D. Jackson, *Classical Electrodynamics*, 3rd edn. (Wiley, New York, 1999)
3. G. Harbeke, *Optical Properties of Semi-Conductors* (North-Holland Publishing Company, Amsterdam, 1972), pp. 21–94
4. F. Abeles, *Optical Properties of Metals* (North-Holland Publishing Company, Amsterdam, 1972), p. 1026
5. C.F. Bohren, D.R. Huffman, *Absorption and Scattering of Light by Small Particles* (Wiley-VCH Verlag GmbH and Co, KGaA, 2004)
6. F. Sagnard, IEEE Trans. Instrum. Meas. **54**(3), 1266 (2005)
7. E. Hecht, *Optics*, 3rd edn. (Addison-Wesley, Boston, 1998)
8. Agilent, *Basics of Measuring the Dielectric Properites of Materials, Application Note* 5989th edn. (Agilent Technologies, Inc, Santa Clara, 2104)
9. J. Leng, J. Opsal, H. Chu, M. Senko, D.E. Aspnes, Thin Film Solids-Elsevier Sci. SA **313–314**, 132 (1998)
10. P.G. Techegoin, E.C. Le Ru, M. Meyer, J. Chem. Phys. **125**(October), 164705 (2006)
11. P.G. Etchegoin, E.C. Le Ru, M. M, J. Chem. Phys. **127**, 189901 (2007)
12. A. Vial, T. Laroche, J. Phys. D: Appl. Phys. **40**, 7152 (2007)
13. A. Vial, J. Opt. A: Pure Appl. Opt. **9**, 745 (2007)
14. V. A, L. T, Appl. Phys. B **93**, 139 (2008)
15. P.B. Johnson, R.W. Christy, Phys. Rev. B **6**(12), 4370 (1972)
16. D.E. Aspnes, A.A. Studna, Phys. Rev. B **27**, 985 (1983). https://doi.org/10.1103/PhysRevB.27.985
17. A.F. McKinley, T.P. White, K.R. Catchpole, J. Appl. Phys. **114**(4), 044317 (2013). https://doi.org/10.1063/1.4816619
18. J.D. Jackson, Am. J. Phys. **70**(9), 917 (2002). https://doi.org/10.1119/1.1491265
19. S.A. Schelkunoff, *Electromagnetic Waves (D)* (Van Nostrand, 1966)
20. R.W.P. King, *The Loop Antenna for Transmission and Reception*, vol. 7, 1st edn., Inter-University Electronic Series (McGraw-Hill, New York, 1969), pp. 458–482
21. T.T. Wu, in *Antenna Theory Part 1*, ed. by R.E. Collin, F.J. Zucker (McGraw-Hill, 1969), pp. 306–351
22. T. Do-Nhat, R.H. Macphie, IEEE Trans. Antennas Propag. **37**(12), 1545 (1989). https://doi.org/10.1109/8.45096
23. P.J. Papakanellos, G. Fikioris, Prog. Electromagn. Res. **69**, 77 (2007)
24. G. Fikioris, P.J. Papakanellos, T.K. Mavrogordatos, N. Lafkas, D. Koulikas, SIAM J. Appl. Math. **71**(2), 559 (2011). https://doi.org/10.1137/100785727
25. I. Tastsoglou, G. Fikioris, IEEE Trans. Antennas Propag. **61**(11), 5517 (2013). https://doi.org/10.1109/TAP.2013.2279423
26. I. Tastsoglou, G. Fikioris, IEEE Trans. Antennas Propag. **61**(11), 5527 (2013). https://doi.org/10.1109/TAP.2013.2279426
27. G. Fikioris, P. Papakanellos, H. Anastassiu, IEEE Trans. Antennas Propag. **56**(1), 151 (2008). https://doi.org/10.1109/TAP.2007.913076
28. G. Fikioris, P. Papakanellos, H. Anastassiu, IEEE Trans. Antennas Propag. **58**(10), 3436 (2010). https://doi.org/10.1109/TAP.2010.2055816
29. G. Zhou, G. Smith, IEEE Trans. Antennas Propag. **39**(8), 1167 (1991). https://doi.org/10.1109/8.97352
30. C.A. Balanis, *Antenna Theory, Analysis and Design*, 4th edn. (Wiley, New York, 2016)
31. G. Fikioris, J. Lionas, C. Lioutas, IEEE Trans. Antennas Propag. **51**(8), 1847 (2003). https://doi.org/10.1109/TAP.2003.815412
32. K. Esselle, S. Stuchly, IEEE Trans. Antennas Propag. **38**(7), 1123 (1990). https://doi.org/10.1109/8.55629

Part II
Standalone Loop Antennas and Rings

Abstract The analytical theory begins with perfectly conducting, thin-wire loops in the radio frequency (RF) regime. The classical solutions elaborated in the 1950s and 1960s are given and the essential equations solved in detail. A new elliptical solution is suggested. The issue of non-solvability of some governing equations is considered. The driving point impedance and admittance admitted by the solution are examined and the modal resonances are discovered. These are contrasted with the anti-resonances. The analysis admits an RLC model for the loop. The current distribution and the radiation characteristics admitted by the solution are then elaborated, ending with characteristics of loops large and small compared with the driving wavelength. Thin-wire, lossy material loops are then analysed, which takes loops into the near infra-red (IR) regime. Resonance saturation due to the loss is examined in some detail. Lossy loops and rings with multiply loaded impedances are then characterised mathematically; their input impedance, current distribution and radiation patterns are examined. The resonant characteristics of single capacitor and multiple capacitor loops are particularly examined. Finally, the governing equation for thick, perfectly conducting loops is derived and partially solved, leaving the remainder to the reader. MATLAB code is available upon request from the author for most mathematical expressions in this volume.

Chapter 3
Thin-Wire Perfectly Conducting Loops and Rings

3.1 The Early History

The closed toroidal loop was first studied analytically at radio frequencies in the early 1900s [1, 2]. Thin-wire and perfect conductor assumptions were applied to make the analyses simple. The techniques used in those pre-computer days tended to be quite creative and ingenious, the result of strong physical intuition, with the goal of solving approximately the equation at hand. A full solution for the impedance of the thin-wire loop was not thought to be ready until the derivations by Storer and Wu [3, 4] in 1956 and 1962 respectively. These results were summarised by King in 1969 [5].

Few measurements supported their results then, but it turns out that modern computer simulations show them to be surprisingly accurate [6]. Their method of solution for the loop current involves a Fourier series expansion over an infinite number of modes.

$$E_\phi = \frac{-j\xi}{2b} \sum_{n=-\infty}^{\infty} a_n I_n e^{jn\phi} \qquad \text{where} \qquad I_n = j\frac{2b}{\xi}\frac{E_n}{a_n} \tag{3.1}$$

Here E_ϕ is the driving $\vec{E}$ field, ξ is the impedance of the medium, b is the radius of the loop, and a_n are coefficients. The series converges or diverges depending on how the ratio E_n/a_n behaves. The numerator E_n depends upon the choice of driving source and the coefficients, a_n, depend upon an integral, the kernel of which is the Green's function $\exp(-jkR)/R$. Hence the behaviour of the series depends on how the integral of the Green's function behaves in relation to the E_n. If it reduces with increasing n more rapidly than does the E_n, the ratio increases and the series may diverge.

A. McKinley, *The Analytical Foundations of Loop Antennas and Nano-Scaled Rings*, Signals and Communication Technology,
https://doi.org/10.1007/978-981-13-5893-7_3

Oseen [1], then Hallen [2, p. 37] used the distance function, Eq. (2.46):

$$R^2 = 4b^2 \mathtt{sin}^2 \frac{(\phi - \phi')}{2} + a^2 \tag{3.2}$$

This function is never singular, as explained in Sect. 2.1.5; indeed, it is finite over all ϕ, even at the driving point (where $\phi \to \phi'$). Hallen, after solving the integral, discovered that the coefficients asymptotically approached the value

$$a_n \approx -\frac{n^2}{\pi k_b}\left(\mathtt{ln}\frac{2b}{a} - \gamma - \mathtt{ln} n\right) \tag{3.3}$$

for all ϕ, where γ is Euler's constant. The a_n become very small, experiencing a singularity near $n = n_o \equiv (2b/a)\mathtt{exp}(\gamma)$. Some 20 years later, Storer indicates that Hallen, using a driven source, found a diverging series for the current. He considered the divergence "non-real" and says: "It must be remembered at this point that current is both bounded and continuous (for physical reasons) and hence the series must converge." [3, p. 608]. Storer dealt with the problem by truncating the series after mode 4 and finding an alternative means of determining the rest of the summation.

Then, in 1962, Wu, referring to similar work he and King had done on linear straight antennas [7], proposed the function Eq. (2.39) for the distance function ?

$$R^2 = 4b^2 \mathtt{sin}^2 \frac{(\phi - \phi')}{2} + 4a^2 \mathtt{sin}^2 \frac{(\psi - \psi')}{2}$$

This modifies the second term of R, making it singular when $\phi \to \phi'$ and $\psi \to \psi'$ simultaneously, at the driving point. The integral of the ensuing Green's function is more difficult to solve, but Wu, on solving the integral, discovered that the current series decayed mildly as n increased, similarly to the linear antenna, and thus the series converged. He says "This difference has the following consequence. When the Fourier series expansion is used,... it is easily verified that the series (for the current) converges for $\phi \neq 0$." [4, p. 1302].

The issue here is one of non-solvability. In some combinations of source, E_ϕ, and distance function, R, the governing equation becomes unsolvable. This essentially means that no function $I(\phi)$ can satisfy the equation. This can happen, for example, when one side of the equation is differentiable at a point of interest, but the other side is not. Such occurs for the governing equation used in much of the literature for linear straight antennas. The left hand side is analytic over z, where z is measured along the length of the antenna, while the derivative on the right hand side becomes discontinuous at the centre where the dipole is driven by a delta-function voltage generator across an infinitesimal gap. Indeed, in 1969 Wu examined the solvability of the equation in much detail in his book chapter *Introduction to Linear Antennas* [8]. The same issue arises with respect to the governing equation of the loop antenna, as used by Hallen, Storer and Wu when driven by a similar delta-function voltage generator across an infinitesimal gap.

The non-solvability of the governing equations of linear and loop antennas are covered in considerable detail in a series of papers spanning the last 20 years, initiated by Wu himself with his student, Fikioris in 2001 [9–16].

The upshot is that the non-singular Green's function, paired with certain types of feeders, leads to an unsolvable equation, while the singular kernel, no matter the feeder, does not. In particular, the singular kernel paired with a delta-function generator, leads to a solution that is finite everywhere except at the origin.

The infinity at the driving point is due to the use of the delta-function source across a infinitesimal gap at $\phi = 0$. This, in fact, acts as an infinite susceptance at the driving point. Wu says in particular, "The two (cross-sectional faces) of the halves of the tubular antenna... are kept at different potentials but separated only infinitesimally. Thus, there is an infinite capacitance between the two circular knife edges." [8, p. 319].

King and Wu were well aware of other types of feeds, besides the delta-function generator; several feeds are examined in Sect. 2.6.

3.2 The Governing Equation and Solutions

The governing equation for the thin-wire, perfectly conducting loop antenna to 150 GHz is given by Eq. (2.124). Repeating them here:

$$\begin{aligned}
E_\sigma &= E_\psi = 0;\\
E_\phi &= \frac{-j\xi}{4\pi}\left[\frac{1}{kb}\frac{\partial}{\partial\phi}\int_{-\pi}^{\pi}\left(\frac{1}{2\pi}\int_{-\pi}^{\pi}\frac{e^{-jkR}}{R}d\psi'\right)\frac{\partial I_{\phi'}}{\partial\phi'}d\phi'\right.\\
&\quad\left.+\,kb\int_{-\pi}^{\pi}\left(\frac{1}{2\pi}\int_{-\pi}^{\pi}\frac{e^{-jkR}}{R}d\psi'\right)I_{\phi'}\cos(\phi-\phi')d\phi'\right]\\
&= \frac{-j\xi}{4\pi}\int_{-\pi}^{\pi}\left[\frac{1}{kb}\frac{\partial I_{\phi'}}{\partial\phi'}\frac{\partial}{\partial\phi}+kbI_{\phi'}\cos(\phi-\phi')\right]\left(\frac{1}{2\pi}\int_{-\pi}^{\pi}\frac{e^{-jkR}}{R}d\psi'\right)d\phi'
\end{aligned}\tag{3.4}$$

Define

$$W(\phi-\phi') \equiv \frac{1}{2\pi}\int_{-\pi}^{\pi}\frac{e^{-jkR}}{R}d\psi' \tag{3.5}$$

$W(\phi-\phi')$ has units of m^{-1}. The term $\partial W(\phi-\phi')/\partial\phi = -\partial W(\phi-\phi')/\partial\phi'$ may be integrated by parts yielding

$$\begin{aligned}\frac{\partial I_{\phi'}}{\partial \phi'}\frac{\partial W(\phi-\phi')}{\partial \phi} &= -\frac{\partial I_{\phi'}}{\partial \phi'}\frac{\partial W(\phi-\phi')}{\partial \phi'}\\ &= -\frac{\partial}{\partial \phi'}\left(I_{\phi'}\frac{\partial W(\phi-\phi')}{\partial \phi'}\right)+I_{\phi'}\frac{\partial^2 W(\phi-\phi')}{\partial \phi'^2}\end{aligned}\tag{3.6}$$

Assuming that the function $I_{\phi'}\frac{\partial W(\phi-\phi')}{\partial \phi'}$ is periodic and harmonic in 2π, the integration over ϕ' causes the first term on the right side to go to zero. Consolidating gives the equation that is to be solved for the ring current:

$$E_{\phi} = \frac{-j\xi}{4\pi}\int_{-\pi}^{\pi} I_{\phi'}\left[kb\cos(\phi-\phi') + \frac{1}{kb}\frac{\partial^2}{\partial \phi'^2}\right]W(\phi-\phi')d\phi' \tag{3.7}$$

The equation may be significantly simplified by expanding $W(\phi-\phi')$, a periodic function defined by Eq. (3.5), in a Fourier series.[1]

$$bW(\phi-\phi') \equiv \sum_{n=-\infty}^{\infty} K_n e^{jn(\phi-\phi')} \tag{3.8}$$

The coefficients are:

$$K_n = \frac{b}{2\pi}\int_{-\pi}^{\pi} W(\phi-\phi')e^{-jn(\phi-\phi')}d(\phi-\phi') \tag{3.9}$$

Using this, note that the definition

$$b\cos(\phi-\phi')W(\phi-\phi') \equiv \sum_{n=-\infty}^{\infty} P_n e^{jn(\phi-\phi')} \tag{3.10}$$

yields the coefficients,

$$\begin{aligned}P_n &= \frac{1}{2\pi}\int_{-\pi}^{\pi} b\cos(\phi-\phi')W(\phi-\phi')e^{-jn(\phi-\phi')}d(\phi-\phi')\\ &= \frac{b}{2\pi}\int_{-\pi}^{\pi} W(\phi-\phi')\frac{(e^{-j(n+1)(\phi-\phi')}+e^{-j(n-1)(\phi-\phi')})}{2}d(\phi-\phi')\end{aligned}\tag{3.11}$$

Therefore,

$$P_n = \frac{1}{2}(K_{n+1}+K_{n-1}) \tag{3.12}$$

Substituting into Eq. (3.7) leads to

[1] A b is added to make K_n unitless, as it is in Storer and Wu. Wu rectified Storer's treatment with respect to the units and we follow him.

$$E_\phi = \frac{-j\xi}{4\pi b}\int_{-\pi}^{\pi} I_{\phi'} \sum_{n=-\infty}^{\infty}\left[kb\left(\frac{K_{n+1}+K_{n-1}}{2}\right)-\frac{n^2K_n}{kb}\right]e^{jn(\phi-\phi')}d\phi' \tag{3.13}$$

Defining

$$a_n \equiv kb\left(\frac{K_{n+1}+K_{n-1}}{2}\right)-\frac{n^2K_n}{kb} \tag{3.14}$$

and rearranging gives

$$E_\phi = \frac{-j\xi}{2b}\sum_{n=-\infty}^{\infty} a_n\left[\frac{1}{2\pi}\int_{-\pi}^{\pi} I_{\phi'}e^{-jn\phi'}d\phi'\right]e^{jn\phi} \tag{3.15}$$

A Fourier series expansion of the current

$$I_{\phi'} = \sum_{n=-\infty}^{\infty} I_n e^{jn\phi'} \tag{3.16}$$

has the coefficients

$$I_n = \frac{1}{2\pi}\int_{-\pi}^{\pi} I_{\phi'}e^{-jn\phi'}d\phi' \tag{3.17}$$

and therefore

$$E_\phi = \frac{-j\xi}{2b}\sum_{n=-\infty}^{\infty} a_n I_n e^{jn\phi} \tag{3.18}$$

Noting that Eq. (3.18) is a Fourier series in its own right, its coefficients are

$$-j\frac{\xi}{2b}a_nI_n = \frac{1}{2\pi}\int_{-\pi}^{\pi} E_\phi(\phi)e^{-jn\phi}d\phi \equiv E_n \tag{3.19}$$

If the source, E_ϕ, is a repeating function, it may be represented by a Fourier Series,

$$E_\phi = \sum_{n=-\infty}^{\infty} E_n e^{jn\phi} \tag{3.20}$$

leading to the current coefficients

$$I_n = j\frac{2b}{\xi}\frac{E_n}{a_n} \tag{3.21}$$

which solves (3.7) for the current once the coefficients, a_n, and the driving source, E_n, are known.

3.3 Determining the Coefficients, a_n and K_n

The coefficients, a_n, depend on the coefficients K_n through (3.14). The latter are determined by solving (3.9), where $W(\phi - \phi')$ comes from Eq. 3.5. Assuming the function is periodic over 2π.

$$K_n = \frac{b}{2\pi}\int_{-\pi}^{\pi} W(\phi - \phi')e^{\pm jn(\phi-\phi')}d(\phi - \phi')$$

$$W(\phi - \phi') = \frac{1}{2\pi}\int_{-\pi}^{\pi} \frac{e^{-jkR}}{R}d\psi' \tag{3.22}$$

The solution depends largely on the form that the distance function R takes. For thin-wire loops, the distance between points on the loop takes the form given in Sect. 2.1.5 by (2.45). This should be positive:

$$R^2 = 4b^2\mathtt{sin}^2\left(\frac{\phi - \phi'}{2}\right) \quad \text{or} \quad R = 2b\left|\mathtt{sin}\left(\frac{\phi - \phi'}{2}\right)\right| \tag{3.23}$$

This rids the integrand in Eq. (3.22) of the variable ψ', but goes to 0 when ϕ approaches ϕ'.

3.3.1 *Storer's Recursive Solution*

$W(\phi - \phi')$ is not a function of ψ'. Using Eq. (3.23)

$$W(\phi - \phi') = \frac{1}{2\pi}\int_{-\pi}^{\pi} \frac{e^{-jkR(\phi-\phi')}}{R(\phi - \phi')}d\psi' = \frac{e^{-jkR(\phi-\phi')}}{R(\phi - \phi')} \tag{3.24}$$

Therefore,

$$K_n = \frac{b}{2\pi}\int_{-\pi}^{\pi} \frac{e^{-jkR}}{R}e^{\pm jn(\phi-\phi')}d(\phi - \phi') \tag{3.25}$$

For ease in a later calculation in Sect. 3.3.3, we select $+n$. We then have

$$K_{n+1} - K_n = \frac{b}{2\pi}\int_{-\pi}^{\pi} e^{-jkR}\left\{\frac{e^{j(n+1)(\phi-\phi')} - e^{jn(\phi-\phi')}}{2b\,|\mathtt{sin}((\phi - \phi')/2)|}\right\}d(\phi - \phi') \tag{3.26}$$

Using $2\theta = (\phi - \phi')$, the limits change and $d(\phi - \phi') = 2d\theta$.

$$K_{n+1} - K_n = \frac{1}{\pi}\int_{-\pi/2}^{\pi/2} e^{-j2k_b|\mathtt{sin}\theta|}\left\{\frac{e^{j2(n+1)\theta} - e^{j2n\theta}}{2\,|\mathtt{sin}\theta|}\right\}d\theta \tag{3.27}$$

Although the denominator goes to zero when $\phi = \phi'$, this kernel does not suffer a pole, because the numerator also goes to zero. Moreover, the kernel is periodic over π, the limits may be shifted by $+\pi/2$ without effect. The absolute sines can be replaced by regular sines, since $|\mathtt{sin}\theta|$ is positive over integral limits. An exponential can be extracted from the internal sum.

$$\begin{aligned}
K_{n+1} - K_n &= \frac{1}{\pi}\int_0^{\pi} e^{-j2k_b\mathtt{sin}\theta} e^{j(2n+1)\theta}\left\{\frac{e^{j\theta} - e^{-j\theta}}{2\mathtt{sin}\theta}\right\}d\theta \\
&= \frac{j}{\pi}\int_0^{\pi} e^{j((2n+1)\theta - 2k_b\mathtt{sin}\theta)}d\theta \\
&= \frac{j}{\pi}\int_0^{\pi}\Big[\mathtt{cos}\,((2n+1)\theta - 2k_b\mathtt{sin}\theta) \\
&\qquad + j\mathtt{sin}\,((2n+1)\theta - 2k_b\mathtt{sin}\theta)\Big]d\theta \\
&= -\mathbf{\Omega}_{2n+1}(2k_b) + j\mathbf{J}_{2n+1}(2kb)
\end{aligned} \tag{3.28}$$

$\mathbf{\Omega}_n(x)$ and $\mathbf{J}_n(x)$ are the Lommel-Weber function and the Bessel function of the first kind respectively.[2]

$$\begin{aligned}
\mathbf{\Omega}_n(x) &= \frac{1}{\pi}\left[\int_0^{\pi} \sin\,(n\theta - x\mathrm{sin}\theta)d\theta\right] \\
\mathbf{J}_n(x) &= \frac{1}{\pi}\left[\int_0^{\pi} \cos\,(n\theta - x\sin\theta)d\theta\right]
\end{aligned} \tag{3.29}$$

Note that: $\mathbf{\Omega}_{-n}(x) = (-1)^n\mathbf{\Omega}_n(x)$ and $\mathbf{J}_{-n}(x) = (-1)^n\mathbf{J}_n(x)$

The result, Eq. (3.28), is a recursion formula, but the first term, K_0 is needed. Using $2\theta = (\phi - \phi')$,

$$\begin{aligned}
K_0 &= \frac{b}{2\pi}\int_0^{2\pi}\frac{e^{-jkR}}{R}d(\phi - \phi') \\
&= \frac{b}{\pi}\int_0^{\pi}\frac{e^{-jkR} - 1}{R}d\theta + \frac{b}{\pi}\int_0^{\pi}\frac{1}{R}d\theta
\end{aligned} \tag{3.30}$$

The first term progresses as follows:

[2] See Appendix A. Storer and Wu both used the Janke-Emde [17] definition where $(x\mathtt{sin}\theta - m\theta)$ is used. This gives rise to a sign difference for $\mathbf{\Omega}_n(z)$ with the standard definition.

$$\begin{aligned}
\frac{b}{\pi}\int_0^{\pi}\frac{e^{-jkR}-1}{R}d\theta &= \frac{b}{\pi}\int_0^{\pi}\frac{e^{-j2kb\mathrm{sin}\theta}-1}{2b\mathrm{sin}\theta}d\theta \\
&= \frac{-j}{2\pi}\int_0^{\pi}\left[\int_0^{2kb}e^{-jx\mathrm{sin}\theta}dx\right]d\theta = \frac{-j}{2}\int_0^{2kb}\left[\frac{1}{\pi}\int_0^{\pi}e^{-jx\mathrm{sin}\theta}d\theta\right]dx \\
&= -\frac{j}{2}\int_0^{2kb}\left[\frac{1}{\pi}\int_0^{\pi}\mathrm{cos}\left(-x\mathrm{sin}\theta\right)+j\mathrm{sin}\left(-x\mathrm{sin}\theta\right)d\theta\right]dx \\
&= \frac{1}{2}\int_0^{2kb}\left[\mathbf{\Omega}_0(x)-j\mathbf{J}_0(x)\right]dx
\end{aligned} \tag{3.31}$$

The second term presents some problems, because, if R is left as $2b\mathrm{sin}\theta$, it goes to infinity at $\phi = 0$. Alternatively therefore, we try other forms for R. We might try $R^2 = 4b^2\mathrm{sin}\theta + a^2$ which is not much different from the original, Eq. (3.2[illegible]), but allows for some radius of the wire and prevents the divergence.

$$\begin{aligned}
\frac{b}{\pi}\int_0^{\pi}\frac{1}{R}d\theta &= \frac{b}{\pi}\int_0^{\pi}\frac{1}{\sqrt{4b^2\mathrm{sin}^2\theta + a^2}}d\theta \\
&= \frac{1}{\pi}\left[\frac{b}{a}\int_0^{\pi}\frac{1}{\sqrt{4(b/a)^2\mathrm{sin}^2\theta+1}}d\theta\right] = \frac{1}{\pi}\mathrm{ln}\frac{8b}{a}
\end{aligned} \tag{3.32}$$

The last step was developed by Oseen [1] and used by Hallen and Storer. A modern numerical engine will verify that it is indeed equal for all practical loops where $5 < \Omega < 15$. Therefore,

$$K_0 = \frac{1}{\pi}\mathrm{ln}\frac{8b}{a} + \frac{1}{2}\int_0^{2kb}\left[\mathbf{\Omega}_0(x) - j\mathbf{J}_0(x)\right]dx \tag{3.33}$$

3.3.2 Storer's Non-recursive Solution

From the recursive solution, Eq. (3.28), use the exponential form rather than the Bessels.

$$\begin{aligned}
K_n &= K_0 + \frac{j}{\pi}\sum_{k=0}^{n-1}\int_0^{\pi}e^{j((2k+1)\theta - 2k_b\mathrm{sin}\theta)}d \\
&= K_0 + \frac{j}{\pi}\int_0^{\pi}e^{-j2k_b\mathrm{sin}\theta}\left[\sum_{k=0}^{n-1}e^{j(2k+1)\theta}\right]d\theta \\
&= K_0 + \frac{1}{2\pi}\int_0^{\pi}e^{-j2k_b\mathrm{sin}\theta}\left[\frac{e^{j2n\theta}-1}{\mathrm{sin}\theta}\right]d\theta
\end{aligned} \tag{3.34}$$

The last statement can be checked on a numerical engine for both real and imaginary parts. Using K_0 from Eq. (3.32) and the exponential form from Eq. (3.31),

$$\begin{aligned}
K_n &= \frac{1}{\pi}\ln\frac{8b}{a} + \frac{1}{2\pi}\int_0^{\pi} \frac{e^{-j2kb\sin\theta}-1}{\sin\theta}d\theta + \frac{1}{2\pi}\int_0^{\pi} e^{-j2k_b\sin\theta}\left[\frac{e^{j2n\theta}-1}{\sin\theta}\right]d\theta \\
&= \frac{1}{\pi}\ln\frac{8b}{a} + \frac{1}{2\pi}\int_0^{\pi} \frac{e^{-j2k_b\sin\theta}e^{j2n\theta}-1}{\sin\theta}d\theta \\
&= \frac{1}{\pi}\ln\frac{8b}{a} + \frac{1}{2\pi}\int_0^{\pi} \left(e^{-j2k_b\sin\theta}-1\right)\frac{e^{j2n\theta}}{\sin\theta}d\theta + \frac{1}{2\pi}\int_0^{\pi} \frac{e^{j2n\theta}-1}{\sin\theta}d\theta
\end{aligned} \tag{3.35}$$

The middle term reduces as in Eq. (3.31), using the Bessel definitions, Eqs. (A.1) and (A.12):

$$\begin{aligned}
\frac{1}{2\pi}\int_0^{\pi}\left(e^{-j2k_b\sin\theta}-1\right)\frac{e^{j2n\theta}}{\sin}\theta d\theta &= -\frac{j}{2\pi}\int_0^{\pi}\left[\int_0^{2k_b} e^{-jx\sin\theta}dx\right]e^{j2n\theta}d\theta \\
&= -\frac{j}{2}\int_0^{2k_b}\left[\frac{1}{\pi}\int_0^{\pi}(\cos(2n\theta-x\sin\theta)+j\sin(2n\theta-x\sin\theta))\,d\theta\right]dx \\
&= \frac{1}{2}\int_0^{2k_b}(\mathbf{\Omega}_{2n}(x)-j\mathbf{J}_{2n}(x))\,dx
\end{aligned} \tag{3.36}$$

The real part of the third term is a summation, while the imaginary part is zero; therefore,

$$\frac{1}{2\pi}\int_0^{\pi}\frac{e^{j2n\theta}-1}{\sin\theta}d\theta = -\frac{2}{\pi}\sum_{k=0}^{|n|-1}\frac{1}{2k+1} \tag{3.37}$$

Combining we have the non-recursive solution:

$$\begin{aligned}
K_n &= \frac{1}{\pi}\ln\frac{8b}{a} + \frac{1}{2}\int_0^{2k_b}(\mathbf{\Omega}_{2n}(x)-j\mathbf{J}_{2n}(x))\,dx - \frac{2}{\pi}\sum_{k=0}^{|n|-1}\frac{1}{2k+1} \\
K_0 &= \frac{1}{\pi}\ln\frac{8b}{a} + \frac{1}{2}\int_0^{2kb}[\mathbf{\Omega}_0(x)-j\mathbf{J}_0(x)]dx
\end{aligned} \tag{3.38}$$

3.3.3 Wu's Solution

Taking the Oseen-Hallen-Storer solution for K_n to its large n approximation shows that it decays quickly to zero near $n = (2b/a)e^{-\gamma}$, where $\gamma = 0.5772$, Euler's constant.[3] This causes a divergence in the current when used in Eq. (3.1) with their delta voltage generator model. This is, of course, not physically possible. In 1962, Wu [4], drawing on work with linear straight antennas, indicated that the reason was the use of Eq. (3.23) for R, and he suggested using Eq. (2.39) instead.

$$R^2 = 4b^2\mathtt{sin}^2\frac{(\phi-\phi')}{2} + 4a^2\mathtt{sin}^2\frac{\psi-\psi'}{2} = b^2\left(4\mathtt{sin}^2\frac{\theta}{2} + A^2\right) \tag{3.39}$$

where $A^2 = (2a/b)^2\mathtt{sin}^2((\psi-\psi')/2)$. For thin wires, this is perfectly acceptable. The method is difficult to follow in his article, but King elaborates in [5]. King's work is expanded here, filling in the gaps and fixing several typographical errors. Essentially, Wu solves Storer's integral using his expanded R above, then examines the difference between his answer and Storer's. This difference he adds as a correction to his answer, which he considers more accurate. Wu then finds that the divergence now does not occur. We focus more critically on the reasons for the divergence in Sect. 3.5.

Wu begins his analysis with Eq. (3.22).

$$K_n = \frac{b}{2\pi}\int_{-\pi}^{\pi}\left[\frac{1}{2\pi}\int_{-\pi}^{\pi}\frac{e^{-jkR}}{R}d\psi'\right]e^{\pm jn(\phi-\phi')}d(\phi-\phi') = K_{-n}$$

Change the variable of integration to A. Consequently,

$$d\psi' = -\frac{1}{(2a/b)\mathtt{cos}((\psi-\psi')/2)}dA = -\frac{1}{\sqrt{(2a/b)^2 - A^2}}dA \tag{3.40}$$

and the limits change to $\pm 2a/b$ and the negative sign disappears with the switch in the limits. Taking $\theta = \phi - \phi'$, $d\theta = d(\phi-\phi')$, then

$$\begin{aligned}
K_n &= \frac{1}{\pi}\int_{-2a/b}^{2a/b}\left[\frac{b}{2\pi}\int_{-\pi}^{\pi}\frac{e^{-jkR}}{R}e^{jn\theta}d\theta\right]\frac{dA}{\sqrt{(2a/b)^2 - A^2}}\\
&= \frac{1}{\pi}\int_{-2a/b}^{2a/b}\left[\hat{K}_n(A)\right]\frac{dA}{\sqrt{(2a/b)^2 - A^2}}\\
\hat{K}_n(A) &= \frac{1}{2\pi}\int_{-\pi}^{\pi}e^{jn\theta}\frac{e^{-jk_b\sqrt{4\mathtt{sin}^2(\theta/2)+A^2}}}{\sqrt{4\mathtt{sin}^2(\theta/2)+A^2}}d\theta
\end{aligned} \tag{3.41}$$

The integral, $\hat{K}_n(A)$, is where Storer began, but without the A. Wu suggests that it may be approximated by setting $\mathtt{sin}^2(\theta/2)$ to $(\theta/2)^2$. This means that the kernel is in error by a considerable amount as ϕ moves past an angle $\pi/2$ from ϕ', regardless of the size of A. Changing variables using $\theta = A\mathtt{sinh}v$, we have $d\theta = A\mathtt{cosh}v\,dv$, the integral reduces to

[3] See Sect. 3.5 for the expansion.

$$\frac{1}{2\pi}\int_{\mathrm{asinh}(-\pi/A)}^{\mathrm{asinh}(\pi/A)} e^{jA(n\sinh v-k_b\cosh v)}dv \tag{3.42}$$

The two equations, Eqs. (3.41) and (3.42), are equal in the real parts. The imaginary parts are approximately zero, since the imaginary part of the kernel is nearly odd. The exponential can be expanded as:

$$(n\sinh v-k_b\cosh v)(n\sinh v+k_b\cosh v)=(n^2-k_b^2)\sinh^2 v-k_b^2 \tag{3.43}$$

Realising that the summations into which these will be inserted will go to infinity, n rapidly becomes much bigger than k_b. The assumption that k_b is small compared to n hurts only when $n < 10$ or so. So assuming this, drop the $-k_b^2$ and substitute into the exponential.

$$\frac{1}{\pi}\int_{0}^{\mathrm{asinh}(\pi/A)} e^{jA(\sqrt{n^2-k_b^2}\sinh v}dv \tag{3.44}$$

The kernel of Eq. (3.44) is even in real and exactly odd in imaginary. The consequence is that the integral matches Eq. (3.42) quite well. In fact, the integral, in which the limits are made variable for 0 to x, converges to Eq. (3.42) by an upper limit of $x = 6$. We can therefore move the upper limit to infinity without harm. The resulting integral is the modified Bessel of the Second Kind Eq. (A.11):

$$\frac{1}{\pi}\int_{0}^{\infty} e^{jA(\sqrt{n^2-k_b^2}\sinh v}dv \equiv \frac{1}{\pi}\mathbf{K}_0\left(A\sqrt{n^2-k_b^2}\right) \quad n > k_b \tag{3.45}$$

An expansion of $\mathbf{K}_0(z)$ around small z gives [18, Eqs. 9.6.13 and 9.6.54]

$$\mathbf{K}_0(z)=-\ln\left(\frac{1}{2}z\right)-\gamma=-\ln\left(\frac{A}{2}\sqrt{n^2-k_b^2}\right)-\gamma \tag{3.46}$$

The difference then between Wu's solution for $\hat{K}_n(A)$ which uses A, and Storer's non-recursive solution, K_n Eq. (3.38), which does not, is approximately

$$\begin{aligned} K_n-\hat{K}_n(A) &= \frac{1}{\pi}\ln\frac{8b}{a}+\frac{1}{2}\int_0^{2k_b}(\mathbf{\Omega}_{2n}(x)-j\mathbf{J}_{2n}(x))\,dx \\ &\quad -\frac{2}{\pi}\sum_{k=0}^{|n|-1}\frac{1}{2k+1}+\frac{1}{\pi}\ln\left(\frac{A}{2}\sqrt{n^2-k_b^2}\right)+\frac{1}{\pi}\gamma \end{aligned} \tag{3.47}$$

The logarithm in the last line can be eliminated with a clever expansion of the first logarithm.

$$
\begin{aligned}
\ln\frac{8b}{a} &= \ln(4|n|) + \ln 2 - \ln(|n|) + \ln\frac{b}{a} \\
&= \ln(4|n|) + \ln\sqrt{n^2 - k_b^2} - \ln(|n|) - \ln\frac{\sqrt{n^2 - k_b^2}}{b/a} + \ln 2 \\
&= \ln(4|n|) + \frac{1}{2}\ln\left(1 - \left(\frac{k_b}{n}\right)^2\right) - \ln\left(\frac{A\sqrt{n^2 - k_b^2}}{2}\right)
\end{aligned}
\tag{3.48}
$$

An A in the last term is substituted for b/a, because it is needed and is about that anyway given its definition. We then have King's intermediate result.[4]

$$
K_n - \hat{K}_n(A) = \frac{1}{2\pi}\left[+\ln\left(1 - \left(\frac{k_b}{n}\right)^2\right) + 2C_n + \pi\int_0^{2k_b} (\boldsymbol{\Omega}_{2n}(x) - j\mathbf{J}_{2n}(x))\, dx\right]
$$

$$
C_n = \gamma - 2\sum_{k=0}^{|n|-1} \frac{1}{2k+1} + \ln(4|n|) \tag{3.49}
$$

If this is the difference between the two of them, then it can be added to Wu's "more accurate" result to give a new answer to Storer's integral

$$
\begin{aligned}
\frac{b}{2\pi}\int_{-\pi}^{\pi} \frac{e^{-jkR}}{R} e^{jn\theta}\, d\theta = \frac{1}{2\pi}\Bigg[&2\mathbf{K}_0\left(A\sqrt{n^2 - k_b^2}\right) \\
&+ \ln\left(1 - \left(\frac{k_b}{n}\right)^2\right) + 2C_n + \pi\int_0^{2k_b} (\boldsymbol{\Omega}_{2n}(x) - j\mathbf{J}_{2n}(x))\, dx\Bigg]
\end{aligned}
\tag{3.50}
$$

Implementing the approximation Eq. (3.46) simplifies a bit. Repeating,

$$
\begin{aligned}
\mathbf{K}_0(A\sqrt{n^2 - k_b^2}) &= -\ln\left(|n|A\sqrt{1 - \frac{k_b^2}{n^2}}\right) - \gamma \\
&= -\ln(|n|A) - \gamma - \frac{1}{2}\ln\left(1 - \frac{k_b^2}{n^2}\right) = \mathbf{K}_0(|n|A) - \frac{1}{2}\ln\left(1 - \frac{k_b^2}{n^2}\right)
\end{aligned}
\tag{3.51}
$$

Substituting,

$$
\frac{b}{2\pi}\int_{-\pi}^{\pi} \frac{e^{-jkR}}{R} e^{jn\theta}\, d\theta = \frac{1}{2\pi}\left[2\mathbf{K}_0(|n|A) + 2C_n + \pi\int_0^{2k_b} (\boldsymbol{\Omega}_{2n}(x) - j\mathbf{J}_{2n}(x))\, dx\right]
$$

[4] It is useful to remember that the sign difference on $\boldsymbol{\Omega}_{2n}(x)$ is due to a difference in its definition as used by Storer and Wu. See Sect. 3.3.1.

Finally, this must integrated in Eq. (3.41).

$$K_n = \frac{1}{\pi}\int_{-2a/b}^{2a/b}\left[\frac{b}{2\pi}\int_{-\pi}^{\pi}\frac{e^{-jkR}}{R}e^{jn\theta}d\theta\right]\frac{dA}{\sqrt{(2a/b)^2 - A^2}}$$
$$= \frac{1}{2\pi^2}\int_{-2a/b}^{2a/b}\left[2\mathbf{K}_0\left(|n|A\right) + 2C_n + \pi\int_0^{2k_b}\left(\mathbf{\Omega}_{2n}(x) - j\mathbf{J}_{2n}(x)\right)dx\right]\frac{dA}{\sqrt{(2a/b)^2 - A^2}} \tag{3.52}$$

The second and third terms are constants in A and the simple integral

$$\int_{-2a/b}^{2a/b}\frac{dA}{\sqrt{(2a/b)^2 - A^2}} = \mathtt{atan}\left(\frac{A}{\sqrt{(2a/b)^2 - A^2}}\right)\Bigg|_{-2a/b}^{2a/b} = \pi \tag{3.53}$$

For the first integral, we use the definition of A from Eq. (3.39), $A^2 = (2a/b)^2\mathtt{sin}^2((\psi - \psi')/2)$

$$\int_{-2a/b}^{2a/b}\mathbf{K}_0\left(|n|A\right)\frac{dA}{\sqrt{2a/b)^2 - A^2}} = \int_{-\pi}^{\pi}\mathbf{K}_0\left((2a/b)|n|\mathtt{sin}(\psi - \psi'/2)\right)d\psi' \tag{3.54}$$

The kernel on the right side is periodic in 2π and therefore we may truncate the limits to $\{0, \pi\}$ if we multiply by 2. Then changing variables to $\theta = (\psi - \psi')/2$, we find the simpler form,

$$4\int_0^{\pi/2}\mathbf{K}_0\left((2a/b)|n|\mathtt{sin}\theta\right)d\theta \tag{3.55}$$

This appears in a text on Bessel functions, cited by King [19] as

$$\int_0^{\pi/2}\mathbf{K}_0(2z\mathtt{sin}\theta)d\theta = \frac{\pi}{2}\mathbf{I}_0(z)\mathbf{K}_0(z) \tag{3.56}$$

The result of integrating fully Eq. (3.41) is

$$K_n = \frac{1}{2\pi^2}\left[2\pi\mathbf{K}_0\left(\frac{|n|a}{b}\right)\mathbf{I}_0\left(\frac{|n|a}{b}\right) + 2\pi C_n + \pi^2\int_0^{2k_b}\left(\mathbf{\Omega}_{2n}(x) - j\mathbf{J}_{2n}(x)\right)dx\right]$$
$$= \frac{1}{\pi}\left[\mathbf{K}_0\left(\frac{|n|a}{b}\right)\mathbf{I}_0\left(\frac{|n|a}{b}\right) + C_n\right] + \frac{1}{2}\int_0^{2k_b}\left(\mathbf{\Omega}_{2n}(x) - j\mathbf{J}_{2n}(x)\right)dx$$
$$\mathtt{where}\quad K_0 = \frac{1}{\pi}\mathtt{ln}\frac{8b}{a} + \frac{1}{2}\int_0^{2kb}\left[\mathbf{\Omega}_0(x) - j\mathbf{J}_0(x)\right]dx$$
$$\mathtt{and\,where}\quad C_n = \gamma - 2\sum_{k=0}^{|n|-1}\frac{1}{2k+1} + \mathtt{ln}(4|n|) \tag{3.57}$$

This is the final answer as presented by Wu and clarified by King. Note that we regain Storer's non-recursive solution if we use the small value approximation Eq. (3.46). The product term reduces by $1/n$ as $n \to \infty$ while the other terms reduce by $1/n^2$.

3.4 An Elliptical Solution

We begin again with the integral that needs to be solved:

$$K_n = \frac{b}{2\pi}\int_{-\pi}^{\pi}\left[\frac{1}{2\pi}\int_{-\pi}^{\pi}\frac{e^{-jkR}}{R}d\psi'\right]e^{jn(\phi-\phi')}d(\phi-\phi') = K_{-n}$$

The critical problem with the solution to this has always been the choice of R. The best choices for thin wires, $R^2 = 4b^2\mathtt{sin}^2(\phi-\phi')/2$ and $R^2 = 4b^2\mathtt{sin}^2(\phi-\phi')/2 + a^2$, as Storer elected, leads to a divergence in the current series. Storer's non-recursive solution is Eq. (3.38)

$$K_n = \frac{1}{\pi}\mathtt{ln}\frac{8b}{a} + \frac{1}{2}\int_0^{2k_b}(\mathbf{\Omega}_{2n}(x) - j\mathbf{J}_{2n}(x))\,dx - \frac{2}{\pi}\sum_{k=0}^{|n|-1}\frac{1}{2k+1}$$

Wu used the form $R^2 = 4b^2\mathtt{sin}^2(\phi-\phi')/2 + 4a^2\mathtt{sin}^2\psi'/2$, which avoids the pole. His solution is Eq. (3.57)

$$K_n = \frac{1}{\pi}\left[\mathbf{K}_0\left(\frac{|n|a}{b}\right)\mathbf{I}_0\left(\frac{|n|a}{b}\right) + C_n\right] + \frac{1}{2}\int_0^{2k_b}(\mathbf{\Omega}_{2n}(x) - j\mathbf{J}_{2n}(x))\,dx$$

$$C_n = \gamma - 2\sum_{k=0}^{|n|-1}\frac{1}{2k+1} + \mathtt{ln}(4|n|) \tag{3.58}$$

Another approach is to use the form $R^2 = 4b^2\mathtt{sin}^2(\phi-\phi')/2 + 4a^2$, which has no pole and which is just as accurate as Wu's form for thin wires. Using $2\theta = \phi - \phi'$, we therefore solve

$$K_n = \frac{b}{\pi}\int_{-\pi/2}^{\pi/2}\frac{e^{-jkR}}{R}e^{j2n\theta}\,d\theta \tag{3.59}$$

This is periodic in π and therefore we may shift the limits. Then, using Hallen's method (see Eq. (3.30))

$$K_n = \frac{b}{\pi}\int_0^{\pi}e^{j2n\theta}\frac{(e^{-jkR}-1)}{R}d\theta + \frac{b}{\pi}\int_0^{\pi}\frac{e^{j2n\theta}}{R}d\theta \tag{3.60}$$

For the first term, we revert to $R = 2b|\mathtt{sin}\theta|$. The kernel remains periodic in π and has no pole. It expands as follows:

$$\begin{aligned}
\frac{b}{\pi}\int_0^{\pi} e^{j2n\theta}\frac{\left(e^{-jkR}-1\right)}{R}d\theta &= \frac{b}{\pi}\int_0^{\pi} e^{j2n\theta}\frac{\left(e^{-jk2b|\mathtt{sin}\theta|}-1\right)}{2b|\mathtt{sin}\theta|}d\theta \\
&= \frac{-j}{2\pi}\int_0^{\pi} e^{j2n\theta}\left[\int_0^{2k_b} e^{-jx|\mathtt{sin}\theta|}dx\right]d\theta \\
&= \int_0^{2k_b}\left[\frac{-j}{2\pi}\int_0^{\pi} e^{j(2n\theta-x|\mathtt{sin}\theta|)}d\theta\right]dx \\
&= \frac{1}{2}\int_0^{2k_b}\left[\mathbf{\Omega}_{2n}(x) - j\mathbf{J}_{2n}(x)\right]dx
\end{aligned} \tag{3.61}$$

For the second term, we use Eq. (3.37) and have

$$\begin{aligned}
\frac{b}{\pi}\int_0^{\pi}\frac{e^{j2n\theta}}{R}d\theta &= \frac{b}{\pi}\left[\int_0^{\pi}\frac{e^{j2n\theta}-1}{R}d\theta + \int_0^{\pi}\frac{1}{R}d\theta\right] \\
&= -\frac{2}{\pi}\sum_{k=0}^{|n|-1}\frac{1}{2k+1} + \frac{1}{2\pi}\int_0^{\pi}\frac{b/a}{\sqrt{(b/a)^2\mathtt{sin}^2\theta+1}}d\theta
\end{aligned} \tag{3.62}$$

The second term on the right is symmetric around $\pi/2$, so the limits may be changed leading to a Complete Elliptical Integral of the First Kind:

$$\int_0^{\pi/2}\frac{1}{\sqrt{(b/a)^2\mathtt{sin}^2\theta+1}}d\theta \equiv F\left(\frac{\pi}{2}\middle| -\frac{b^2}{a^2}\right) \equiv K\left(-\frac{b^2}{a^2}\right) \tag{3.63}$$

This is a tabulated function, like the Bessels. It can of course be numerically computed from the integral itself. The effect of the aspect ratio of the loop is apparent. The end result is

$$\begin{aligned}
K_n &= \frac{1}{2}\int_0^{2k_b}\left[\mathbf{\Omega}_{2n}(x) - j\mathbf{J}_{2n}(x)\right]dx - \frac{2}{\pi}\sum_{k=0}^{|n|-1}\frac{1}{2k+1} + \frac{b/a}{\pi}K\left(-\frac{b^2}{a^2}\right) \\
K_n &= \frac{1}{2\pi}\left[\pi\int_0^{2k_b}\left[\mathbf{\Omega}_{2n}(x) - j\mathbf{J}_{2n}(x)\right]dx - 4\sum_{k=0}^{|n|-1}\frac{1}{2k+1} + \frac{2b}{a}K\left(-\frac{b^2}{a^2}\right)\right]
\end{aligned} \tag{3.64}$$

To develop K_0, we return to Eq. (3.32), and rejecting Oseen's derivation, recognize the integral from Eq. (3.63). Using $R^2 = 4b^2\mathtt{sin}^2(\phi - \phi')/2 + 4a^2$

$$
\begin{aligned}
K_0 &= \frac{b}{\pi}\int_0^{\pi}\frac{1}{R}d\theta + \frac{1}{2}\int_0^{2kb}\left[\boldsymbol{\Omega}_0(x) - j\mathbf{J}_0(x)\right]dx \\
&= \frac{b/a}{\pi}\int_0^{\pi/2}\frac{1}{\sqrt{(b/a)^2\sin^2\theta + 1}}d\theta + \frac{1}{2}\int_0^{2kb}\left[\boldsymbol{\Omega}_0(x) - j\mathbf{J}_0(x)\right]dx \\
&= \frac{b/a}{\pi}K\left(-\frac{b^2}{a^2}\right) + \frac{1}{2}\int_0^{2kb}\left[\boldsymbol{\Omega}_0(x) - j\mathbf{J}_0(x)\right]dx
\end{aligned} \tag{3.65}
$$

The difference between Storer's non-recursive solution for K_n and this solution is

$$
\Delta_S = \frac{1}{\pi}\ln\frac{8b}{a} - \frac{b/a}{\pi}K\left(-\frac{b^2}{a^2}\right) \tag{3.66}
$$

That between Wu's solution and the new one is

$$
\Delta_W = \frac{1}{\pi}\left[\mathbf{K}_0\left(\frac{|n|a}{b}\right)\mathbf{I}_0\left(\frac{|n|a}{b}\right) + \gamma + \ln(4|n|)\right] - \frac{b/a}{\pi}K\left(-\frac{b^2}{a^2}\right) \tag{3.67}
$$

Δ_W reduces to Δ_S when the $\mathbf{K}_0$ and $\mathbf{I}_0$ are taken to first order approximations. The maximum difference is on the order of 11% for $\Omega = 12$ thin-wire loops.

3.5 Determining Non-solvability and Convergence

Recalling the current coefficients of the Fourier Series, Eq. (3.1):

$$
I_n = j\frac{2b}{\xi}\frac{E_n}{a_n} \tag{3.68}
$$

The coefficients, a_n are given by Eq. (3.14)

$$
a_n = k_b\left(\frac{K_{n+1} + K_{n-1}}{2}\right) - \frac{n^2 K_n}{k_b} \tag{3.69}
$$

with K_n given by Eq. (3.57) just above.

It is important that the coefficients, a_n, decrease with n as slowly as or more slowly than E_n, since, if it does not, a divergence in the summation then occurs. The decrease in a_n depends upon the behaviour of K_n as $n \to \infty$. In Storer's non-recursive solution, Eq. (3.38), the integral of the Lommel-Weber and Bessel functions reduce to 0 as $n \to \infty$ [18, use Eqs. 9.3.1 and 12.3.5]. As $n \to \infty$.

$$
\lim_{n\to\infty}\left(\sum_{k=0}^{|n|-1}\frac{1}{(2k+1)}\right) = \lim_{n\to\infty}\left(-\sum_{k=1}^{n}\frac{(-1)^k}{k}\right) + \lim_{n\to\infty}\left(\sum_{k=1}^{n}\frac{1}{2k}\right) \tag{3.70}
$$

The first term on the right side is $\mathtt{ln}(2)$. The second term is a harmonic series related to Euler's constant $\gamma = 0.5772$, as

$$\gamma \equiv \lim_{n\to\infty}\left(-\mathtt{ln}(|n|) + \sum_{k=1}^{n}\frac{1}{k}\right) \tag{3.71}$$

Substituting this into Eq. (3.38) yields for the coefficients

$$K_{n\to\infty} \approx \frac{1}{\pi}\left[\mathtt{ln}\frac{2b}{a} - \gamma - \mathtt{ln}(|n|)\right] \tag{3.72}$$

The large n asymptotic for of the K_n coefficients decay to zero near $n = n_o \equiv (2b/a)e^{-\gamma}$. At this value of n then, the a_n decay rapidly and the current diverges. For an $\Omega = 10$ thin loop, the current diverges at mode $n_o = 26$. Storer's response was to truncate the summation at term 5 and estimate the value of the rest of the summation.

Wu's formulation leads to

$$K_{n\to\infty} \approx \frac{1}{\pi}\left[\mathbf{K}_0\left(\frac{|n|a}{b}\right)\mathbf{I}_0\left(\frac{|n|a}{b}\right) + \frac{\gamma}{2} + \mathtt{ln}(4|n|)\right] \tag{3.73}$$

The K_n decay mildly avoiding a divergence in the current series.

The elliptical solution has the same behaviour as Storer's formulation, albeit at a different mode number. The elliptical integral is a constant dependent on b/a. Specify $K(-(b/a)^2) \equiv k$, then

$$K_{n\to\infty} \approx \frac{1}{\pi}\left[\frac{b}{a}k + \mathtt{ln}\frac{4}{|n|} - \gamma\right] \tag{3.74}$$

and the divergence occurs as $n = n_o = 4e^{-(\gamma-(b/a)\mathbf{k})}$. For $\Omega = 10$ this is mode 202.

In an illuminating paper on the solvability problem, [13], Fikioris et al. suggested that the Hallen/Storer K_n decay very rapidly in large n:

$$bK_n\Big|_{n\to\infty} \propto \frac{(f(2b/a))^{|n|}}{\sqrt{|n|}}$$
$$\text{where } f\left(\frac{2b}{a}\right) = 1 - \frac{2(\sqrt{(2b/a)^2+1}-1)}{(2b/a)^2} \tag{3.75}$$

Following [13], substitution into Eq. (3.14) yields the following form for the a_n:

$$a_n \propto \frac{|n|^3(f(2b/a))^{|n|}}{|n|^{3/2}} \tag{3.76}$$

This rises initially to a peak at about $n = 100$ then falls beyond that to approach 0 rapidly. This gives the current

$$I_n\Big|_{n\to\infty} \propto \frac{|n|^{3/2}E_n}{|n|^3(f(2b/a))^{|n|}} \tag{3.77}$$

If E_n is constant in n, the current diverges above $n = 100$. The numerator E_n must offset this behaviour by decaying faster than the denominator falls in large n or else the current will diverge. We examine now the voltage sources arising from the methods of coupling examined in Sect. 2.6.

Direct Coupling. The voltage Fourier coefficients for a loop modelled using delta function source (that is, a Dirac Delta Distribution - see Eq. 2.134) are constant:

$$E_n = E_{\delta n} = -\frac{V_0}{2\pi b} \tag{3.78}$$

The current in Eq. (3.77) therefore diverges and the governing equation, Eq. (3.7), is not solvable. Similarly, the voltage coefficients for a loop modelled using a finite gap are constant if $n = 0$ and inversely proportional to n for $n \neq 0$:

$$E_n = E_{\Delta n} = \begin{Bmatrix} -V_0\mathtt{sin}(n\gamma/2)/(\pi b n\gamma) \;\; \mathtt{if} \;\; n \neq 0 \\ -V_0/(2\pi b), \;\; \mathtt{if} \;\; n = 0 \end{Bmatrix} \tag{3.79}$$

In the latter case, the additional linearity in the denominator cannot offset the exponential behaviour of the K_n, and the current again diverges somewhat after $n > 200$, again making the governing equation unsolvable. The essential reason for this difficulty is that the delta distribution implies an infinitesimal gap width across which the voltage source is placed, and this further implies an infinite capacitance at the source. Such an argument gives rise to an infinite input susceptance, which has caused problems with the analysis of cylindrical dipole antennas, of which there is a long history in the literature [8, 9, 12, 15, 16, 20, 21]. Identical problems arise for the analysis of the loop [13, 22]. See Tastsoglou's serial papers [16, 21] for a history of the gap problem.

Inductive Coupling. For the voltage Fourier coefficients of a loop modelled using a frill generator, Fikioris gives

$$E_n = E_{fn}(\phi) = \frac{V}{2\pi b\,\mathtt{ln}\,(a_o/a)}\Big[(K_{0,n+1} + K_{0,n-1}) - (K_{n+1} - K_{n-1})\Big] \tag{3.80}$$

where $K_{0,n}$ is equal to K_n, but with a replaced by the radius of the frill, a_0. Since for large n, $K_{0,n} \ll K_n$, the large n asymptote is proportional to

$$E_{fn}(\phi) \to \frac{(f(2b/a))^{|n|}}{\sqrt{|n|}} \tag{3.81}$$

yielding for the current a form

$$I_n \rightarrow j\frac{1}{|n|^2} \tag{3.82}$$

The current therefore converges, making the governing equation solvable.

Illuminated Coupling. The $\vec{E}$ field at the point Q on the wire due to an illuminating TEM wave is given in Sect. 2.6.3. The principal equation is given by Eq. (2.150) in general for the closed loop of any thickness. An argument can be made that under the thin-wire approximation, where $\sigma_q \rightarrow 0$ only the $\hat{\phi}$ component affects the currents in the loop, then

$$\vec{E}_Q = E_0 e^{-j\vec{k}\cdot\vec{R}}\left[(\cos\gamma\cos\alpha + \sin\gamma\cos\theta_p\sin\alpha)\hat{\phi}\right] \tag{3.83}$$

where $\vec{k}\cdot\vec{R}$ takes an abbreviated form of Eq. (2.146), namely $\vec{k}\cdot\vec{R} = z\cos\alpha - kd_0$ where $z = k_b\sin\theta_p$. The $-kd_0$ is necessary, since this plane wave is not approaching the wire broadside (that is, $\theta_p \neq 0$). When $d_0 = 0$, the wave is on top of the loop but at angle to the plane of the loop. These were found by Kanda in 1984 [23].[5] The Fourier expansion of this function yields Fourier's coefficients:

$$E_Q = \sum_{-\infty}^{\infty} E_{Q,n}e^{jn\alpha} \quad E_{Q,n} = \frac{1}{2\pi}\int_{-\pi}^{\pi} E_Q e^{-jn\alpha}d\alpha$$

$$\begin{aligned} E_{Q,n} &= E_0\cos\gamma\left[\frac{1}{2\pi}\int_{-\pi}^{\pi}\cos\alpha e^{jz\cos\alpha}e^{-jn\alpha}d\alpha\right] \\ &+ E_0\sin\gamma\cos\theta_p\left[\frac{1}{2\pi}\int_{-\pi}^{\pi}\sin\alpha e^{jz\cos\alpha}e^{-jn\alpha}d\alpha\right] \end{aligned} \tag{3.84}$$

These are related to the Bessel function Eq. (A.3) and their derivatives, Eq. (A.8)

$$\mathbf{J}_n(z) \equiv \frac{j^{-n}}{2\pi}\int_{-\pi}^{\pi} e^{jz\cos\alpha}e^{-jn\alpha}d\alpha$$

$$\frac{2n}{z}\mathbf{J}_n(z) = \mathbf{J}_{n+1}(z) + \mathbf{J}_{n-1}(z) \quad \text{and} \quad \mathbf{J}_n'(z) = \frac{n}{z}\mathbf{J}_n(z) - \mathbf{J}_{n+1}(z)$$

Using

$$\cos\alpha e^{-jn\alpha} = \frac{1}{2}\left(e^{-j(n-1)\alpha} + e^{-j(n+1)\alpha}\right)$$

$$\sin\alpha e^{-jn\alpha} = \frac{1}{2j}\left(e^{-j(n-1)\alpha} - e^{-j(n+1)\alpha}\right)$$

[5]This is true, except for a sign change on the second term of $\hat{\phi}$ in Eq. (3.83) compared with his Eq. (12). This appears to be due to Kanda's coordinate system twisted by 180° compared with Eq. (2.142). In that case, our γ is π radians out of phase with his. Kanda also has an extra phase factor of $e^{-jn\phi_p}$, which is unnecessary here since ϕ_p is assimilated into α.

we have on substituting,

$$\begin{aligned} E_{Q,n} =& \frac{E_0 \cos\gamma}{2}\left[\frac{\mathbf{J}_{n-1}(z)}{j^{-(n-1)}} + \frac{\mathbf{J}_{n+1}(z)}{j^{-(n+1)}}\right] \\ &+ \frac{E_0 \sin\gamma \cos\theta_p}{2j}\left[\frac{\mathbf{J}_{n-1}(z)}{j^{-(n-1)}} - \frac{\mathbf{J}_{n+1}(z)}{j^{-(n+1)}}\right] \\ =& j^{(n-1)} E_0 \cos\gamma \mathbf{J}_n'(z) - j^n E_0 \sin\gamma \cos\theta_p \frac{n}{z}\mathbf{J}_n(z) \end{aligned} \tag{3.85}$$

These were found by Kanda [23]. The current becomes

$$I_n = j\frac{2b}{\xi}\frac{E_{Q,n}}{a_n} \tag{3.86}$$

Fikioris et al. [13] note that both $\mathbf{J}_n(z)$ and its derivative decay much faster than does a_n, leading the current to decay rapidly and the series to converge.

There may be one exception to this general rule. We establish a straight-on broadside illumination onto the plane of the loop, as in Figs. 2.16 and 2.17, by setting $\theta_p = 0$, $\phi_p = 0$ and $\gamma = 0$. This gives us a reduction of Eq. (3.85). We have:

$$E_{Q,n} = j^{n-1} E_0 \mathbf{J}_n'(0) = E_0 \tag{3.87}$$

for all n, as it should. This is a constant and does not decay as the Bessels decay. Indeed, if we assert Eq. (3.78), $E_0 = V_0/(2\pi b)$, the current becomes

$$I_n = j\frac{2b}{\xi a_n} E_{Q,n} = \frac{V_0}{Z_n} \tag{3.88}$$

as for the driven loop.

Conclusions. These conclusions[6] on solvability limit models of the driving source to inductive coupling and illuminated plane waves. However, the nature of the divergences in the case of gap driven models is such that the summation may be truncated before the rapid divergence occurs. This leads to some error, of course, but "waiting too long" causes unnatural, non-physical oscillations in the results near the driving point when using moment methods. Considerable study has focused on the issue and shows that truncation should occur before n reaches $2\pi b/a$ or, in terms of the thickness measure, $n < e^{\Omega/2}$. For thin wires $\Omega > 12$, this is on the order of $n < 400$. For the thickest rings used in meta-materials, $\Omega = 7$, this is on the order of $n < 33$.

[6] Much more detail about convergence of these equations appears in a series of articles by Fikioris et al. specifically [9, 10, 13, 14, 22].

3.6 Summary of Solutions to the Thin-Wire PEC Loop

The solution to the governing equation Eq. (3.7) of the thin-wire PEC loop (for example, a metal loop with thickness $\Omega \geq 10$ up to 150 GHz) is the current

$$I_\phi(\phi) = \sum_{n=-\infty}^{\infty} I_n e^{jn\phi} \quad \texttt{where} \quad I_n = j\frac{2b}{\xi}\frac{E_n}{a_n} \tag{3.89}$$

where

$$a_n = k_b\left(\frac{K_{n+1} + K_{n-1}}{2}\right) - \frac{n^2 K_n}{k_b} \tag{3.90}$$

There are three solutions for the K_n, all giving very nearly the same answer but with different effects on the convergence problem. The three are:
Storer's Solution:

$$\begin{aligned} K_n &= \frac{1}{\pi}\texttt{ln}\frac{8b}{a} + \frac{1}{2}\int_0^{2k_b} (\mathbf{\Omega}_{2n}(x) - j\mathbf{J}_{2n}(x))\,dx - \frac{2}{\pi}\sum_{k=0}^{|n|-1}\frac{1}{2k+1} \\ K_0 &= \frac{1}{\pi}\texttt{ln}\frac{8b}{a} + \frac{1}{2}\int_0^{2kb} [\mathbf{\Omega}_0(x) - j\mathbf{J}_0(x)]dx \end{aligned} \tag{3.91}$$

Wu's Solution:

$$\begin{aligned} K_n &= \frac{1}{\pi}\left[\mathbf{K}_0\left(\frac{|n|a}{b}\right)\mathbf{I}_0\left(\frac{|n|a}{b}\right) + \gamma - 2\sum_{k=0}^{|n|-1}\frac{1}{2k+1} + \texttt{ln}(4|n|)\right] \\ &\quad + \frac{1}{2}\int_0^{2k_b} (\mathbf{\Omega}_{2n}(x) - j\mathbf{J}_{2n}(x))\,dx \end{aligned} \tag{3.92}$$

with the same K_0 as Storer's solution.

Elliptical Solution:

$$\begin{aligned} K_n &= \frac{1}{2\pi}\left[\pi\int_0^{2k_b} [\mathbf{\Omega}_{2n}(x) - j\mathbf{J}_{2n}(x)]\,dx - 4\sum_{k=0}^{|n|-1}\frac{1}{2k+1} + \frac{2b}{a}K\left(-\frac{b^2}{a^2}\right)\right] \\ K_0 &= \frac{b/a}{\pi}K\left(-\frac{b^2}{a^2}\right) + \frac{1}{2}\int_0^{2kb} \left[\mathbf{\Omega}_0(x) - j\mathbf{J}_0(x)\right]dx \end{aligned} \tag{3.93}$$

See above sections for details. If the loop is embedded in free space, then $\xi = \xi_0 = 377\,\Omega$. The coefficients E_n depend on the model used for the driving source. Delta distribution generators across infinitesimal or finite gaps cause non-physical oscillations in the results when using moment methods due to a diverging current summation. The frill generator model and the illuminating plane wave model yield converging current solutions and are recommended. However the vast amount of literature on the loop continues to use the Delta distribution across an infinitesimal gap and this is permissible with the thin loop since the divergence begins to be noticeable around $n = 100$ for an $\Omega = 10$ loop using Wu's solution and higher for the elliptical solution. The errors compared with results from simulations are small, as we shall see.

Given this, we shall forego use of inductive coupling models, since they require a rather arbitrary definition for the parameter a_0 and yield nearly the same information as do the two gap models using the Delta function generator. We shall forego the finite gap model in preference to the easier infinitesimal gap model, and we shall return to the illuminating plane wave model in Chap. 9 when we study meta-materials. Therefore, we shall use Eq. (3.78) with the current coefficients:

$$I_n = j\frac{2b}{\xi}\frac{E_n}{a_n} = \frac{V_0}{j\pi\xi a_n} \tag{3.94}$$

for the thin-wire loop antenna up to 150 GHz.

3.6.1 *The Symmetry of the Closed Loop Coefficients, I_n, Around Mode $n = 0$*

The Fourier representation of the closed loop's current is given by Eq (3.89), where the summation is taken from $-\infty$ to $+\infty$. The summation includes the negative modes. The question is whether the coefficients are symmetric in n; that is, does $I_{-n} = I_n$?

The coefficients will be symmetric if the a_n are symmetric, and these will be symmetric if the K_n are symmetric. In all three solutions of the K_n, the only terms that are possibly asymmetric are the Bessels and these are always of the form:

$$\mathbf{B}_{2n} = \mathbf{\Omega}_{2n}(x) - j\mathbf{J}_{2n}(x) \tag{3.95}$$

By the Bessel definitions given in Eq. (3.29), namely,

$$\mathbf{\Omega}_{-n}(x) = (-1)^n \mathbf{\Omega}_n(x) \quad \text{and} \quad \mathbf{J}_{-n}(x) = (-1)^n \mathbf{J}_n(x) \tag{3.95}$$

we have

$$\begin{aligned}\mathbf{B}_{-2n} &= \mathbf{\Omega}_{-2n}(x) - j\mathbf{J}_{-2n}(x) = (-1)^{2n}\left[\mathbf{\Omega}_{2n}(x) - j\mathbf{J}_{2n}(x)\right] \\ &= \mathbf{\Omega}_{2n}(x) - j\mathbf{J}_{2n}(x) = \mathbf{B}_{2n}\end{aligned} \tag{3.97}$$

The current coefficients for the closed loop are therefore always symmetric around $n = 0$.

References

1. C.W. Oseen, Ark. Mat. Astr. Fys. **9**, 1 (1913)
2. E. Hallen, Nova Actae Regiae Soc. Sci. Ups. Ser. IV **11**(4), 1 (1938)
3. J.E. Storer, Trans. AIEE **75**, 606 (1956)
4. T.T. Wu, J. Math. Phys. **3**(6), 1301 (1962)
5. R.W.P. King, in *Antenna Theory, part 1, Inter-University Electronic Series*, vol. 7, ed. by R.E. Collin, F.J. Zucker, 1st edn. (McGraw-Hill, New York, 1969), Chap. 11, pp. 458–482
6. A.F. McKinley, T.P. White, K.R. Catchpole, J. Appl. Phys. **114**(4), 044317 (2013). https://doi.org/10.1063/1.4816619, http://scitation.aip.org/content/aip/journal/jap/114/4/10.1063/1.4816619
7. T. Wu, R.W.P. King, J. Appl. Phys. **30**, 76 (1959)
8. T.T. Wu, in *Antenna Theory Part 1*, ed. by R.E. Collin, F.J. Zucker (McGraw-Hill, 1969), Chap. 8, pp. 306–351
9. G. Fikioris, T.T. Wu, I.E.E.E. Trans, Antennas Propag. **49**(3), 383 (2001)
10. G. Fikioris, J. Lionas, C. Lioutas, IEEE Trans, Antennas Propag. **51**(8), 1847 (2003). https://doi.org/10.1109/TAP.2003.815412
11. H. Anastassiu, I.E.E.E. Trans, Antennas Propag. **54**(3), 860 (2006). https://doi.org/10.1109/TAP.2006.869929
12. P.J. Papakanellos, G. Fikioris, Prog. Electromagn. Res. **69**, 77 (2007)
13. G. Fikioris, P. Papakanellos, H. Anastassiu, IEEE Trans, Antennas Propag. **56**(1), 151 (2008). https://doi.org/10.1109/TAP.2007.913076
14. P. Papakanellos, G. Fikioris, A. Michalopoulou, IEEE Trans, Antennas Propag. **58**(5), 1635 (2010). https://doi.org/10.1109/TAP.2010.2044319
15. G. Fikioris, P.J. Papakanellos, T.K. Mavrogordatos, N. Lafkas, D. Koulikas, SIAM, J. Appl. Math. **71**(2), 559 (2011). https://doi.org/10.1137/100785727
16. I. Tastsoglou, G. Fikioris, I.E.E.E. Trans, Antennas Propag. **61**(11), 5517 (2013). https://doi.org/10.1109/TAP.2013.2279423
17. E. Janke, F. Emde, F. Loesch, *Tafeln Hoherer Funktionen* (B. G. Verlagsgesellschaft, Stuttgart, 6th edition, 1960)
18. I. Stegun, M. Abramowitz, *Handbook of Mathematical Functions with Formulas, Graphs and Mathematical Tables*. Applied Mathematics Series 55 (US. Government Printing Office, WDC, 1964)
19. G.N. Watson, *Theory of Bessel Functions* (The MacMillan Company, New York, 1945)
20. T. Do-Nhat, R.H. Macphie, IEEE Trans, Antennas Propag. **37**(12), 1545 (1989). https://doi.org/10.1109/8.45096
21. I. Tastsoglou, G. Fikioris, IEEE Trans, Antennas Propag. **61**(11), 5527 (2013). https://doi.org/10.1109/TAP.2013.2279426
22. G. Fikioris, P. Papakanellos, H. Anastassiu, IEEE Trans, Antennas Propag. **58**(10), 3436 (2010). https://doi.org/10.1109/TAP.2010.2055816
23. M. Kanda, IEEE Trans. Electromagn. Compat. **EMC-26**(3), 102 (1984)

Chapter 4
The Driving Point Impedance and Admittance of Thin, PEC Loops and Rings

4.1 Formation of the Input Impedance

It is immediately evident by looking at Eqs. (3.89) and (3.94) that for the closed loop

$$I_\phi(\phi) = \sum_{n=-\infty}^{\infty} I_n e^{jn\phi} = \sum_{n=-\infty}^{\infty} \frac{V_0}{Z_n} e^{jn\phi} \quad \texttt{where} \quad Z_n \equiv j\pi\xi a_n = \frac{1}{Y_n} \tag{4.1}$$

The input impedance of the loop is defined in the standard way as the ratio of the source voltage to the resulting current at the driving point, $\phi = 0$. The closed loop impedance, then, is

$$Z_{cl} \equiv \frac{V_0}{I_\phi(0)} = \left[\sum_{n=-\infty}^{\infty} \frac{1}{Z_n}\right]^{-1} \equiv \frac{1}{Y_{cl}} \tag{4.2}$$

Note that if V_0 is set to 1, the current at the driving source is given by the closed loop admittance Y_{cl}.

From Sect. 3.6.1, $a_n = a_{-n}$, and therefore $Z_n = Z_{-n}$ and $I_n = I_{-n}$ for the closed loop; that is, both the input impedance and the Fourier current coefficients of the closed loop are symmetrical around mode 0. Therefore,

$$\begin{aligned} I_\phi(\phi) &= \sum_{n=-\infty}^{\infty} I_n e^{jn\phi} = I_0 + 2\sum_{n=1}^{\infty} I_n \texttt{cos}(n\phi) = 2\sum_{n=0}^{\infty} \varepsilon_n I_n \texttt{cos}(n\phi) \\ &\texttt{where } I_0 = \frac{V_0}{Z_0}, \quad I_n = \frac{V_0}{Z_n}, \; n > 0 \\ &\text{and where} \quad \varepsilon_n = 1/2 \text{ for } n = 0 \text{ and } 1 \text{ for } n \neq 0 \end{aligned} \tag{4.3}$$

A. McKinley, *The Analytical Foundations of Loop Antennas and Nano-Scaled Rings*, Signals and Communication Technology,
https://doi.org/10.1007/978-981-13-5893-7_4

Inverting,

$$Y_{cl} = \frac{1}{Z_{cl}} \equiv \frac{I_\phi(0)}{V_0} = 2\sum_{n=0}^{\infty} \varepsilon_n \frac{I_n}{V_0} = 2\sum_{n=0}^{\infty} \frac{\varepsilon_n}{Z_n} \tag{4.4}$$

These results arise from thin-wire and perfectly conducting material assumptions. Practically, this implies a loop antenna at frequencies below 150 GHz using metals, such as gold, silver, copper or aluminium.[1] The input impedance is complex, $Z = R + jX$; the real part indicating the resistance of the loop, the imaginary part indicating the reactance of the loop.

The most important feature of these curves is where the ring resonates and anti-resonates. These occur when the reactance crosses through $k_b = 0$. These "zero-crossings (ZC)" are marked in Fig. 4.1a. The resonances and anti-resonances of the loop are distinguishable by looking at the form of the impedance locally near the zero-crossings and comparing it to the forms for series resonant RLC circuits and for parallel RLC circuits. Series resonant circuits are capacitive below the resonant frequency, inductive above and act like a low resistance on resonance; current flows readily from source to the resistive load. In the case of the loop, current flows best between source and the loop at the resonances, making resonance the best transmitting and receiving frequency. Parallel resonant circuits are inductive below the resonant frequency, capacitive above, and highly resistive on resonance; current does not flow between source and load and therefore little current flows in the loop. Some authors call this resonance the "open-circuit" resonance.

Note that the first and second resonances occur near $k_b = 1$ and $k_b = 2$. These are the fundamental and second harmonic of the loop. At these resonances, the current wave closes on itself as it traverses the circumference of the loop. These harmonic resonances are often associated with the "Bohr orbits", the first six of which are shown in Fig. 4.2. The associated current distributions are shown in Fig. 4.3. The anti-resonances occur at about half the resonance values because the circumference is $1/2(n + 1)$ times the wavelength of the current. The current wave effectively cancels itself on each trip around the loop. The loop displays very high input resistance at the anti-resonances.

Another important feature of these curves is the position of peaks of the real part of the admittance in Fig. 4.1b. If V_0 is taken to be 1 volt, Eq. (4.4) indicates that the admittance and the input current at the input ($I(0)$) are equivalent in both real and imaginary parts; that is, the admittance curves may also be considered current curves. The real part, called the "conductance", shows peaks just to the left of the resonances. The imaginary part, called the "susceptance", also shows the resonances and anti-resonances at the zero-crossings. Both conductance and susceptance are measured in "milli-Siemens" (mS).

Figure 4.4 shows a comparison between theory, as shown in Fig. 4.1, given by Eq. (4.4), and simulations for $\Omega = 10$ and 12. The analytical theory weakens as the loop thickens.

---

[1] See Sect. 2.3.2.

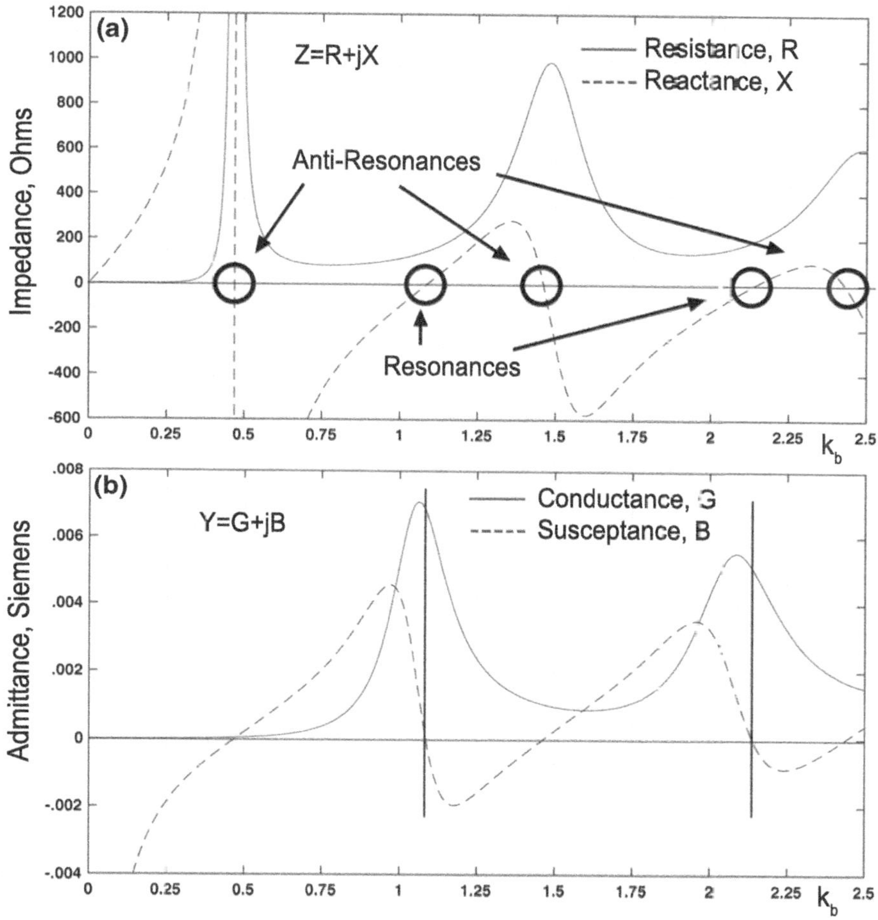

Fig. 4.1 The impedance (**a**) and admittance (**b**) of a size $\Omega = 12$ ring; $b/a \approx 64.2$. Resonances and anti-resonances are marked by circles in **a**. Notice that the peaks of the admittance are to the left of the true resonances. Notice too that these apply to any thin-wire, PEC loop antenna. These are plotted using the model, Eq. (4.4)

4.2 The Circuit Element Representation of the Loop

It is evident from the form of the impedance in Eq. (4.2), that the loop antenna displays itself as a parallel set of impedances. Even more interesting, plots of the individual modal impedances, Z_n, show forms consistent with series resonant circuits.[2] As an example, Fig. 4.5 shows the impedances Z_1 and Z_2. In fact, all of the modal impedances display themselves with the form:

[2]This was also noticed in [1].

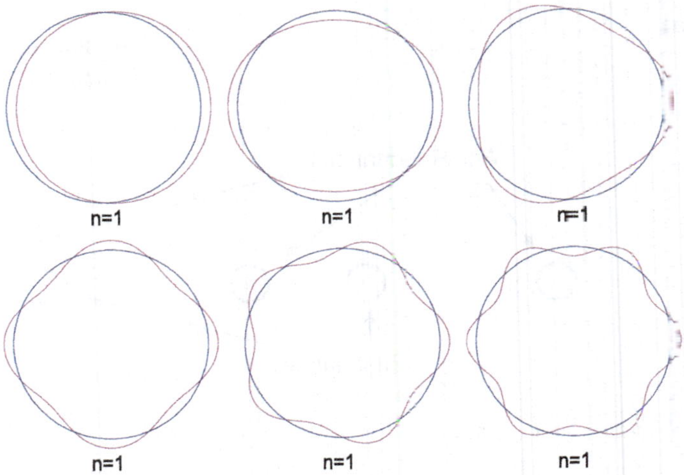

Fig. 4.2 The proper modal resonances, ω_n of a circular loop, commonly called the "Bohr Orbits"

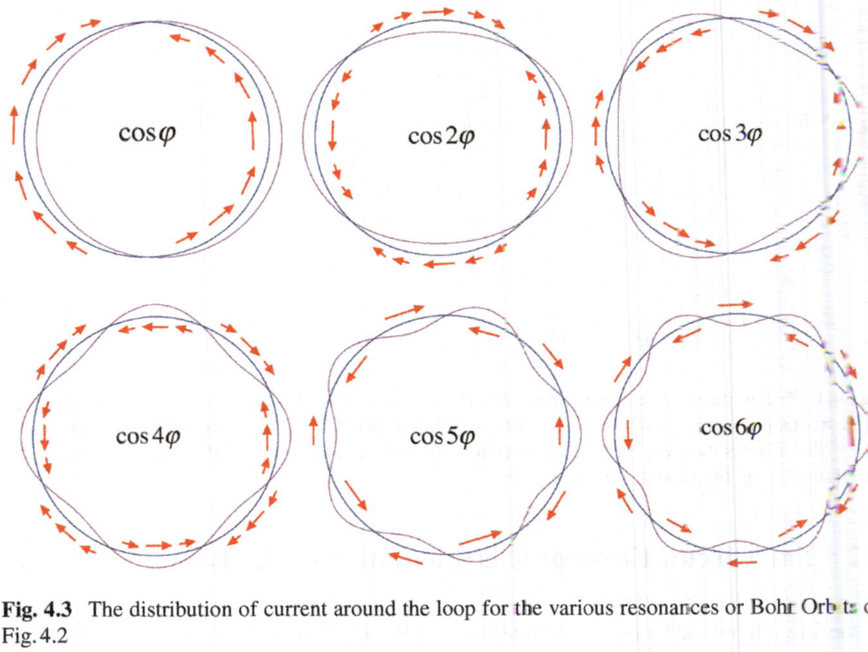

Fig. 4.3 The distribution of current around the loop for the various resonances or Bohr Orbits of Fig. 4.2

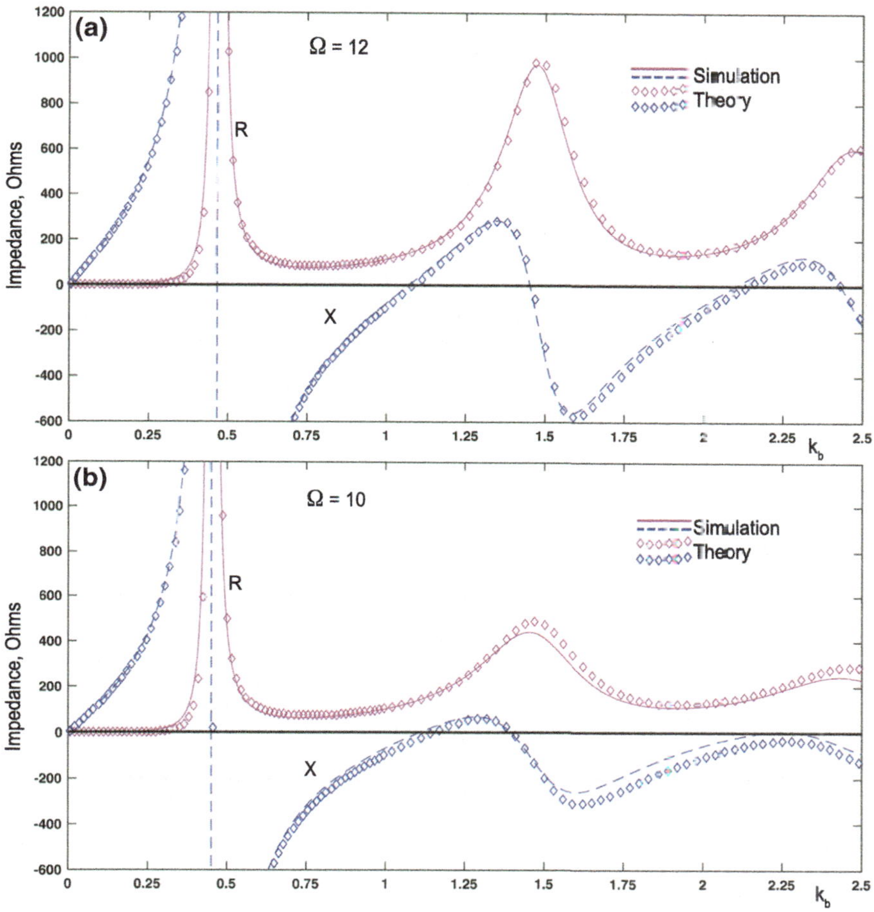

Fig. 4.4 A comparison of theory given by Eq. (4.4) and numerical simulation results using MWS for **a** $\Omega = 12$, and **b** $\Omega = 10$ thin wire PEC loop antennas. The match with simulation results grows weaker as the aspect ratio, b/a, decreases, due to assumptions of the thin-wire approximation

$$Z_n = R_n + j\left(\omega L_n - \frac{1}{\omega C_n}\right) = R_n\left(1 + jQ_n\left(\frac{\omega}{\omega_n} - \frac{\omega_n}{\omega}\right)\right)$$
$$\text{where } Q_n = \sqrt{L_n/C_n}/R_n. \qquad (4.5)$$

This suggests that the loop may be modelled as a set of parallel series resonant circuits, as shown in Fig. 4.6.

Assuming the loop is embedded in free space, $\xi = \xi_0$ and substituting Eqs. (3.90) and (3.92) into Eqs. (4.1) and (4.5),

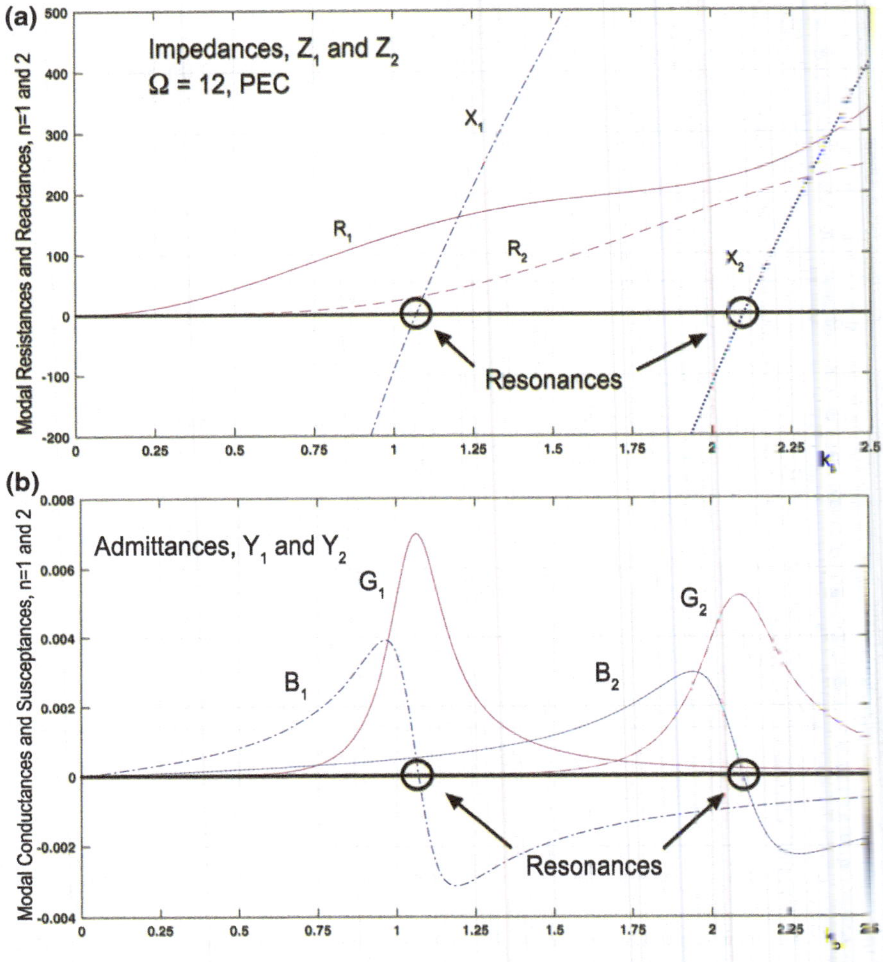

Fig. 4.5 The impedances (**a**) and admittances (**b**) for modes $n = 1$ and $n = 2$ of a practical thin-wire loop antenna. Notice that the resonances (circled) have the same form as series resonant circuits. Notice too that the modes tend to overlap, thus affecting the total impedance and admittance, seen in Fig. 4.1. The reason for this is the thickness of the wire. Less thick wires have higher Q resonances, but are impractical to make

Fig. 4.6 The proper *RLC* model of a circular loop antenna looking in from the source

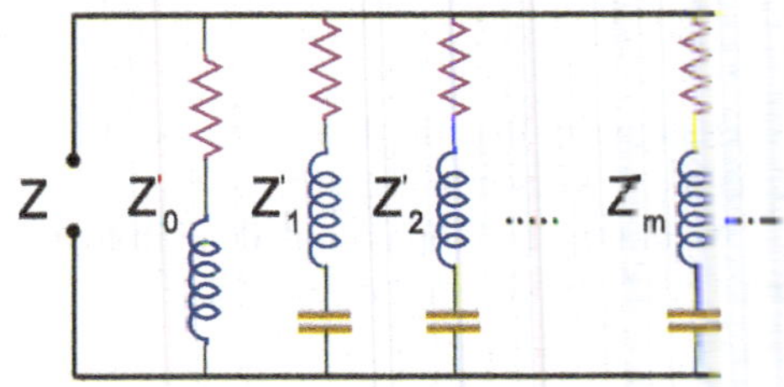

$$R_n \equiv \xi_0 \left[k_b r_n - n^2/(k_b g_n)\right]$$
$$L_n \equiv \mu_o b l_{\mu n}$$
$$C_n \equiv \varepsilon_o b l_{\varepsilon n}/n^2 \tag{4.6}$$

where r_n, g_n, $l_{\mu n}$ and $l_{\varepsilon n}$ are unit-less functions of k_b. They are, in detail[3]:

$$r_n = \begin{cases} \frac{\pi}{4}\int_0^{2k_b} \left(\mathbf{J}_{2(n+1)}(x) + \mathbf{J}_{2(n-1)}(x)\right) dx & \text{if } n > 0, \\ \frac{\pi}{2}\int_0^{2k_b} \mathbf{J}_2(x) dx & \text{if } n = 0. \end{cases} \tag{4.7}$$

$$g_n = \begin{cases} 1/\left[\frac{\pi}{2}\int_0^{2k_b} \mathbf{J}_{2n}(x) dx\right] & \text{if } n > 0, \\ \infty & \text{if } n = 0. \end{cases} \tag{4.8}$$

$$l_{\mu n} = \begin{cases} \ln\left(\frac{8b}{a}\right) - \sum_{k=0}^{n}(2k+1)^{-1} - \sum_{k=0}^{n-2}(2k+1)^{-1} \\ -\frac{\pi}{4}\int_0^{2k_b}\left[\mathbf{\Omega}_{2(n+1)}(x) + \mathbf{\Omega}_{2(n-1)}(x)\right] dx & \text{if } n > 0, \\ \ln\left(\frac{8b}{a}\right) - 2 - \frac{\pi}{2}\int_0^{2k_b}\left[\mathbf{\Omega}_2(x)\right] dx & \text{if } n = 0. \end{cases} \tag{4.9}$$

$$l_{\varepsilon n} = \begin{cases} 1/\left[\ln\left(\frac{8b}{a}\right) - 2\sum_{k=0}^{n-1}(2k+1)^{-1} - \frac{\pi}{2}\int_0^{2k_b}\left[\mathbf{\Omega}_{2n}(x)\right] dx\right] & \text{if } n > 0 \\ \infty & \text{if } n = 0. \end{cases} \tag{4.10}$$

The elements r_n, g_n, $l_{\mu n}$, and $l_{\varepsilon n}$ in Eq. (4.6) allow a general alternative way of specifying resistance, conductance, inductance and capacitance for a loop of radius b at a given incident wavelength. For example, an inductance specified as $l_\mu = 1.0$ has, on a loop of radius b with incident wavelength equal to the circumference ($k_b = 1.0$), the value $L = \mu_0 b$ henries and the reactance $X_L = \xi_0 = 377$ ohms. Each of these individual modal elements in Eq. (4.6) are therefore given by complicated functions of the variable k_b; some with Bessel functions and some with integrals of Bessel functions.

4.2.1 The Closed Loop as an R, L and C Circuit at Any k_b

A system with many resonances will have a number of inductors and capacitors, some in series, some in parallel, but when the analysis is complete, the system will consist of some mathematical combination of the inductors to produce one inductance, and

[3]These differ from the results given in [2] by a factor of 2 for $n > 0$, because in that paper the factor of 2 in Eq. (4.4) was assimilated into the Z_n. We do not do that here.

of some mathematical combination of the capacitors to produce one capacitance. The same is true of the resistors. Therefore, the complicated infinity of series resonant circuits in parallel, illustrated by Fig. 4.6 can be reduced to a single R, L, and C in series. From Eq. (4.4)

$$Y_{cl} = \frac{1}{Z_{cl}} = 2\sum_{n=0}^{\infty}\frac{\varepsilon_n}{Z_n} = 2\sum_{n=0}^{\infty}\frac{\varepsilon_n}{R_n + jX_n} = \frac{1}{R_{cl} + jX_{cl}}$$
$$\text{where} \quad \varepsilon_n = 1/2 \text{ for } n = 0 \text{ and } 1 \text{ for } n \neq 0 \tag{4.11}$$

Equation (4.12) gives the total resistance and reactance of the loop, taking into account all of the modal impedances. R_{cl} and X_{cl} are shown in Figs. 4.1 and 4.4. They are calculated by taking the real and imaginary parts of Z_{cl}.

$$\begin{aligned} Z^*_{cl} &= \frac{Z^*_{cl}Z_{cl}}{Z_{cl}} = 2(Z^*_{cl}Z_{cl})\sum_{n=0}^{\infty}\frac{\varepsilon_n}{Z_n} = 2(Z^*_{cl}Z_{cl})\sum_{n=0}^{\infty}\varepsilon_n\frac{Z^*_n}{Z^*_nZ_n} \\ &= 2(Z^*_{cl}Z_{cl})\sum_{n=0}^{\infty}\varepsilon_n\frac{R_n - jX_n}{(Z^*_nZ_n)} = 2(Z^*_{cl}Z_{cl})\left[\sum_{n=0}^{\infty}\frac{\varepsilon_nR_n}{(Z^*_nZ_n)} - j\sum_{n=0}^{\infty}\frac{\varepsilon_nX_n}{(Z^*_nZ_n)}\right] \\ &= R_{cl} - jX_{cl} \end{aligned}$$
$$\text{where} \quad \varepsilon_n = 1/2 \text{ for } n = 0 \text{ and } 1 \text{ for } n \neq 0 \tag{4.12}$$

In a similar fashion, the reactance can be expanded to give the total inductance and capacitance of the loop. Remembering that the closed loop has no mode 0 capacitive reactance ($C_0 = \infty$).

$$\begin{aligned} X_{cl} &= 2(Z^*_{cl}Z_{cl})\sum_{n=0}^{\infty}\frac{\varepsilon_nX_n}{(Z^*_nZ_n)} \\ X_L = \omega L &= 2(Z^*_{cl}Z_{cl})\sum_{0}^{\infty}\frac{\varepsilon_n\omega L_n}{(Z^*_nZ_n)} \\ X_C = \frac{1}{\omega C} &= 2(Z^*_{cl}Z_{cl})\sum_{1}^{\infty}\frac{1/(\omega C_n)}{(Z^*_nZ_n)} \end{aligned} \tag{4.13}$$

Reducing, we have for L and C:

$$L = \mu_0 bl_\mu = 2(Z^*_{cl}Z_{cl})\sum_{0}^{\infty}\frac{\varepsilon_nL_n}{(Z^*_nZ_n)} \quad \text{and} \quad \frac{1}{C} = \frac{1}{\varepsilon_0 bl_\varepsilon} = 2(Z^*_{cl}Z_{cl})\sum_{1}^{\infty}\frac{1/C_n}{(Z^*_nZ_n)} \tag{4.14}$$

Table 4.1 Details of the first two harmonic resonances and the anti-resonances for two sizes of thin-wire PEC loop antennas, as calculated from Eq. (4.12) through Eq. (4.15). The values of the resonances vary slightly depending on code implementation. The conductance peak is slightly off the resonance due to the thin wire approximation. Note that $X_L - X_C \neq 0$ – see Sect. 4.2.2 for explanation

Ω	Resonances							G		Anti-resonances
	k_{br} at $X = 0$	R (ohms)	L (μH)	X_L (ohms)	C (pF)	X_C (ohms)	Q	Peak		k_b at $X = 0$
								k_b	mS	
12	1.0877	148.36	1.26	857.1	1.70	862.7	5.80	1.06	7.0	0.46
	2.1545	205.78	1.16	1572	0.47	1577	7.65	2.09	5.5	1.43
10	1.15	165	0.98	710.4	1.93	718.6	4.3	1.09	7.0	0.45
	NA	NA	NA	NA	NA	NA	NA	2.12	5.7	1.40

and consequently,

$$l_\mu = 2(Z_{cl}^* Z_{cl}) \sum_0^\infty \frac{\varepsilon_n l_{\mu n}}{(Z_n^* Z_n)} \quad \text{and} \quad \frac{1}{l_\varepsilon} = 2(Z_{cl}^* Z_{cl}) \sum_1^\infty \frac{1/l_{\varepsilon n}}{(Z_n^* Z_n)} \tag{4.15}$$

These are functions of k_b. The quality factor of the loop can now be obtained: $Q = (L/C)/R$.

The first two sets of ZC resonances and anti-resonances for thin rings of size $\Omega = 10$ and 12 are shown in Table 4.1. Note that the resistance at every k_b is the radiation resistance of the ring. This is because the rings are perfectly conducting and therefore have no heat losses. Quality factors, Q, for the first two resonances are also given. Note that the resonance peaks of the admittance do not occur at the resonance point itself (where the reactance goes to zero); it is slightly lower in frequency.

Notice also that as the ring thickens, the resonances slightly blue shift. A very thin loop, say with $\Omega = 40$, has its first resonance very nearly at $k_b = 1.0$. But thicker loops vary slightly from that. This is because the current finds a path around the inner circumference, $2\pi(b - a)$ rather than along the thin-wire circumference, $2\pi b$.

Evidently these component elements are not constant over frequency, as they would be in typical RLC circuit theory. An important effect of the variation over frequency is the placement of the resonances.

$$\omega_n = \frac{1}{\sqrt{L_n C_n}} \quad \text{and} \quad k_{bn} = \frac{n}{\sqrt{l_{\mu n} l_{\varepsilon n}}} \tag{4.16}$$

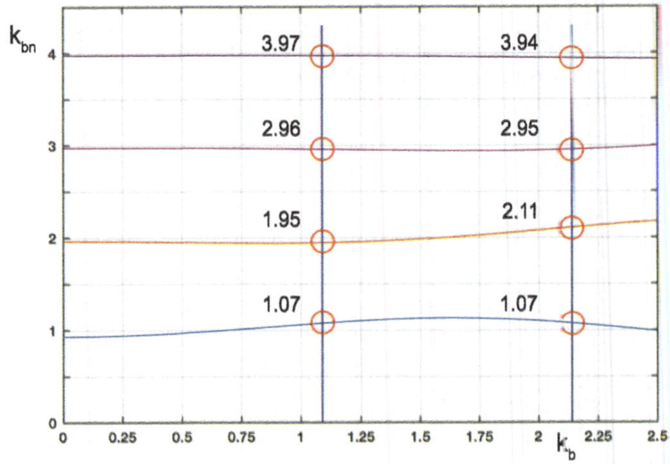

Fig. 4.7 The resonance functions, $\tilde{k}_{bn}$ from Eq. (4.16), for the first four modes, n, of the fundamental and second harmonic of a size $\Omega = 12$ loop. The blue lines mark the fundamental ($k_{br} = 1.09$) and the second harmonic ($k_{br} = 2.14$), of the $\Omega = 12$ loop. Different sized loops will have different resonances. The thinner the loop, the closer will the loop resonate to $k_{br} = 1.0$ and $k_{br} = 2.0$

A set of rolling curves over k_b results, as shown in Fig. 4.7 for the first four modes.[4] For example, for the fundamental resonance an intersection of the line $y = 1.09$ with the function k_{bn} gives the resonance of each mode. These resonances, with associated values of the unit-less functions, are given in Table 4.2 for two thin loop antennas.[5] When referring to a value that occurs at a modal resonance, a tilde above the symbol will be used henceforth; for example, $\tilde{R}_2$ refers to the resistance of the loop at the modal resonance, $\tilde{k}_{b2}$. It is important to note that the inductance and capacitance depend directly on the loop radius, but the resistance does not. Apparently, all thin-wire, PEC loop antennas, no matter the size, have the same radiation resistance.

4.2.2 *Difficulties with the Series Resonant Model of the Loop*

The fact that the peak of the conductance curve does not match the zero-crossing of the reactance, and that $X_L - X_C \neq 0$, points to the need for some critical thinking about the model. It can be shown easily enough, by increasing Ω to a very large number that the input reactance crosses zero closer and closer to 1.0 as Ω increases. This suggests that the culprit for the difficulties is the thin-wire approximation

[4] A corresponding Fig. 4 in [2] shows the primary resonances for the first four modes. This figure contains more information.

[5] This table corrects Table 1 in [2] for $\Omega = 12$.

Table 4.2 Elemental circuit values at the two fundamental resonances for two differently sized thin-wire PEC loops. These give a sense of the effect of each of the first four modes on the resonances. Each mode is a series resonant RLC circuit, and their effects summed together create the harmonic resonances k_{br}, as calculated from Eq. (4.12)

Ω k_{br}	n	$\tilde{k}_{brn}$	$\tilde{r}_n$	$\tilde{g}_n$	$\tilde{R}_n$ (ohm)	$\tilde{l}_{\mu n}$	$\tilde{L}_n$ (μH)	$\tilde{X}_{Ln}$ (ohm)	$\tilde{l}_{\varepsilon n}$	$\tilde{C}_n$ (pF)	$\tilde{X}_{Cn}$ (ohms)
	0	0	0.53	∞	217	4.7	2.8	1931	∞	∞	0
12	1	1.07	1.16	1.89	292	4.1	2.5	1680	0.21	0.90	1632
1.0877	2	1.95	0.27	29.7	62	4.0	2.4	1632	0.27	0.28	5228
	3	2.96	0.02	1016.0	4	3.4	2.0	1375	0.31	0.14	10150
	4	3.97	0.00	60,359.0	0	3.0	1.8	1221	0.34	0.09	16277
Ω k_{br}	n	$\tilde{k}_{brn}$	$\tilde{r}_n$	$\tilde{g}_n$	$\tilde{R}_n$ (ohm)	$\tilde{l}_{\mu n}$	$\tilde{L}_n$ (μH)	$\tilde{X}_{Ln}$ (ohm)	$\tilde{l}_{\varepsilon n}$	$\tilde{C}_n$ (pF)	$\tilde{X}_{Cn}$ (ohms)
	0	0	0.61	∞	264	3.7	2.2	1618	∞	∞	0
10	1	1.11	1.17	1.6	310	3.0	1.8	1316	0.27	1.1	1223
1.1504	2	1.93	0.31	23	75	3.0	1.8	1304	0.36	0.4	3678
	3	2.93	0.02	698	5	2.4	1.4	1035	0.44	0.2	6729
	4	3.96	0.00	36935	0	2.0	1.2	873	0.51	0.1	10319

The individual modal resonances also suffer from this problem, the conductance peaks do not line up precisely with the zero-crossings. This suggests that the model is not as accurate as we might want it to be; but we know this. The thin-wire approximation means something, and it means precisely this: that the model grows weaker the thicker the wire. The simulations however suggest that we can trust the values in these tables as long as we don't become dogmatic about them. These are approximations that are good to a thickness of about $\Omega = 9$ or 10. We drop the thin-wire approximation in Chap. 8, but this is currently the cutting edge of loops and rings; those thick wire equations have not yet been solved. Equation (4.7) through Eq. (4.10) and therefore the values given in two Tables 4.1 and 4.2 can be trusted up to a point. The latter table uses Eq. (4.12) through Eq. (4.15) and that derivation is straightforward, offering no surprises. Since any configuration of passive elements can be molded by math to form a single RLC series circuit, those values must then be given by the RLC values found in Table 4.2. They are the best we have.

This latter idea yields still another way of deriving the circuit elements of the closed loop, but it does not rely on a summation of the modal elements, and as such turns out to give values quite wide of those in the previous tables. Suppose we assume that Eq. (4.5) holds for this molded RLC circuit, but instead of calculating R, L and C using the modal summations we use

$$Z_{cl} = R_{cl} + j\left(\omega L_{cl} - \frac{1}{\omega C_{cl}}\right) = R_{cl}\left(1 + jQ_{cl}\left(\frac{\omega}{\omega_n} - \frac{\omega_n}{\omega}\right)\right)$$

$$\text{where } Q_{cl} = \sqrt{L_{cl}/C_{cl}}/R_{cl}. \tag{4.17}$$

Take the derivative of the reactance as follows:

$$X_{cl} = RQ\left(\frac{\omega}{\omega_r} - \frac{\omega_r}{\omega}\right) = RQ\left(\frac{k_b}{k_{br}} - \frac{k_{br}}{k_b}\right)$$
$$\left.\frac{\partial X}{\partial k_b}\right|_{k_{br}} = RQ\left.\left(\frac{1}{k_{br}} + \frac{k_{br}}{k_b^2}\right)\right|_{k_{br}} = \frac{2RQ}{k_{br}} \tag{4.18}$$

It suggests that the slope of the reactance at resonance is proportional to the Q of the loop. Simply by measuring the slope of the reactance at k_{br} and inserting the value for R at k_{br}, we ought to know the Q of the loop. The problem with this is that both R and Q are functions of k_b and have slopes through k_{br}, whereas in a true RLC resonant circuit, this is not the case; there, R and Q are constants. Even if we do the calculation properly:

$$\left.\frac{\partial X}{\partial k_b}\right|_{k_{br}} = \left.\left[RQ\left(\frac{1}{k_{br}} + \frac{k_{br}}{k_b^2}\right) + Q\left(\frac{k_b}{k_{br}} - \frac{k_{br}}{k_b}\right)\frac{\partial R}{\partial k_b} + R\left(\frac{k_b}{k_{br}} - \frac{k_{br}}{k_b}\right)\frac{\partial Q}{\partial k_b}\right]\right|_{k_{br}} \tag{4.19}$$

we have problems. Suppose, for the moment, that we were able to measure the slope of Q at k_{br} and substituted that on the right hand side. When we then take the result at k_{br} the second and third terms go to zero and we are left with the simple answer above in Eq. (4.18), $2RQ/k_{br}$, which does not give the same answers as do the set of equations based on the modal summations.

It is wise then to avoid too much of a reliance on the notion that a loop can be represented by a single RLC circuit unless those circuit elements have been found using the modal summations, and then be humble enough to allow for the discrepancy between the expected $X = 0$ and $X_L - X_C \neq 0$.

4.3 The Subwavelength Anti-resonance

Table 4.1 indicates that in the region near $k_b \approx 0.46$, a strong anti-resonance appears in the impedance of the loop. The region below $k_b = 0.50$ is called "sub-wavelength", because the circumference of the loop is about one-half the wavelength of the incident or driving wavelength. In other words, the loop is small compared with the driving wavelength; indeed, about half the length and hence the incident wave tends to cancel itself out on its travel around the loop. At this point the loop has little to no current flow and acts as if it were nearly an infinite resistance; for all practical purposes, it acts like as an open circuit.

An anti-resonance, like a true resonance, occurs when the reactance of the loop goes to zero. There are two parts to the reactance: a positive inductive part and a negative capacitive part. At the anti-resonance, the parts are equal and the sum goes to zero. This leaves the loop looking purely resistive. How is it then that the

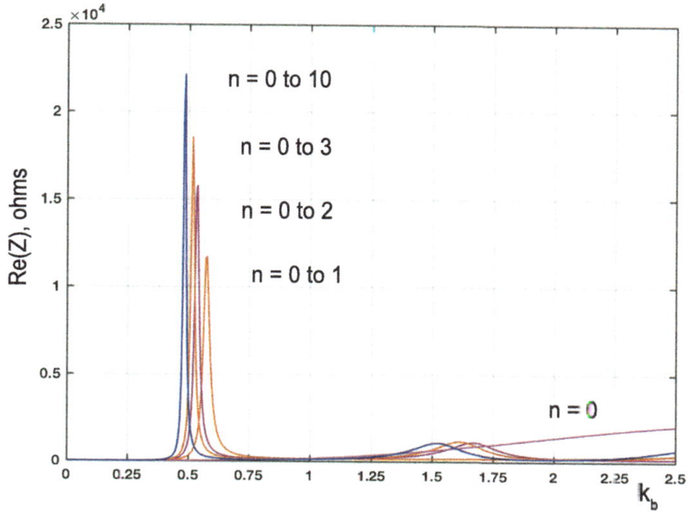

Fig. 4.8 The real part of the input impedance as the sum of modes near $k_b = 0.46$. The mode 0 term has no influence on the anti-resonance, because mode 0 has no capacitive reactance to offset the mode 0 inductive reactance and hence cannot create a anti-resonance. However, all modes 1 and higher influence the formation of the anti-resonance, although the lower modes influence it stronger than do the higher modes because of their distance from the sub-wavelength region

individual modal resistances, the R_1 and R_2, show no interest in forming an infinite peak at $k_b = 0.46$ in Fig. 4.5? The answer does not lie with R_0, which is also quite small in this region. Indeed, remember that the total impedance is **NOT** given by the sum $\sum Z_n$, but by the inverse of the sum of inverses: examine Eq. (4.4) carefully. Figure 4.8 shows the real part of sum of the first few terms of Eq. (4.4). The first term (the $n = 0$ term) shows no influence on the anti-resonance. However, the anti-resonance is already quite well stated with the sum of the first two terms, $n = 0$ and 1. It is because the inductive reactance of the loop and the capacitive reactance of the loop are equal and opposite at precisely the point of the anti-resonance. The anti-resonance is as strong as it is because the gradient of the difference between these two reactive parts is quite large.

References

1. K. Esselle, S. Stuchly, IEEE Trans. Antennas Propag. **38**(7), 1123 (1990), https://doi.org/10.1109/8.55629
2. A.F. McKinley, T.P. White, I.S. Maksymov, K.R. Catchpole, J. Appl. Phys. **112**(9), 094911 (2012), https://doi.org/10.1063/1.4764104, http://link.aip.org/link/?JAP/112/094911/1

Chapter 5
Current Distribution and Radiation Characteristics of Thin, PEC Loops and Rings

5.1 Current Characteristics

The current for PEC loop antennas is given in summary in Sect. 3.6:

$$I_{\phi}(\phi) = \sum_{n=-\infty}^{\infty} I_n e^{jn\phi} \tag{5.1}$$

It is plotted in Fig. 5.1 at the fundamental resonance and at the second harmonic. The modal currents are symmetric ($I_n = I_{-n}$), as they should be for the closed loop.

5.2 Characterizing Radiation of the PEC Loop

The quality factor, Q, is given in circuit theory by the definition

$$Q = 2\pi \frac{\texttt{Energy Stored}}{\texttt{Energy lost or dissipated}} \quad \texttt{over one cycle} \tag{5.2}$$

Consequently, since the resistive loss in the PEC loop is zero, the quality of the loop antennas in this section should be taken as infinite. This seems to pose some problem, since, evidently from plots of the admittance, resonances appear with a determined height and bandwidth, allowing Q also to be defined by its alternative:

$$Q = \frac{\texttt{Resonant frequency}}{\texttt{Bandwidth}} = \frac{\omega_r}{\Delta\omega} = \frac{k_{br}}{\Delta k_b} \tag{5.3}$$

How are these reconciled in the case of the PEC loop? The question arises because of the unfortunate use of the word "resistance" together with the word "radiation".

A. McKinley, *The Analytical Foundations of Loop Antennas and Nano-Scaled Rings*, Signals and Communication Technology,
https://doi.org/10.1007/978-981-13-5893-7_5

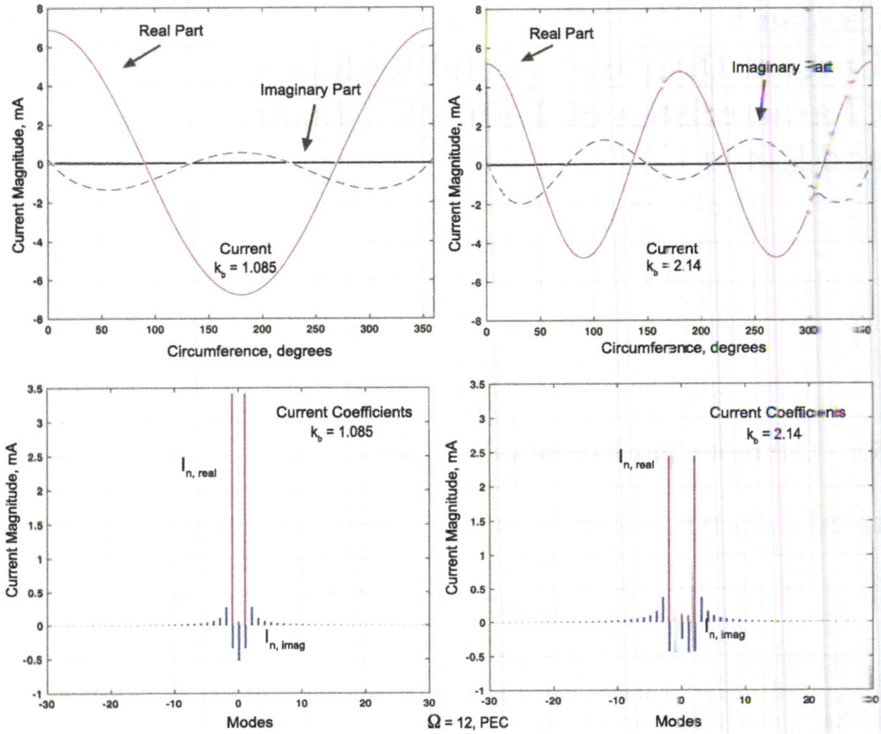

Fig. 5.1 The current and current coefficients for the first two harmonic resonances of a thin-wire PEC loop antenna occurring at $k_b = 1.085$ (left side) and $k_b = 2.14$ (right side). The real parts are strong in the proper modes. The imaginary parts are small, as they should be at resonances. The dipole and quadrupole associated with these two harmonics are clear. The real and imaginary parts are placed side by side, so that in case of overlap both can be easily distinguished

In antenna design one wants as high a radiation "resistance" as one can get, because it is defined as the ratio of power radiated to the square of the input current producing the radiation [1]:

$$2P_{rad} = |I(0)|^2 R_{rad} \tag{5.4}$$

Typically, then, in an antenna, one wants both the current and the radiation resistance to be as large as possible, so the power radiated is large. The current is maximized in a circuit equivalent model of the antenna by minimizing resistive losses, which is automatic in a PEC loop, since the resistive loss is zero. The radiation resistance is given by the real part of the input impedance at the driving point. Therefore, when Q is calculated using $Q = \sqrt{L/C}/R$, as in (4.5), **it gives the quality of the circuit as seen from the driving point, not the "quality" of the radiation**. That is, this Q is not a measure of how well the antenna radiated; that's a different thing. Indeed measures of the resonant frequency and the bandwidth of the admittance curve with (5.3) will give the same value for the Q, **but it will not be Q_{rad}, the quality of the antenna associated with the radiated energy**.[1] The definition of Q in the latter case must be changed to

$$Q_{rad} = 2\pi \frac{\texttt{Energy Stored}}{\texttt{Energy radiated}} \quad \texttt{over one cycle} \tag{5.5}$$

and the denominator can only be known by calculating the radiated power and energy.

The radiation efficiency of the antenna, which is defined as the ratio of the radiation resistance to the sum of the resistive loss plus the radiation resistance, is a better way of understanding how well an antenna radiated:

$$\varepsilon_{ff} = \frac{R_{rad}}{R_{loss} + R_{rad}} \tag{5.6}$$

In the case of the PEC loop antenna, the efficiency is, of course, 100%; all the power at the input radiated. Note that since there may be standing waves on the transmission cable between transmitter and the input point, the power from the transmitter is not necessarily all radiated.

5.2.1 *Radiating $\vec{E}$ Field Patterns, Near and Far*

Closed form expressions for the far-field radiation patterns of large, thin, PEC loop antennas, those for which $k_b > 0.5$, were first derived by Martin in 1960 [2] using a straightforward integration technique on the vector potential $\vec{A}$ that results from any given current distribution on the loop. The vector potential is important because both the $\vec{E}$ and $\vec{H}$ can be determined directly from it. Using (2.92) and (2.52c) with $\vec{J} = 0$, assuming no charge in the medium, and proceeding as in Sect. 2.4,

$$\begin{aligned} \vec{\nabla} \times \vec{B} &= \mu\varepsilon_b \frac{\partial \vec{E}}{\partial t} \\ \vec{\nabla} \times \vec{\nabla} \times \vec{A} = \vec{\nabla}(\vec{\nabla} \cdot \vec{A}) - \vec{\nabla}^2 \vec{A} = \vec{\nabla}(\vec{\nabla} \cdot \vec{A}) + k^2 \vec{A} &= j\omega\mu\varepsilon_b \vec{E} \end{aligned} \tag{5.7}$$

These yield

$$\vec{H} = \frac{1}{\mu} \nabla \times \vec{A} \quad \text{and} \quad \vec{E} = -j \frac{1}{\omega\mu\varepsilon_b} \left[\nabla(\nabla \cdot \vec{A}) + k^2 \vec{A} \right] \tag{5.8}$$

[1] See [1] Sect. 11.5, where Q_{rad} is discussed in terms of radiated spherical mode waves.

The magnetic potentials are

$$
\begin{aligned}
A_r(r,\theta,\phi) &\approx \frac{\mu b \sin\theta}{4\pi}\int_0^{2\pi} I(\phi')\sin(\phi-\phi')\frac{e^{-jkR'}}{R'}d\phi' \\
A_\theta(r,\theta,\phi) &\approx \frac{\mu b \cos\theta}{4\pi}\int_0^{2\pi} I(\phi')\sin(\phi-\phi')\frac{e^{-jkR'}}{R'}d\phi' \\
A_\phi(r,\theta,\phi) &\approx \frac{\mu b}{4\pi}\int_0^{2\pi} I(\phi')\cos(\phi-\phi')\frac{e^{-jkR'}}{R'}d\phi'
\end{aligned} \tag{5.9}
$$

where

$$R' = \sqrt{r^2 + b^2 - 2br\sin\theta\cos(\phi-\phi')} \tag{5.10}$$

This corresponds to the geometry given in Fig. 5.2.

Werner [3, 4] was able to develop closed form solutions for the integrals in (5.9) and then showed that these were applicable as well to the near field. The term "closed form expression" in this case means a solution to the integral using summation expansions of well-known functions. Werner solved the above potential integrals using a wonderful trick. Note that the vector potential components can also be written

$$
\begin{aligned}
A_r(r,\theta,\phi) &= -\frac{\mu}{j2kr}\frac{d}{d\phi}\mathscr{F}(r,\theta,\phi) \\
A_\theta(r,\theta,\phi) &= -\frac{\mu}{j2kr\tan\theta}\frac{d}{d\phi}\mathscr{F}(r,\theta,\phi) \\
A_\phi(r,\theta,\phi) &= \frac{\mu}{j2kr\cos\theta}\frac{d}{d\theta}\mathscr{F}(r,\theta,\phi)
\end{aligned} \tag{5.11}
$$

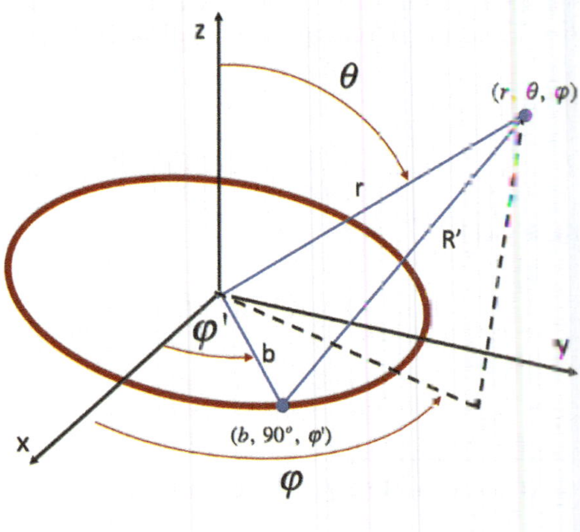

Fig. 5.2 Far-field geometry for the loop antenna. The current element, dI, which occupies the section of the loop, $bd\phi$, generates the field at (r, θ, ϕ) and is located at $(b, 90^o, \phi')$

where

$$\mathscr{F}(r,\theta,\phi) = \frac{1}{2\pi}\int_0^{\infty} I(\phi')e^{-jkR'}d\phi' \tag{5.12}$$

This integral is then common to all three components and only it needs to be solved. With some doing, the solution is

$$\mathscr{F}(r,\theta,\phi) = G_0 e^{-jkR} + \sum_{m=1}^{\infty} G_m(\phi) \times \frac{(k^2 br\,\mathtt{sin}\theta)^m}{m!}\frac{h_{m-1}^{(2)}(kR)}{(kR)^{m-1}} \tag{5.13}$$

where

$$G_0 = \frac{1}{2\pi}\int_0^{2\pi} I(\phi')d\phi'; \quad G_m = \frac{1}{2\pi}\int_0^{2\pi} I(\phi+\phi')\mathtt{cos}^n(\phi')d\phi'$$
$$\mathtt{and} \quad R = \sqrt{r^2-b^2} \tag{5.14}$$

The functions, $h_{m-1}^{(2)}(kr)$, are Hankel functions of the second kind. The problem boils down to solutions for G_0 and G_m for the given assumed current distribution. They are solvable for the sinusoidal configuration of (5.15). The components of $\vec{A}$ are then found using (5.13). Werner derived the near and far fields for general current distributions of this form,

$$I(\phi) = \sum_0^{\infty} I_{wn}\mathtt{cos}(n\phi) \tag{5.15}$$

We shall use ours from (4.3)

$$I(\phi) = 2\sum_{n=0}^{\infty}\varepsilon_n I_n \mathtt{cos}(n\phi)$$
$$\text{where } \varepsilon_n = 1/2 \text{ for } n=0 \text{ and } 1 \text{ for } n \neq 0 \tag{5.16}$$

The far field potential patterns, transformed using our current, are:

$$A_r(r,\theta,\phi) \approx \frac{\mu}{j}\frac{e^{-jkr}}{kr}\sum_{n=1}^{\infty} nj^n I_n \mathtt{sin}(n\phi)\mathbf{J}_n(k_b\mathtt{sin}\theta)$$

$$A_\theta(r,\theta,\phi) \approx \frac{\mu}{j\mathtt{tan}\theta}\frac{e^{-jkr}}{kr}\sum_{n=1}^{\infty} nj^n I_n \mathtt{sin}(n\phi)\mathbf{J}_n(k_b\mathtt{sin}\theta)$$

$$A_\phi(r,\theta,\phi) \approx \frac{\mu k_b}{j}\frac{e^{-jkr}}{kr}\sum_{n=0}^{\infty}\varepsilon_n j^n I_n \mathtt{cos}(n\phi)\mathbf{J}_n'(k_b\mathtt{sin}\theta) \tag{5.17}$$

where $\mathbf{J}_n'(k_b \mathtt{sin}\theta)$ is the derivative of the Bessel with respect to its argument. After substantial calculations based on (5.8) and (5.17), the far field is:

$$\begin{aligned}
H_r &= E_r \approx 0 \\
H_\theta &\approx \frac{k_b e^{-jkr}}{r} \sum_{n=0}^{\infty} j^n \varepsilon_n I_n \mathtt{cos}(n\phi) \mathbf{J}_n'(k_b \mathtt{sin}\theta) \\
H_\phi &\approx -\frac{\mathtt{cot}\theta e^{-jkr}}{r} \sum_{n=1}^{\infty} n j^n I_n \mathtt{sin}(n\phi) \mathbf{J}_n(k_b \mathtt{sin}\theta) \\
E_\theta &\approx -\frac{\xi \mathtt{cot}\theta e^{-jkr}}{r} \sum_{n=1}^{\infty} n j^n I_n \mathtt{sin}(n\phi) \mathbf{J}_n(k_b \mathtt{sin}\theta) \\
E_\phi &\approx -\frac{\xi k_b e^{-jkr}}{r} \sum_{n=0}^{\infty} j^n \varepsilon_n I_n \mathtt{cos}(n\phi) \mathbf{J}_n'(k_b \mathtt{sin}\theta)
\end{aligned} \tag{5.18}$$

Substituting the current coefficients of (3.94) into these expressions yields the final far-field expressions for the thin-wire PEC closed loop. These are all functions of r, θ, ϕ and k_b. The derivatives of the Bessels, namely, $\mathbf{J}_n'(k_b \mathtt{sin}\theta)$, are taken with respect to the argument.

It is worth noting that an alternative derivation of the near and far-fields were accomplished using Legendre polynomials by Li et al. [5]. The method uses the Dyadic Green's function.

5.2.2 *Radiated Power, Radiation Intensity, Directivity and Gain*

The time averaged Poynting vector gives the average power density (power per unit area) as a function of the radius, the angles θ and ϕ, and of course k_b:

$$P(r,\theta,\phi,k_b) = \frac{1}{2}\mathtt{Re}\left[\vec{E} \times \vec{H}^*\right] = \frac{1}{2\xi}(E_\theta^* E_\theta + E_\phi^* E_\phi) \tag{5.19}$$

In the far field, the Poynting vector is normal to the surface of the sphere. Consequently, integrating this over the surrounding sphere gives the time averaged, total power radiated [1, Sect. 2.3] for a given radius and k_b:

$$P_{rad}(k_b) = \int_0^{2\pi} \int_0^{\pi} \frac{(E_\theta^* E_\theta + E_\phi^* E_\phi)}{2\xi} r^2 \mathtt{sin}\theta d\theta d\phi \tag{5.20}$$

Substituting (5.18) and keeping the counters distinguishable leads to integrals of the form

$$Q^{(p)}_{m,n}(k_b) = \int_0^{\pi/2} \mathbf{J}_m(k_b \mathrm{sin}\theta)\mathbf{J}_n(k_b \mathrm{sin}\theta)\mathrm{sin}^p\theta d\theta \tag{5.21}$$

These are a class of auxiliary functions introduced in 2002 by Savov [6] specifically for expressions of radiated power on some types of antennas. Lu et al. [7] applied these integrals,[2] using (4.1).[3] The radiated power becomes

$$\begin{aligned} P_{rad}(k_b^2) &= \xi\pi k_b^2|V_0|^2 \\ &\times \sum_{n=0}^{\infty} \varepsilon_n |Y_n|^2 \left[\frac{1}{2}Q^{(1)}_{n-1,n-1}(k_b) + \frac{1}{2}Q^{(1)}_{n+1,n+1}(k_b) - \frac{n^2}{k_b^2}Q^{(1)}_{n,n}(k_b)\right] \\ &\text{where } \varepsilon_n = 1/2 \text{ for } n=0 \text{ and } 1 \text{ for } n \neq 0 \end{aligned} \tag{5.22}$$

This makes use of the helpful recurrence relation when $n = m$:

$$Q^{(-1)}_{n,n}(k_b) = \frac{k_b^2}{4n^2}\left[Q^{(1)}_{n-1,n-1}(k_b) + 2Q^{(1)}_{n-1,n+1}(k_b) + Q^{(1)}_{n+1,n+1}(k_b)\right] \tag{5.23}$$

An alternative series representation comes directly from the integral definition of the Q's:

$$Q^{(1)}_{n,n}(k_b) = \int_0^{\pi/2} \mathbf{J}_n^2(k_b \mathrm{sin}\theta)\mathrm{sin}(\theta)d\theta = \frac{1}{k_b}\sum_{k=0}^{\infty}\mathbf{J}_{2n+2k+1}(2k_b) \tag{5.24}$$

Here Y_n are the closed loop input modal admittances. Note that the power radiated is a function of k_b^2, indicating that given equal currents, the higher the frequency the larger will be the power radiated. That factor is found in the E_ϕ component (see (5.29)). It is very interesting to note that the dependency on the radius has disappeared. The reason is quite straightforward: the power radiated outward in the far field for the forms of the $\vec{E}$ and $\vec{H}$ fields in (5.18) fall off with $1/r^2$ as the surface area grows with r^2. The kernel of (5.20) can be written so that

$$\begin{aligned} &E^0_\theta(\theta,\phi,k_b) \equiv re^{jkr}E_\theta \quad \text{and} \quad E^0_\phi(\theta,\phi,k_b) \equiv re^{jkr}E_\phi \\ &P_{rad}(k_b) = \int_0^{2\pi}\int_0^{\pi}\frac{1}{2\xi}\frac{(E^{0*}_\theta E^0_\theta + E^{0*}_\phi E^0_\phi)}{r^2}r^2 \mathrm{sin}\theta d\theta d\phi \end{aligned} \tag{5.25}$$

and the radius cancels. This leaves us with the nice definition

$$U(\theta,\phi,k_b) \equiv \frac{1}{2\xi}\left[(E^{0*}_\theta E^0_\theta + E^{0*}_\phi E^0_\phi)\right] = r^2 P(r,\theta,\phi,k_b) \tag{5.26}$$

---

[2]See their appendix for fast computational methods.

[3]Lu et al. also included the lossy metal terms, given in (6.3).

at least for the fields here, but indeed for *all* antennas in the far-field, since the fields are always outward flowing normal to the surface of the surrounding sphere

$P_{rad}(k_b)$ is a measure of the power radiated, and is independent of radius. $U(\theta, \phi, k_b)$ is also a measure of the power radiated, but contains the angular fluctuations as well, and therefore is also a measure of how directional the antenna is. It is called the "Radiation Intensity". Dividing $U(\theta, \phi, k_b)$ by $P_{rad}(k_b)$ normalizes it across k_b. Typically though, it is normalized to $P_{rad}(k_b)/(4\pi)$, an average value, since the full solid angle of the sphere is

$$\int_0^{2\pi}\int_0^{\pi} \sin\theta d\theta d\phi = 4\pi \tag{5.27}$$

The typical measure for Directivity then is

$$D(\theta, \phi, k_b) = 4\pi \frac{U(\theta, \phi, k_b)}{P_{rad}(k_b)} \tag{5.28}$$

The directivity indicates how intense the radiation is in some direction relative to the total power emitted. Using again (5.18) and watching the counters,

$$\begin{aligned} E_\theta^{0*}E_\theta^0 &= |E_\theta^0|^2 = |V_0|^2\xi^2\cot^2\theta \times \\ &\left[\sum_{m=1}^{\infty}\sum_{n=1}^{\infty}(-1)^n mnj^{m+n}Y_mY_n^*\sin(m\phi)\sin(n\phi)\mathbf{J}_m(k_b\sin\theta)\mathbf{J}_n(k_b\sin\theta)\right] \\ E_\phi^{0*}E_\phi^0 &= |E_\phi^0|^2 = |V_0|^2\xi^2k_b^2 \times \\ &\left[\sum_{m=0}^{\infty}\sum_{n=0}^{\infty}(-1)^n j^{m+n}Y_mY_n^*\cos(m\phi)\cos(n\phi)\mathbf{J}'_m(k_b\sin\theta)\mathbf{J}'_n(k_b\sin\theta)\right] \end{aligned} \tag{5.29}$$

The gain simply includes the efficiency of the antenna itself:

$$G(\theta, \phi, k_b) = \varepsilon_{ff}D(\theta, \phi, k_b) \tag{5.30}$$

which means that, since the efficiency of the PEC loop is 1.0, the Gain and the Directivity are the same.

5.3 Characterising the Large Closed Loop

The radiation patterns, given above, apply to all sizes of PEC loop antennas at low frequencies. The admittance coefficients that appear in the equations above need to be extended to include the material characteristics of the wire. That is done in Part II of this volume. (see Chap. 6). The PEC loop therefore can be considered "ideal", and

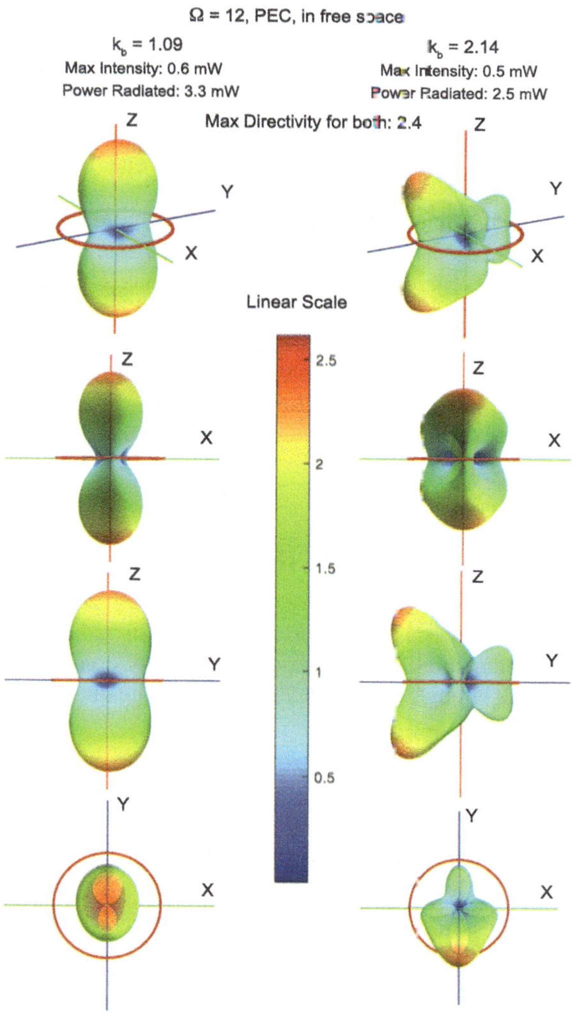

Fig. 5.3 Linearly scaled directivity patterns for the thin-wire PEC loop at the first two resonances. The red circle represents the loop. All axes are equal. The directivity determines the size as well as the surface colouring. Patterns are shown along the main axes, x, y, and z, except the top pattern which is randomly viewed. The bottom patterns show the dipole and quadrupole character of the first and second harmonics. The maxima for power radiated, intensity and directivity are shown at the top

that makes it worthwhile to examine the radiation characteristics more carefully; they become the standard against which loops made of real materials can be compared.

The power radiated, the maximum intensity and the maximum directivity of the PEC loop are functions of k_b, meaning that we can plot 3D visualisations of the radiation patterns,[4] at any k_b desired. Two good choices would include the first two resonances, $k_b = \{1.09, 2.14\}$.

---

[4]Appreciation to Jogender Nagar for use of variations in his MATLAB code that produced these patterns and others in this volume.

Figure 5.3 shows the patterns for $k_b = \{1.09, 2.14\}$, the first two resonances. The patterns are linearly scaled to show true gain; that is, they follow (5.28). Decibel scaling, in which $10 \times \texttt{log}_{10}(\texttt{D})$ is plotted, emphasises the side lobes, which are not important in this case. The axes are the same in all figures and the maximum intensity determines both the size of the pattern and the colouring.

Several points of interest:

1. The resonance at the fundamental harmonic, $k_b = 1.09$, shows strong directivity along the z-axis normal to the plane of the loop.
2. The resonance at the second harmonic, $k_b = 2.14$, shows standard quadrupole behaviour.
3. In the XY cuts, the first resonance shows a stronger lobe at $\phi = 90^o$ to where the driving source is located, while the second resonance shows a stronger lobe in the direction of $\phi = 270^o$ to the driving source.
4. The maximum directivity for both is 4.8.

5.4 Characterising the Small Closed Loop

The small loop is easily characterised in its radiation patterns because the current is quite plausibly, and indeed quite practically, constant around the periphery, if the loop is small enough [1, Sect. 5.2][3, Fig. 2]. The question is "When does this assumption break down".

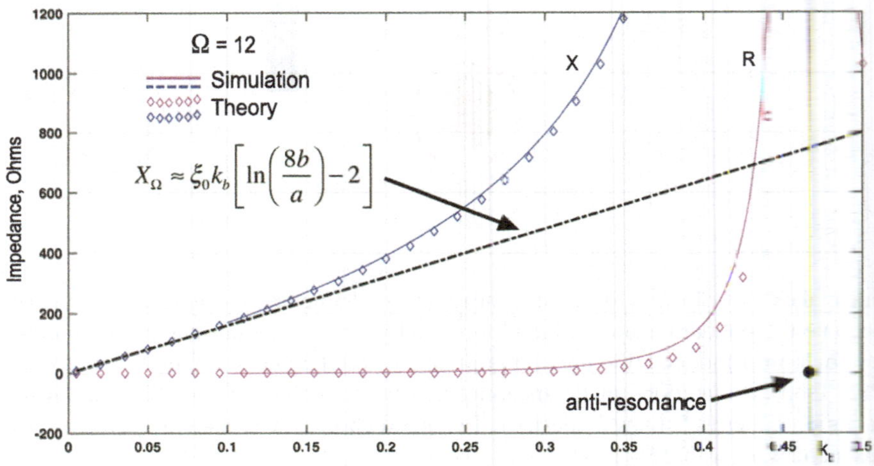

Fig. 5.4 An expansion of Fig. 4.4 around the low k_b region, showing the full resistance and impedance of a thin PEC loop antenna. The reactance is mainly inductive and the slope of the curve rises quite sharply after $k_b \approx 0.20$. The slope may be given by the often used small loop approximation $X_\Omega = \xi_0 k_b[\texttt{ln}(8b/a) - 2]$. A second order and even third approximation derived from the circuit model takes the estimate out only to about 0.25

Based on the constancy of current in an $\Omega = 10$ loop, Balanis suggests that a loop larger than $k_b \approx 0.20$ cannot be considered small. We now examine this suggestion from another point of view, that of the RLC circuit model. Figure 5.4 is an expansion of the low k_b region of Fig. 4.4a, showing the resistance and capacitance of the PEC loop. Immediately evident is that while the resistance is quite small and constant, the reactance is not. It is in fact highly inductive with a large linear slope out to about $k_b \approx 0.20$, moving into an exponential slope beyond. The estimate of inductance as $L = \mu_0 b[\mathtt{ln}(8b/a) - 2]$, which comes from the simple argument of constant current, is shown overlayed. It appears in the Eq. (4.9), and therefore explains the correspondence. As Balanis suggests, the estimate is good out to $k_b \approx 0.20$, but certainly not any further. A second order approximation turns out to be no help. The reason for the sudden and dramatic rise in the reactance beyond 0.20 is the rather strong influence of the anti-resonance that appears near $k_b \approx 0.46$. Indeed, there is no easy approximation to the curve based on the circuit model in Sect. 4.2.

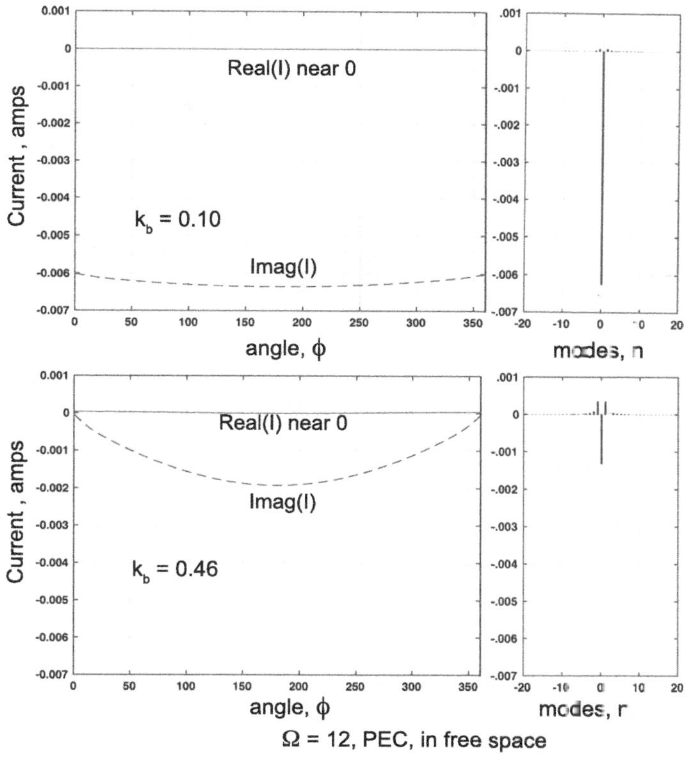

Fig. 5.5 The current and current coefficients for two small thin-wire PEC loop at $k_b = 0.10$ and $k_b = 0.46$. The second is very near the anti-resonance, where no current should be flowing. Even at four decimals, it is difficult to determine exactly. The strong negative imaginaries indicate leading currents in both cases and therefore strongly inductive loops

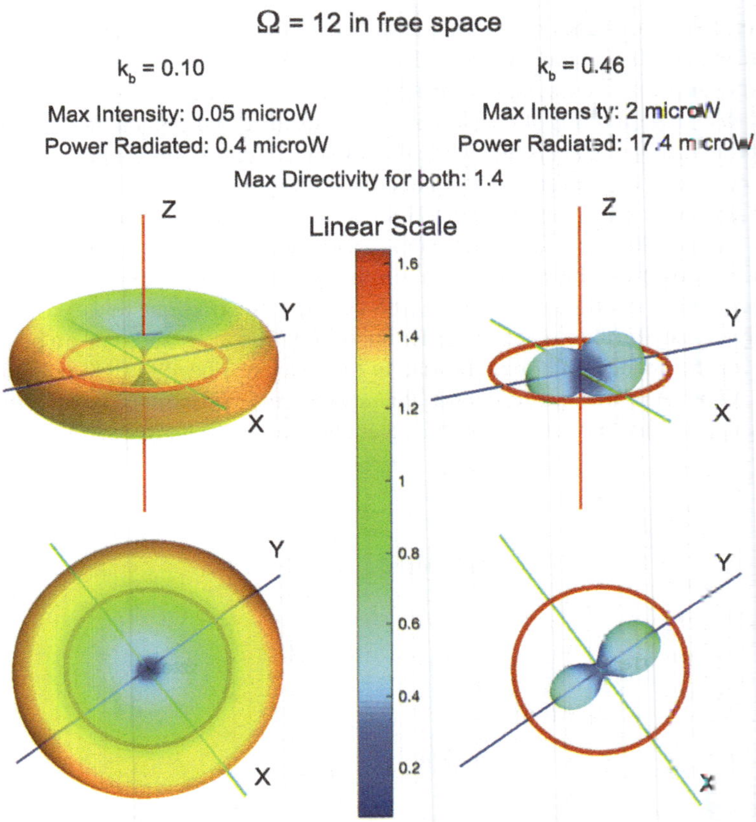

Fig. 5.6 Linearly scaled directivity patterns for small loops on the same scale as in Fig. 5.3. "Small" loops are those below $k_b = 0.10$. The corresponding current is shown in Fig. 5.5. Note that the pattern is quite uniform in the azimuth direction. A comparison is made with a loop at the anti-resonance, $k_b = 0.46$, which has much lower current. The power radiated from the loop at anti-resonance is larger than the small loop, however, due to the k_b^2 factor in (5.22)

Figure 5.5 shows the current and current coefficients for the small circular loop at $k_b = 0.10$ and 0.46. The loop acts highly inductive in both cases and therefore the current at the input lags the voltage substantially. Figure 5.6 shows the directivity patterns for a small loop at $k_b = 0.10$. It is compared with the pattern at the anti-resonance, when a much smaller current flows. The maximum directivities are about the same, but the powers radiated are substantially higher. This is due to the k_b^2 term in E_ϕ.

References

1. C.A. Balanis, *Antenna Theory, Analysis and Design*, 4th edn. (Wiley, New York, 2016)
2. E.J. Martin, IRE Trans. Antennas Propag. 106–107 (1960)
3. D.H. Werner, IEEE Trans. Antennas Propag. **44**(2), 157 (1996)
4. D.H. Werner, in *Lommel Expansions in Electromagnetics, Chapter in Frontiers in Electromagnetics*, ed. by D.H. Werner, R. Mittra (IEEE Press, 2000), pp. 474–522
5. L.W. Li, M.S. Leong, P.S. Kooi, T.S. Yeo, IEEE Trans. Antennas Propag. **45**(12), 1741 (1997)
6. S. Savov, IEEE Antennas Propag. Mag. **44**(5), 98 (2002), https://doi.org/10.1109/MAP.2002.1077781
7. B.Q. Lu, J. Nagar, T. Yue, M.F. Pantoja, D.H. Werner, IEEE Trans. Antennas Propag. **65**(1), 121 (2017)

Chapter 6
Lossy Thin Loops and Rings

6.1 The Effect of Surface Impedance on Loop Characteristics

The models developed in Part Two were limited by two approximations.

(1) The thin wire approximation says that the circumference of the wire is quite a bit less than the wavelength of the source frequency: $2\pi a \ll \lambda$, or equivalently, $ka \ll 1$. This ultimately limits the models to $\Omega \geq 10$.[1]

(2) The perfectly conducting metal approximation limits the usefulness of the models to frequencies below 150–200 GHz.[2]

We wish now to lift the second of these restrictions so that the models also reflect the behaviour of loops at higher frequencies, most notably in the THz and optical regions. Such rings are therefore more appropriately called "nano-scaled rings" rather than "loop antennas". The thin-wire approximation will continue to hold.

The behavioural change at about 150 GHz occurs due to dispersive properties of the material from which the ring is made. The effect is simply taken care of by adding a dispersive term to the governing equation, (2.94)

$$\vec{E} = -\vec{\nabla}\Phi - \frac{\partial \vec{A}}{\partial t} \tag{6.1}$$

The driving voltage has been taken as a delta function across an infinitesimal gap (see Sect. 3.6 and previous). In a PEC loop, this driving voltage appears only across the gap, but in a dispersive material, the voltage appears also across the full circumference of the ring. This can be taken care of by modifying the governing equation to:

---

[1] See Sect. 2.5.3.

[2] See Sects. 2.3.1 through 2.3.3.

A. McKinley, *The Analytical Foundations of Loop Antennas and Nano-Scaled Rings*, Signals and Communication Technology,
https://doi.org/10.1007/978-981-13-5893-7_6

$$\vec{E} - J_\phi(\phi)Z_s = -\vec{\nabla}\Phi - \frac{\partial \vec{A}}{\partial t} \tag{6.2}$$

and finding a relevant expression for Z_s. The impedance Z_s is a surface impedance with units of 'ohms/m'. The analysis of skin depth in Sect. 2.3.3 indicates that thin metallic wires are fully penetrated by the far infra-red. Lower than 2 THz, current flows on the surface only; above 48 THz, it flows across the entire cross-sectional area. The thin-wire approximation is used no matter how the current flows and it limits current density to the circumferential direction only. The limiting factor is the b/a ratio; that is, $k_b \ll b/a$. Too small a ratio makes the assumption invalid. The field calculated at the center of the wire is distributed uniformly around the circumference, $2\pi a$.

Following the same strategy, then, as in Sect. 4.1, the results lead directly to

$$I_\phi(\phi) = \sum_{n=-\infty}^{\infty} I_n e^{jn\phi} \quad \texttt{where} \quad I_n = \frac{V_0}{j\pi\xi a_n + (b/a)Z_s} \tag{6.3}$$

and a_n is given by (3.89). In other words, a complex lossy term involving both the aspect ratio of the wire and the surface impedance is added to each modal impedance. This surface impedance term, $(b/a)Z_s$, is a constant for every mode at a given k_b. It may be added to any non-lossy model Z_n term in any equation expressed in earlier chapters. Hence, the modal impedance of the lossy metal loop antenna transforms to

$$Z_n \rightarrow Z_n + (b/a)Z_s \tag{6.4}$$

6.1.1 The Functional Dependencies of the Surface Impedance

A cylindrical wire, made of an imperfectly conducting material, propagates an electromagnetic field inside that can be described well using Bessel functions of the first kind. The field external to the wire uses Hankel functions. Derivations are shown in Stratton [1] and Hanson [2], but Hallen [3, pg 9] knew of the term. Novotny [4] uses the same approach to derive a theory of wavelength scaling in dipoles. Stratton and Hanson give the surface impedance of a cylindrical wire as

$$z_s = \frac{\gamma}{2\pi a\sigma}\frac{\mathbf{J}_0(\gamma a)}{\mathbf{J}_1(\gamma a)} \equiv \frac{1}{2\pi a}Z_s \tag{6.5}$$

in ohms/m. γ is the transverse propagation constant in the wire, the same as k in Eq. (2.69), $\gamma = k = \beta + j\alpha$; α is the damping coefficient. a is the radius of the wire, so γa expresses the total phase shift and damping over the radius of the wire. $\mathbf{J}_0(\gamma a)$ and $\mathbf{J}_1(\gamma a)$ are the zero and first order Bessel functions of the First Kind.

The form of the conductivity of the material now depends upon frequency and whether the material is dielectric or metal. We shall work only with metal; work with dielectrics is, at the moment, at the cutting edge. These behavioural characteristics were discussed in detail in Sect. 2.3.2. The conductivity was given in (2.84) for the mid-range and optical frequencies as

$$\sigma(\omega) = j\omega\varepsilon_0(\varepsilon_r - 1) = j\omega\varepsilon_0(\eta^2 - 1) \tag{6.6}$$

where ε_r is the relative permittivity of the metal and since the permittivity is related to the square of the index of refraction by (2.65). We have

$$\begin{aligned} \gamma &= \frac{\omega}{c}\eta \\ \frac{\gamma}{\sigma} &= -i\xi_0\frac{\eta}{\eta^2 - 1} \\ Z_s &= -i\xi_0\left(\frac{\eta}{\eta^2 - 1}\right)\frac{\mathbf{J}_0(\gamma a)}{\mathbf{J}_1(\gamma a)} \end{aligned} \tag{6.7}$$

where ξ_0 is the characteristic impedance of free-space, $\sqrt{\mu_0/\varepsilon_0}$. This surface impedance applies to any metal for which the complex index of refraction is known. The effects of the extra term on the current and impedance differ from metal to metal, but this term accurately handles these differences.

6.1.2 Modelling the Index of Refraction

A full spectrum response model was developed in Sect. 2.3.2 for the index of refraction of any material. By "full spectrum" is meant the frequency spectrum from Radio Frequency (RF) through the Optical Region (OR). The model for metals, given by (2.89), is repeated here:

$$\begin{aligned} \eta^2 = \varepsilon_r = 1 - f_0\frac{\omega_p^2}{\omega}\left(\frac{1}{\omega - j2\Gamma_0} + \frac{\alpha}{\omega - j2\beta\Gamma_0}\right) \\ + \sum_{n=1}^{\infty}\frac{f_n\omega_p^2}{2\omega_n}\left[\frac{e^{j\pi/\gamma_n}}{(\omega_n - \omega + j\Gamma_n)^{a_n}} + \frac{e^{-j\pi/\gamma_n}}{(\omega_n + \omega - j\Gamma_n)^{a_n}}\right] \end{aligned}$$

The values for all of the coefficients are given in Table 2.2 for the metals gold, silver, and copper. The DC value of the conductivity is found by taking $\omega \to 0$ and realising that the resonances do influence the DC conductivity. Then,

$$\sigma_0 = \frac{\omega_p^2\varepsilon_0}{2\Gamma_0}\left[1 + \frac{\alpha}{\beta}\right] \tag{6.8}$$

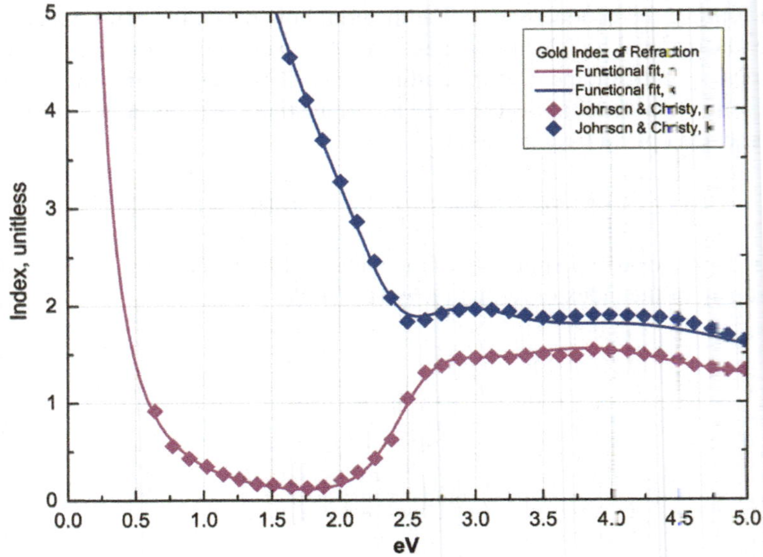

Fig. 6.1 Gold's index of refraction, as given by [5] and by the function Eq. (6.8)

The index is shown in Fig. 6.1 for gold in the optical region, using diamonds to indicate the data measured by [5] and using a solid line to indicate the fit. Note that the original Johnson and Christy data assumes a $\exp(-j\omega t)$ dependence, and therefore the complex conjugate of their original data has been taken to show the functional fit of Fig. 6.1.

Figure 6.2 shows the real and imaginary parts of the wire impedance of a thin gold ring ($\Omega = 12$) given by Eq. (6.10) for three different circumference lengths in the low THz, the near infra-red region (NIR) and the optical region. The principal reason for the sudden increases in the curves is the effect of the transition bands on the Bessel function ratio, $\mathbf{J}_0(\gamma a)/\mathbf{J}_1(\gamma a)$. The figure shows that the degree to which a given ring is affected by the transitions depends on its circumference and the driving wavelength. Different sized rings respond to the same driving wavelength differently. For example, a 600 nm circumference ring will be affected by the transitions when the driving wavelength reaches the length of the circumference, at 600 nm. But a 350 nm ring will be affected when the driving wavelength reaches double its circumference, at 700 nm. In any case, resonance saturation puts all of the modal resonances of such rings lower in frequency than the transition effects. Section 6.4 will show that at optical circumferences these rings resonate in the region $k_b \leq 0.2$.

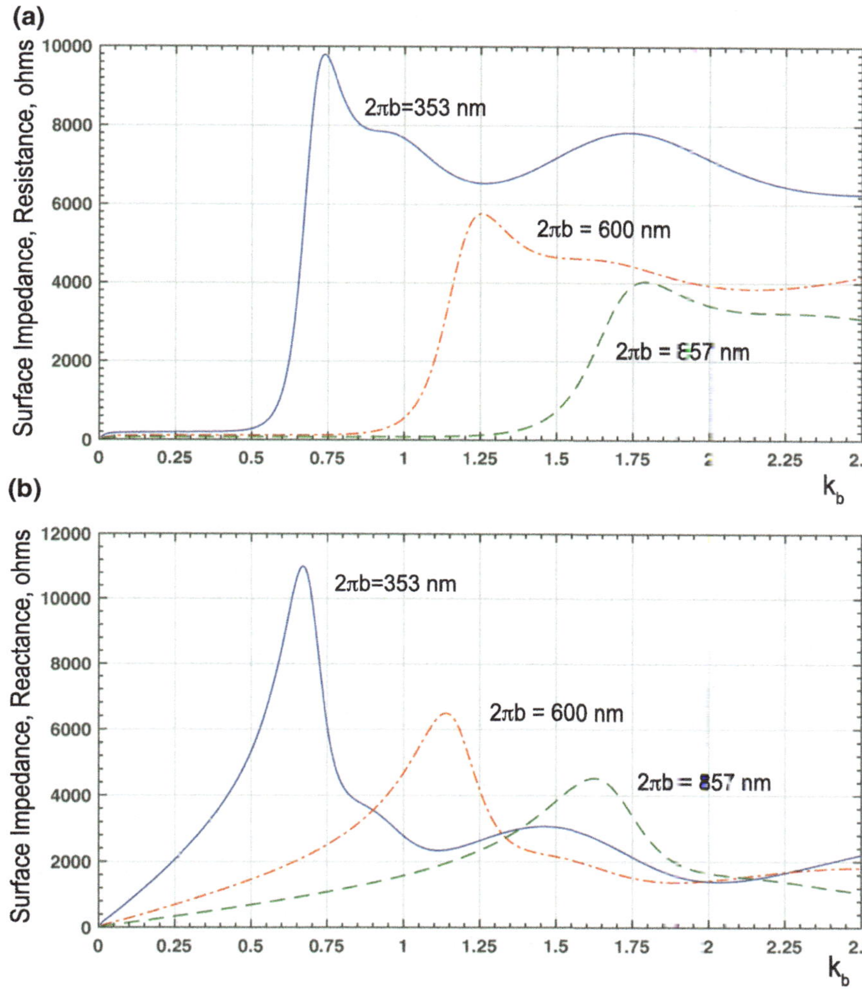

Fig. 6.2 The real and imaginary parts of the surface impedance for an $\Omega = 12$ gold loop in the low THz, the infra-red and the optical region: **a** surface resistance and **b** surface reactance. The rise in all cases is due to the inter-band transitions

6.2 The Driving Point Impedance and Admittance

The transformation from closed loop modal impedances to lossy modal impedances, is given in (6.4).

$$Z_n \rightarrow Z_n + (b/a)Z_s \tag{6.9}$$

There is really no need for additional notation to show this, since the lossy term is constant at a given k_b and added to every mode. Hence, whenever one sees Z_n in this volume, consider it to have the lossy term involved automatically. Y_{in} and Z_{in} will be used for the general input admittance and impedance, regardless of construction. The admittance becomes, from (4.4)

$$Y_{in} = \frac{1}{Z_{in}} \equiv \frac{I_\phi(0)}{V_0} = 2\sum_{n=0}^{\infty} \frac{\varepsilon_n}{Z_n + (b/a)Z_s}$$
$$\text{where} \quad \varepsilon_n = 1/2 \text{ for } n = 0 \text{ and } 1 \text{ for } n \neq 0 \tag{6.10}$$

Figures 4.1 and 6.3 show the input impedances for three gold nano-rings with different circumference lengths, one in the GigaHertz (GHz) region, one in the FIR and one in the OR, as given by Eq. (6.3) and by numerical simulation using CST's Microwave Studio™(MWS) [6]. The match between the analytical model and numerical simulation is remarkably close for all three rings and supports substantially the correctness of the model.

An examination of Figs. 4.1, 6.3 show that the extra lossy term in Eq. (6.3) causes the input impedance function to compress; all resonances migrate toward wavelengths longer than the circumference. Three points are important: (1) the shape of the functions are essentially the same; (2) the resonances migrate; and (3) the magnitude of the impedance increases with shorter circumferences. The first point means that all rings respond similarly to incident energy. The second means that the resonances move toward smaller values of k_b as the circumference decreases; the resonances must therefore eventually saturate. The third means that smaller rings carry smaller current magnitudes and lag the driving field.

In Fig. 6.3b notice that although the ring circumference matches a wavelength in the mid-visible region, the ring's first resonance is at $600/0.095 = 6316$ nm, well outside the OR. In Fig. 6.3c the ring circumference matches a wavelength in the UV region, the ring's first resonance is at $353/0.057 = 6192$ nm, not too far from the first resonance of the previous ring, which is 70% longer in circumference.

Note that in Fig. 6.3, the optical region begins to the right of the x-axis. For the 3000 nm ring, $k_b = 0.50$ corresponds to 6 μm, and in the 600 nm ring, it corresponds to 1.2 μm.

6.3 The RLC Model for Lossy Metals

An exact RLC model for the perfectly conducting closed ring was presented in Sect. 4.2. That model may be applied to rings in the region $0 < k_b < 2.5$. The goal now is to produce an exact RLC model for the lossy rings. This is difficult in the model's present form, because the Bessel functions in Eq. (6.7) cannot be reframed to show R, L, C elements. However, by assuming $\gamma a \ll 1$ (that is, by again asserting the thin-wire approximation), the Bessel functions can be approximated to first order

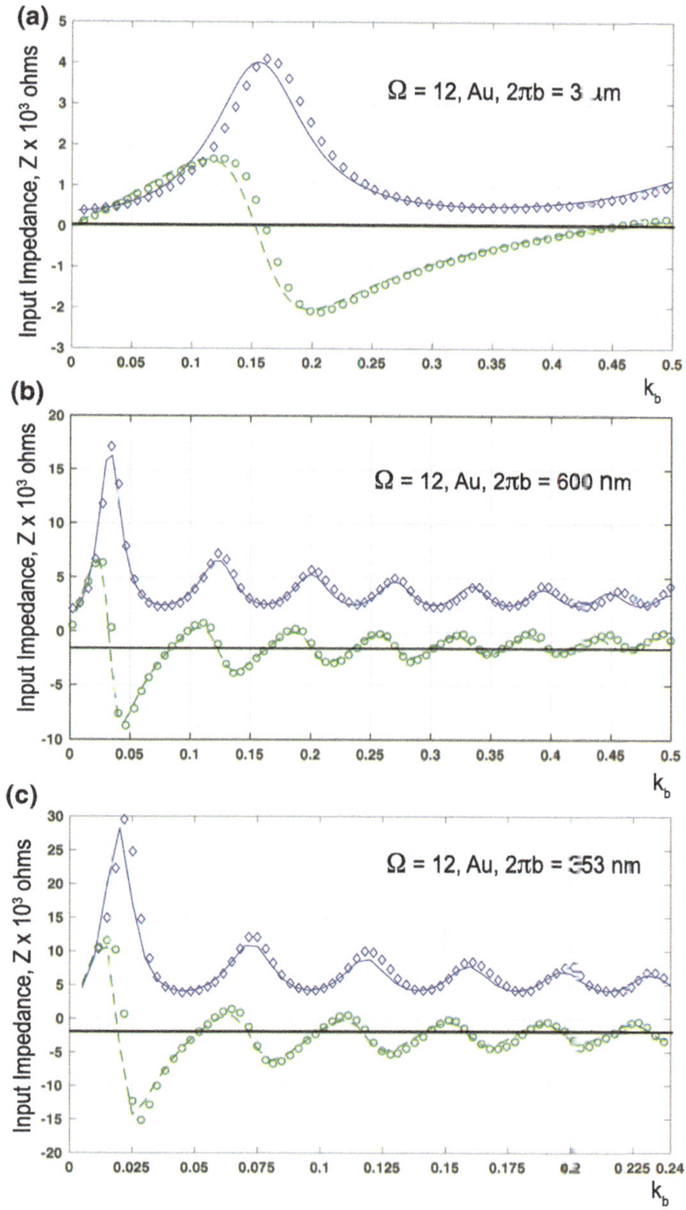

Fig. 6.3 The impedance for three circular thin gold rings, calculated from Eq. (6.3) compared with numerical simulations. **a** $2\pi b = 3000\,\text{nm}$, **b** $2\pi b = 600\,\text{nm}$ and $k_b = 353\,\text{nm}$. Note that the optical region begins far to the right of the x-axis cutoff given here. In c the first resonance is at $353/0.057 = 6192\,\text{nm}$ and the x-axis ends at $353/0.24 = 1240\,\text{nm}$

in γa [7]. This will be the only approximation for the following RLC model:

$$\frac{\mathrm{J}_0(\gamma a)}{\mathrm{J}_1(\gamma a)} \approx \frac{1-(\gamma a)^2/4}{(\gamma a)/2} \tag{6.11}$$

Substituting into Eq. (6.7), the lossy term from (6.5), $2\pi b \times z_s = (b/a)Z_s$, can be approximated as:

$$\begin{aligned}
\frac{b}{a}Z_s &= -i\xi_0\left(\frac{b}{a}\right)\left(\frac{\eta}{\eta^2-1}\right)\frac{\mathrm{J}_0(\gamma a)}{\mathrm{J}_1(\gamma a)} \\
&= \frac{\xi_0}{\left|\eta^2-1\right|^2}\times\left[\left(\frac{k_b}{2}+\frac{2(b/a)^2}{k_b}\right)\mathrm{Im}(\eta^2)+i\frac{k_b}{2}\left(\left|\eta^2\right|^2-\mathrm{Re}(\eta^2)\right)\right] \\
&+ i\frac{\xi_0}{\left|\eta^2-1\right|^2}\left[\frac{2(b/a)^2}{k_b}\left(1-\mathrm{Re}(\eta^2)\right)\right] \\
&= \xi_0\left[\left(k_b r_s-\frac{1}{k_b g_s}\right)+i\left(k_b l_{\mu s}-\frac{1}{k_b l_{\varepsilon s}}\right)\right]
\end{aligned} \tag{6.12}$$

Using Eq. (6.3) and the definitions,

$$\begin{aligned}
R_n' &\equiv \xi_0\left[k_b(r_n+r_{sn})-\frac{1}{k_b}\left(\frac{n^2}{g_n}+\frac{1}{g_{sn}}\right)\right] \\
L_n' &\equiv \mu_o b(l_{\mu n}+l_{\mu sn}) \\
C_n' &\equiv \varepsilon_o b\frac{1}{(n^2/l_{\varepsilon n}+1/l_{\varepsilon sn})}
\end{aligned} \tag{6.13}$$

the modal input impedance in Eq. (6.3) becomes:

$$\begin{aligned}
Z_n' &= R_n' + j\left(\omega L_n' - \frac{1}{\omega C_n'}\right) \\
&= \xi_0\left(k_b(r_n+r_{sn})-\frac{1}{k_b}\left(\frac{n^2}{g_n}+\frac{1}{g_{sn}}\right)\right) \\
&+ i\xi_0\left(k_b(l_{\mu n}+l_{\mu sn})-\frac{1}{k_b}\left(\frac{n^2}{l_{\varepsilon n}}+\frac{1}{l_{\varepsilon_s n}}\right)\right)
\end{aligned} \tag{6.14}$$

where

$$r_{sn} = \begin{cases} \mathrm{Im}(\eta^2)/(4|\eta^2-1|^2) \text{ if } n>0 \\ \mathrm{Im}(\eta^2)/(2|\eta^2-1|^2) \text{ if } n=0 \end{cases}$$

$$g_{sn} = \begin{cases} -|\eta^2-1|^2/((b/a)^2\mathrm{Im}(\eta^2)) \text{ if } n>0 \\ -|\eta^2-1|^2/(2(b/a)^2\mathrm{Im}(\eta^2)) \text{ if } n=0 \end{cases}$$

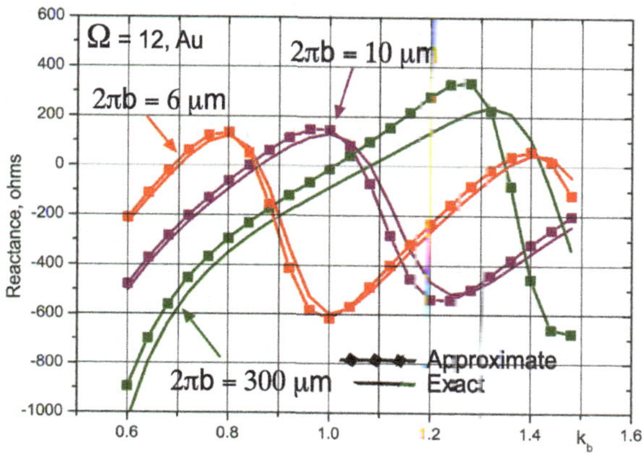

Fig. 6.4 The reactance for three thin gold rings, comparing the effect of the exact lossy term Eq. (6.5) with the approximate Eq. (6.12). The smaller the ring, the more exact the lossy term approximation becomes

$$
\begin{aligned}
l_{\mu sn} &= \begin{cases} (|\eta^2|^2 - \mathrm{Re}(\eta^2))/(4|\eta^2 - 1|^2) \text{ if } n > 0 \\ (|\eta^2|^2 - \mathrm{Re}(\eta^2))/(2|\eta^2 - 1|^2) \text{ if } n = 0 \end{cases} \\
l_{\varepsilon sn} &= \begin{cases} -|\eta^2 - 1|^2/((b/a)^2(1 - \mathrm{Re}(\eta^2))) \text{ if } n > 0 \\ -|\eta^2 - 1|^2/(2(b/a)^2(1 - \mathrm{Re}(\eta^2))) \text{ if } n = 0 \end{cases}
\end{aligned}
\tag{6.15}
$$

Since the index of refraction is wavelength dependent, it is not sufficient to specify only a value for k_b in order to calculate the modal RLC elements. One must specify a circumference as well. Figure 6.4 compares the reactance of Eq. (6.3) using the exact Z_s term, Eq. (6.5), and using the approximate Z_s term, Eq. (6.14), for various circumferences of a thin ring, $\Omega = 12$ ($b/a = 64$). For circumferences larger than 20–30 μm, the approximate reactance in Eq. (6.14) is a poor fit with the exact result given by Eq. (6.5). This is due to that $\gamma a \ll 1$ approximation, made so that the Bessels could be reduced to first order. But for 6 μm and shorter, the agreement is very good. **Therefore, any calculations with the approximation Eq. (6.11) can be applied to rings with circumferences shorter than about 10 μm.**

6.3.1 *The Total R, L and C of the Lossy Metal Loop at Any k_b*

The closed loop represents itself as an RLC circuit over all k_b, as described in Sect. 4.2.1. At the anti-resonances it shows as an open circuit and at the true resonances as a short circuit. The values of these circuit elements for the closed loop may be calculated using (4.12) through (4.14)).

These equations do not change in form when the lossy term is added to Z_n.

$$Z_n \rightarrow R_n + jX_n + (b/a)Z_s \tag{6.16}$$

The other equations follow accordingly.

6.4 Resonance Saturation of Rings in the Optical Region

Several authors have noticed that as dipoles and rings are made shorter, they tend to resonate at wavelengths longer than their lengths, or circumferences, would suggest [4, 8–11]. Indeed, they eventually reach a minimum wavelength below which they will not resonate. The model given above yields an accurate functional curve for this behavior in rings. The reason is the extra lossy term in Eq. (6.3).

Using the approximation Eq. (6.14), an equivalent expression to Eq. (4.16) which gives the modal resonances in the OR, is:

$$k_{bnr} = \frac{2\pi b}{\lambda_{nr}} = \sqrt{\frac{n^2 + l_{\varepsilon n}/l_{\varepsilon sn}}{l_{\varepsilon n} l_{\mu n} + l_{\varepsilon n} l_{\mu sn}}}. \tag{6.17}$$

Since this assumes the small γa approximation, it applies only to rings with circumferences less than 10 μm. $l_{\mu n}$ and $l_{\varepsilon n}$ are the low frequency, unit-less, functions related to the ring inductance and capacitance, respectively, found in (4.9) and (4.10). $l_{\mu sn}$ and $l_{\varepsilon sn}$ are the high frequency additions coming from the new lossy term. Figure 6.5 shows a plot of (6.17) for rings of various thicknesses. The model is less accurate for thick rings, because of the thin-wire approximation. Even so, the analytic model matches numerical simulations down to thicknesses of $\Omega = 8$ $(b/a = 8.7)$. We see immediately a resonance saturation for each ring. The cutoff for the thinnest ring

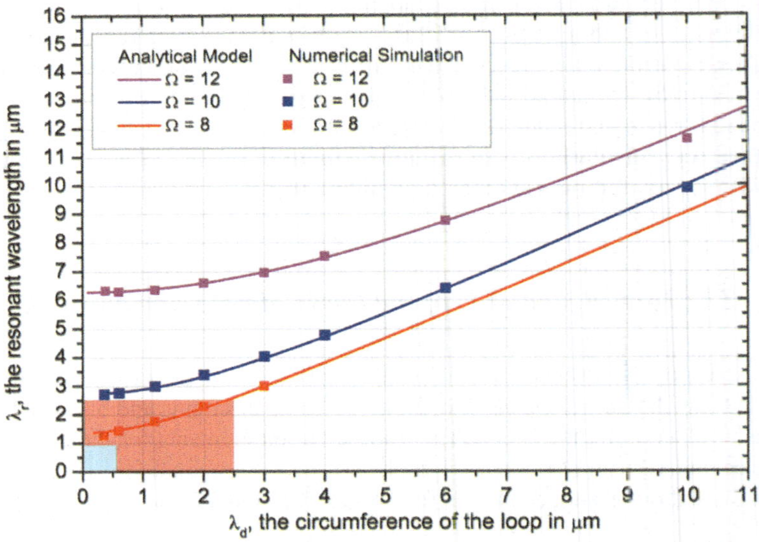

Fig. 6.5 The first modal resonance ($n = 1$) as a function of the circumference for four gold rings of various thicknesses, given by (6.17). Only the thickest rings show cutoff in the near infrared. Red indicates the near infra-red and light blue the optical region

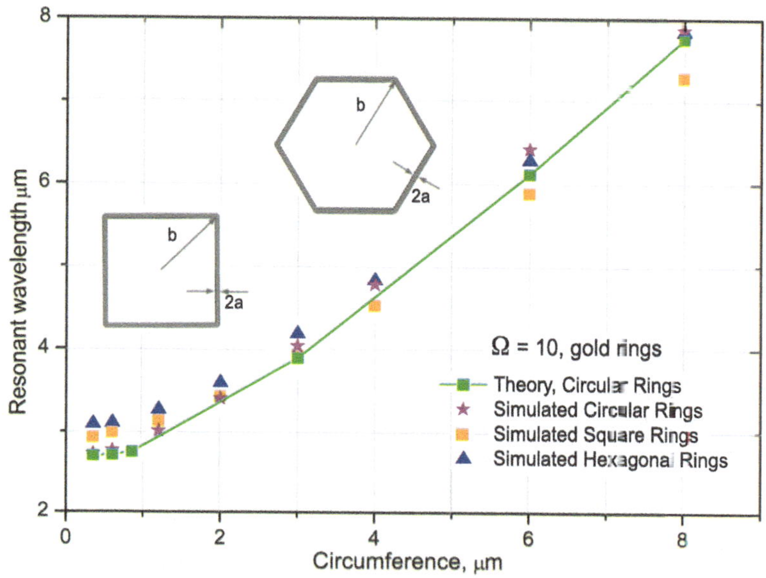

Fig. 6.6 The first resonances of simulated gold circular, hexagonal and square nano-rings of size $\Omega = 10$ $(b/a = 24)$ compared with resonances measured from plots using Eq. (6.3). The line joins the points calculated from Eq. (6.3) as a guide to the eye

is about 6.2 µm, below which closed, gold nano-rings will have no first harmonic $(n = 1)$ resonance. The analytical model predicts a cutoff for a thick $\Omega = 6$ ring of about 600 nm, but the simulations put it at about 1.0 µm. This is because the thin-wire approximation makes theoretical results for thicker rings less accurate.

Resonance saturation occurs for hexagons and squares as well. Figure 6.6 shows resonances for simulated circular, hexagonal and square rings, compared with the resonances identified from numerical input impedance plots of circular rings using Eq. (6.3). The radius of the wire, a, for these rings is taken to be the radius of a circumscribing circle through the middle of the wire at the corners. The circle circumscribing the hexagon or square itself has radius b. We conclude that the shape of the ring is not an important parameter for saturation.

An interesting discussion of saturation in split-rings (rings with one gap) appears in [12].

6.5 Radiation and Radiation Losses in the Thin-Wire Loop

The radiation characteristics of the thin-wire, perfectly conducting loop or ring is discussed in Sect. 5.2. The only required change to all that has preceded this chapter due to the surface impedance is to use (6.4) for the modal impedances.

The radiation efficiency of a loop is given by (5.6), repeated here

$$\varepsilon_{ff} = \frac{R_{rad}}{R_{loss} + R_{rad}} \tag{6.18}$$

R_{rad} is the radiation resistance and is given by (5.4). Then using $Y_{el} = I_\phi(0)/V_0$, from (4.4) and (5.22)

$$R_{rad} = \frac{2P_{rad}}{|I_\phi(0)|^2} = |Z_{in}|^2 \frac{\xi \pi k_b^2}{2} T(k_b) \tag{6.19}$$

where $Z_{in} = 1/Y_{in}$ as given in (6.10). This assumes that the current is symmetrical around the loop. Asymmetrical currents for loaded loops are handled in Chap. 7.

The current should now include the surface impedance in all the modal terms. The power loss is related to the loss resistance, R_{loss}, which depends on that surface impedance term. It is given by [13–15]

$$|I_\phi(0)|^2 R_{loss} = \frac{b}{a} \mathrm{Re}(Z_s) \frac{1}{2\pi} \int_0^{2\pi} |I(\phi)|^2 \, d\phi \tag{6.20}$$

Using (4.3)

$$\left|I_\phi(\phi)\right|^2 = 4V_0^2 \sum_{n=0}^{\infty} \varepsilon_n^2 |Y_n|^2 \cos^2(n\phi) \quad \text{and} \quad |I_\phi(0)|^2 = |I_{in}|^2 = V_0^2 |Y_{in}|^2 \tag{6.21}$$

we have

$$R_{loss} = 2|Z_{in}|^2 \mathrm{Re}(Z_s) \frac{b}{a} \sum_0^{\infty} \varepsilon_n^2 |Y_n|^2 \tag{6.22}$$

The directivity is given by (5.28) and the gain by (5.30).

References

1. J.A. Stratton, *Electromagnetic Theory* (McGraw-Hill Book Company, New York, 1941)
2. G.W. Hanson, IEEE Trans. Antennas Propag. **54**(12), 3677 (2006)
3. E. Hallen, Nova Actae Regiae Soc. Sci. Upsaliensis Ser. IV, **11**(4), 1 (1938)
4. L. Novotny, Phys. Rev. Lett. **98**, 266802 (2007). https://doi.org/10.1103/PhysRevLett.98.266802
5. P.B. Johnson, R.W. Christy, Phys. Rev. B **6**(12), 4370 (1972)
6. Computer Simulation Technology (2018). https://www.cst.com
7. I. Stegun, M. Abramowitz, *Handbook of Mathematical Functions With Formulas, Graphs and Mathematical Tables*, vol. 55, Applied Mathematics Series (US. Govenent Printing Office, Washington, D.C., 1964)

8. T. Okamoto, T. Otsuka, S. Sato, T. Fukuta, M. Haraguchi, Opt. Express **20**(21), 24059 (2012). https://doi.org/10.1364/OE.20.024059
9. J.B. Khurgin, G. Sun, Appl. Phys. Lett **99**(21), 211106 (2011). https://doi.org/10.1063/1.3664105, http://link.aip.org/link/?APL/99/211106/1
10. C.M. Soukoulis, T. Koschny, J. Zhou, M. Kafesaki, E.N. Economou, Phys. Status Solidi (b) **244**(4), 1181 (2007). https://doi.org/10.1002/pssb.200674503
11. M.W. Klein, C. Enkrich, M. Wegener, C.M. Soukoulis, S. Linden, Opt. Lett. **31**, 1259 (2006)
12. D.Ö. Güney, T. Koschny, C.M. Soukoulis, Phys. Rev. B **80**(12), 125129 (2009). https://doi.org/10.1103/PhysRevB.80.125129
13. V. Savov, IEEE Antennas Propag. Mag. **45**(3), 129 (2003)
14. B.Q. Lu, J. Nagar, T. Yue, M.F. Pantoja, D.H. Werner, IEEE Trans. Antennas Propag. **65**(1), 121 (2017)
15. M.F. Pantoja, J. Nagar, B. Lu, T. Yue, D.H. Werner, in *2016 10th European Conference on Antennas and Propagation (EuCAP)* (2016), pp. 1–4. https://doi.org/10.1109/EuCAP.2016.7481565

Chapter 7
Lossy Thin Loops and Rings with Multiple Impedance Loads

7.1 The Usefulness of Loop Antennas with Multiple Loads

In 1965, Iizuka [1] provided the first calculation of current and input impedance on a loop loaded at various points on the periphery with impedances. At the time, Iizuka was interested in the effects that negative resistive diodes (Esaki diodes) had on current distribution around the loop and on resonances. It turned out that negative resistors caused stronger resonances than was capable with the simpler closed loop antenna. These diodes are, of course, active devices, rather than passive devices, but their use in loop antennas does not seem to have survived in practice. A bit later that year, Harrington [2] looked at the problem of multiply-loaded loops in an arbitrary field. At the same time, however, the amateur radio community and the United States Army explored loop antennas loaded with a single capacitor [3, 4]. These caused much stronger resonances than did the negative resistance loads. There was also some experimentation with loop antennas loaded with as many as six capacitors, but this work was never published [5].

Equation (4.6) through (4.10) indicate that loops of small circumference compared to the incident wavelength will display only the $n = 0$ mode and will therefore have small resistance (since the upper limit of the $\mathbf{J}_2$ integral, namely $2k_b$, is near zero), infinite conductance, infinite capacitance and an inductance of about $L_0 \approx \mu_0 b\,[\mathtt{ln}(8b/a) - 2]$ (see also Sect. 5.4). The inductance changes with the aspect ratio of the loop, b/a. This value is constant up to about $k_b \approx 0.18$, as is seen in Fig. 5.4 for a thin $\Omega = 12$ loop. There is no capacitive reactance, since the charge is the same at every point along the length of the ring. Therefore, resonances simply do not occur. If, therefore, a capacitor were added of the proper value, one could have perhaps a significant resonance in the vicinity of very low, sub-wavelength k_b.

This phenomenon will now be studied in some detail by extending the thin-wire PEC theory to loaded loop antennas in such a way that any number and variety of impedances can be added in the periphery of the loop. This allows us to see the generation of significant resonances and radiation patterns due to the addition of

A. McKinley, *The Analytical Foundations of Loop Antennas and Nano-Scaled Rings*, Signals and Communication Technology,
https://doi.org/10.1007/978-981-13-5893-7_7

capacitors and inductors placed variously around the periphery. These results are not confined then to the sub-wavelength region but at every region of the k_b spectrum between 0 and 2.5, and are applicable to all thin wire PEC loop antennas. With the addition of the material characteristics, as suggested in Chap. 6, they are also applicable through the optical region.

7.2 Current in Multiply Loaded Loops

Iizuka's original derivation [1] begins, as it did for the closed loop antenna by establishing a broadband spectrum voltage source across a gap at $\phi = 0$. The gap is reduced to infinitesimal separation and a Dirac Delta distribution, $V\delta(\phi)$, is used as the source. This leads to the current distribution given in (4.1), repeated here

$$I_\phi(\phi) = \sum_{n=-\infty}^{\infty} I_n e^{jn\phi} = \sum_{n=-\infty}^{\infty} \frac{V_0}{Z_n} e^{jn\phi} \quad \texttt{where} \quad Z_n \equiv j\pi\xi a_n = \frac{1}{Y_n} \tag{7.1}$$

Z_n are the modal input impedances of the PEC closed loop. With loss, the modal impedances are replaced with (6.9), namely

$$Z_n \to Z_n + \frac{b}{a} Z_s \quad \text{where} \quad Z_s = -i\xi_0 \left(\frac{\eta}{\eta^2 - 1}\right) \frac{\mathbf{J}_0(\gamma a)}{\mathbf{J}_1(\gamma a)} \tag{7.2}$$

Define

$$Z_n' \equiv Z_n + \frac{b}{a} Z_s \tag{7.3}$$

Note that the input impedance for a closed lossy loop is, from (4.2) with the loss included. The prime on Z_n indicates that loss is included.

$$Z_{cl} \equiv \frac{V_0}{I_\phi(0)} = \left[\sum_{n=-\infty}^{\infty} \frac{1}{Z_n'}\right]^{-1} \equiv \frac{1}{Y_{cl}} \tag{7.4}$$

We now place in the periphery M similar voltage sources, evenly spaced, separated by an angle $\Delta\phi = 2\pi/M$, as shown in Fig. 7.1. Any given source q is then placed at the angle $\phi_q = 2\pi(q-1)/M$ and the current in the periphery arises from a superposition of the currents due to each source, V_q. The resulting current is then given by

$$I(\phi) = \sum_{q=1}^{M} I_q(\phi) = \sum_{q=1}^{M} \left(\sum_{n=-\infty}^{\infty} \frac{e^{jn(\phi - 2\pi(q-1)/M)}}{Z_n'} \right) V_q \equiv \sum_{q=1}^{M} Y(\phi, q) V_q \tag{7.5}$$

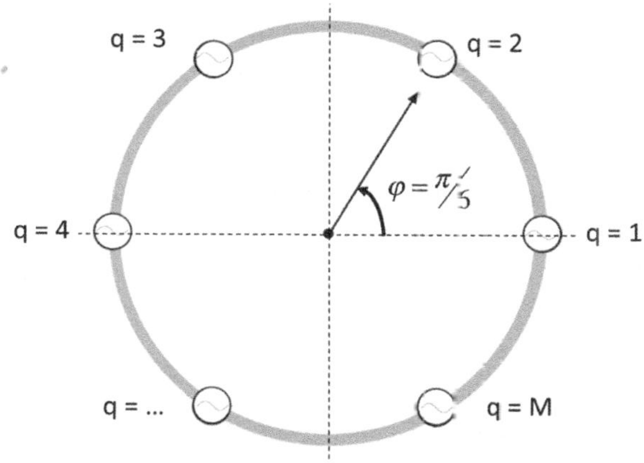

Fig. 7.1 The geometry associated with multiply loaded sources, numbered 1 through M. The sources are distributed evenly around the periphery of the loop

This notation suggest two vectors, where $Y(\phi, q)$ and the voltage sources, V_q, becomes a column vector, so that the summation of the current becomes more simply written in matrix notation as

$$I(\phi) = \mathbf{Y}^{\mathrm{T}}(\phi)\mathbf{V} \tag{7.6}$$

$\mathbf{Y}^{\mathrm{T}}(\phi)$ is the transpose of $\mathbf{Y}(\phi)$. Here,

$$Y(\phi, q) = \sum_{n=-\infty}^{\infty} \frac{e^{jn(\phi-2\pi(q-1)/M)}}{Z'_n} \tag{7.7}$$

At a given source, for example, source p, located at ϕ_p,

$$Y(\phi, q) = \sum_{n=-\infty}^{\infty} \frac{e^{jn2\pi(p-q)/M}}{Z'_n} \equiv [Y_{pq}] \equiv \mathbb{Y} \tag{7.8}$$

This notation $[Y_{pq}]$ suggests a square matrix, $\mathbb{Y}$, with $M \times M$ members; p being the row counter and q the column counter.

An alternative Tensor index notation has no notion of row and column. Instead, it suppresses the summation and assumes that whenever an index is repeated, a summation over that index is implied. The indices must be raised and lowered and cannot be used again in another summation in the same term. In other words, the same index may not appear more than once in either the superscript position or the subscript position. The notation has some advantages over matrix notation. So we will use them both.

For example, writing

$$Y(\phi, q) \equiv Y_q(\phi) \tag{7.9}$$

we have

$$I(\phi) = \sum_{q=1}^{M} Y_q(\phi)V_q \equiv Y_q(\phi)V^q = Y^q(\phi)V_q \tag{7.10}$$

Notice that Y^q and V_q may be switched without detriment to the meaning. In similar manner,

$$[Y_{pq}] \to Y_q^p \tag{7.11}$$

indicates a two rank tensor, since q and p indicate the loads, each up to M.

One might think that $\mathbb{Y}$ or Y_q^p is Hermitian in p and q, but it is not since Z_n' is complex and does not change when p and q flip places. From (7.8), these may be written

$$\begin{aligned}
\mathbb{Y} \equiv Y_q^p &= \sum_{n=-\infty}^{\infty} \frac{e^{jn2\pi(p-q)/M}}{Z_n'} \\
&= \frac{1}{Z_0} + \sum_{n=1}^{\infty} \frac{1}{Z_n'}\left(e^{jn2\pi(p-q)/M} + e^{-jn2\pi(p-q)/M}\right) \\
&= \frac{1}{Z_0} + 2\sum_{n=1}^{\infty} \frac{\cos(2\pi n(p-q)/M)}{Z_n'}
\end{aligned} \tag{7.12}$$

It is therefore symmetric around $n = 0$.

Impedances can be added at each source under the realisation that impedances are voltage sinks: $-Z_k^q I^k$. Since the voltage across each load depends on the current generated by each of the voltage sources spread around the loop, these voltage sinks need to be added. Hence the summation over k. When constructed as a matrix Z_k^q is diagonal with the individual loads, Z^q, on the diagonal. We have for the total voltage at each angle ϕ_q:

$$V^q - Z_k^q I^k \equiv \mathbf{V} - \mathbb{Z}\mathbf{I} \tag{7.13}$$

Here the vector $\mathbf{I}$ has elements that refer to the current due to the sources. These are not the modal currents. The matrix, $\mathbb{Z}$ is diagonal square with $M \times M$ members. When V^q is set to 0, the impedance appears alone at ϕ_q. Setting ϕ_p in (7.10) and substituting the transformation (7.13)

$$\begin{aligned}
I^p = Y_q^p V^q \to Y_q^p\left(V^q - Z_k^q I^k\right) &= Y_q^p V^q - Y_q^p Z_k^q I^k \\
I^p + Y_q^p Z_k^q I^k &= Y_q^p V^q \\
\left(\mathbb{I}_k^p + Y_q^p Z_k^q\right) I^k = Y_q^p V^q \quad &\Leftrightarrow \quad (\mathbb{I} + \mathbb{Y}\mathbb{Z})\,\mathbf{I} = \mathbb{Y}\mathbf{V}
\end{aligned} \tag{7.14}$$

The matrix $\mathbb{I}_k^p \equiv \mathbb{I}$ is the identity matrix and allows isolation of I^k. Now define

$$F_k^p \equiv \mathbb{I}_k^p + Y_q^p Z_k^q \quad \Leftrightarrow \quad \mathbb{F} = \mathbb{I} + \mathbb{Y}\mathbb{Z} \tag{7.15}$$

Equation (7.14) becomes

$$I^k = [F^{-1}]_p^k Y_q^p V^q \quad \Leftrightarrow \quad \mathbf{I} = \mathbb{F}^{-1}\mathbb{Y}\mathbf{V} \tag{7.16}$$

where $[F^{-1}]_p^k$ is the inverse of the matrix F_k^p. Using (7.13) and (7.16) in (7.10), the current as a function of angle results for the loop. Watching the counters:

$$\begin{aligned} I(\phi) &= Y_q(\phi)\left(V^q - Z_k^q I^k\right) = Y_q(\phi)V^q - Y_q(\phi)Z_k^q I^k \\ &= Y_l(\phi)V^l - Y_q(\phi)Z_k^q[F^{-1}]_p^k Y_l^p V^l \\ &= (Y_l(\phi) - Y_q(\phi)Z_k^q[F^{-1}]_p^k Y_l^p)V^l \end{aligned}$$

$$\Leftrightarrow \quad \boxed{I(\phi) = \left(\mathbf{Y}^{\mathrm{T}}(\phi) - \mathbf{Y}^{\mathrm{T}}(\phi)\mathbb{Z}\mathbb{F}^{-1}\mathbb{Y}\right)\mathbf{V}} \tag{7.17}$$

7.2.1 A Simplification

One may notice that there is no real reason to place the loads evenly around the loop periphery, since that idea never plays a role in the subsequent development of the theory.[1] One may simply set

$$Y_q(\phi) = \sum_{n=-\infty}^{\infty} \frac{e^{jn(\phi-\phi_q)}}{Z_n'} \quad \text{and} \quad Y_q^p = \sum_{n=-\infty}^{\infty} \frac{e^{jn(\phi_p-\phi_q)}}{Z_n'} \tag{7.18}$$

where ϕ_q is any angle whatsoever, as in Fig. 7.2. In this case, $M = 3$ only, but $\phi_2 \neq 2\pi(2-1)/3$; it is an angle free of M. The rest of the theory follows as normal. Note from (7.4) that when $p = q$,

$$Y_q^p = \sum_{n=-\infty}^{\infty} \frac{1}{Z_n'} = Y_{cl} \quad \text{when} \quad p = q \tag{7.19}$$

[1] Appreciation to Mario Pantoja for this insight.

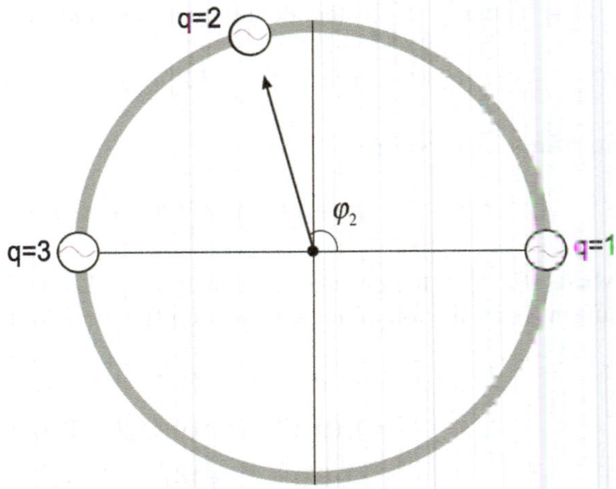

Fig. 7.2 The simplified Iizuka configuration. Here $M = 3$, but $\phi_2 \neq 2\pi(2-1)/3$; it is an angle free of M

7.2.2 Current Coefficients

The current coefficients can be determined readily. From (7.17):

$$I(\phi) = \sum_{n=-\infty}^{\infty} I_n e^{jn\phi} = (Y_l(\phi) - Y_q(\phi) Z_k^q [F^{-1}]_p^k Y_l^p) V^l$$

$$\Leftrightarrow \boxed{I(\phi) = \left(\mathbf{Y}^{\mathbf{T}}(\phi) - \mathbf{Y}^{\mathbf{T}}(\phi) \mathbb{Z}\mathbb{F}^{-1}\mathbb{Y}\right) \mathbf{V}} \tag{7.20}$$

The I_n here are not the same as I^k given by (7.16), which is instead the current at each load, k. I_n are the Fourier current coefficients for each mode, n. The current given by (7.17) can be expanded by substituting (7.5).

$$Y_l(\phi) \equiv \sum_{n=-\infty}^{\infty} \frac{e^{jn(\phi-\phi_l)}}{Z_n'} = \sum_{n=-\infty}^{\infty} \frac{e^{-jn\phi_l}}{Z_n'} e^{jn\phi} \equiv \sum_{n=-\infty}^{\infty} y_{nl} e^{jn\phi}$$

$$\Leftrightarrow \mathbf{Y}(\phi) \tag{7.21}$$

Substituting,

$$I(\phi) = \sum_{n=-\infty}^{\infty} \left[(y_{nl} - y_{nq} Z_k^q [F^{-1}]_p^k Y_l^p) V^l \right] e^{jn\phi}$$

$$= \sum_{n=-\infty}^{\infty} y_{nq} \left[(\mathbb{I}_l^q - Z_k^q [F^{-1}]_p^k Y_l^p) \right] V^l e^{jn\phi}$$

$$\equiv \sum_{n=-\infty}^{\infty} Y_{nl} V^l e^{jn\phi} \tag{7.22}$$

Therefore the current coefficients are, in Tensor and matrix notation

$$I_n = Y_{nl} V^l \quad \Leftrightarrow \quad I_n = \mathbf{Y_n^T V} \tag{7.23}$$

where $Y_{nl} \Leftrightarrow \mathbf{Y_n}$ are the modal admittances at each of the load sources; they are related to the modal impedances

$$(\mathbf{Y_n})^{-1} \equiv \mathbf{Z_n} = \mathbf{R_n} + j\mathbf{X_n} \tag{7.24}$$

The modal reactances are complicated enough to make a separation into inductive and capacitive elements impossible.

7.2.3 Asymmetries Due to Loads

We now examine the current coefficients I_n for symmetry across mode 0. The modal impedances, $Z'_n = Z_n + (b/a)Z_s = Z_{-n} + (b/a)Z_s$, are symmetrical across mode 0, as discussed in Sect. 3.6.1. The loss term added to each mode is independent of the mode number, and therefore even under loss, these impedances are symmetrical across mode 0. The matrix coefficients, F_k^p given in (7.15), are symmetrical because the matrix coefficients, Y_q^p, given by (7.8), and the diagonal load impedances, Z_k^q, are symmetrical across $n = 0$. The set of coefficients, y_{nl}, are not symmetrical:

$$y_{nl} = \frac{e^{-jn\phi_l}}{Z'_n} \quad \text{and} \quad y_{(-n)l} = \frac{e^{jn\phi_l}}{Z'_n} \tag{7.25}$$

Consequently, the modal currents are, in general, asymmetrical across $n = 0$, unlike the modal currents for the closed loop. All physical variables derived from the current must take this asymmetry into account. Since the imaginary parts are asymmetrical, the modes differ in their phasing around the loop. Therefore,

$$I_{(-n)} \neq I_n \tag{7.26}$$

7.3 The Input Impedance and Circuit Representation of Multiply Loaded Loops

The asymmetry in the current coefficients propagates through to the current at the input terminals; i.e., at $\phi = 0$. The input impedance is defined as usual, with only one voltage source. Taking it at $\phi = 0$, $V^1 \equiv V_0$, we have from (7.22):

$$Y_{in} \equiv \frac{I(0)}{V_0} = \sum_{n=-\infty}^{\infty} Y_{n1} = \sum_{n=-\infty}^{\infty} y_{nq} \left(\mathbb{I}_1^q - Z_k^q [F^{-1}]_p^k Y_1^p \right) \tag{7.27}$$

The subscript 1 indicates, of course, the first column of the matrix; that is, the position of the source at $\phi = 0$, which, by definition of input impedance, is the only voltage source in the loop. Note that due to asymmetry across n,

$$Y_{n1} \neq Y_{(-n)1} \tag{7.28}$$

7.4 Radiation from Multiply Loaded Loops

Radiation from the closed, unloaded loop is discussed in detail in Sects. 5.2 and 6.5. The asymmetry of the current coefficients in loaded loops affects their radiation pattern significantly. The new patterns can be determined by again solving (5.14) for G_0 and G_m but this time using for the current

$$I(\phi) = \sum_{n=-\infty}^{\infty} I_n e^{jn\phi} = \sum_{n=0}^{\infty} \varepsilon_n (I_n e^{jn\phi} + I_{-n} e^{-jn\phi})$$
$$\text{and where } \varepsilon_n = 1/2 \text{ for } n = 0 \text{ and } 1 \text{ for } n \neq 0 \tag{7.29}$$

When the current is symmetric, $I_n = I_{-n}$, this reduces to

$$I(\phi) = 2 \sum_{n=0}^{\infty} \varepsilon_n I_n \cos(n\phi) \tag{7.30}$$

which what we used in the closed loop case, (5.16). The fields become [6]

$$H_r = E_r \approx 0$$

$$H_\theta \approx \frac{k_b e^{-jkr}}{2r} \sum_{n=0}^{\infty} j^n \varepsilon_n \left(I_n e^{jn\phi} + I_{-n} e^{-jn\phi} \right) \mathbf{J}'_n(k_b \sin\theta)$$

$$H_\phi \approx j \frac{\cot\theta e^{-jkr}}{2r} \sum_{n=1}^{\infty} n j^n \left(I_n e^{jn\phi} - I_{-n} e^{-jn\phi} \right) \mathbf{J}_n(k_b \sin\theta)$$

$$E_\theta \approx j\frac{\xi \cot\theta e^{-jkr}}{2r}\sum_{n=1}^{\infty} nj^n \left(I_n e^{jn\phi} - I_{-n} e^{-jn\phi}\right) \mathbf{J}_n(k_b \sin\theta)$$

$$E_\phi \approx -\frac{\xi k_b e^{-jkr}}{2r}\sum_{n=0}^{\infty} j^n \varepsilon_n \left(I_n e^{jn\phi} + I_{-n} e^{-jn\phi}\right) \mathbf{J}_n'(k_b \sin\theta) \quad (7.31)$$

The expressions will reduce to the expressions given by (5.18) for symmetric currents. To obtain the radiated power, these are to be integrated according to the prescription in (5.20), leading to

$$P_{rad}(k_b) = \frac{\xi\pi k_b^2}{2}|V_0|^2$$
$$\times \sum_{n=0}^{\infty} \varepsilon_n \left[|Y_{n1}|^2 + Y_{(-n)1}|^2\right]\left[\frac{1}{2}Q_{n-1,n-1}^{(1)}(k_b) + \frac{1}{2}Q_{n+1,n+1}^{(1)}(k_b) - \frac{n^2}{k_b^2}Q_{n,n}^{(1)}(k_b)\right] \quad (7.32)$$

where Y_{nl} is for the loaded loop, given in (7.27) and is asymmetric around $n = 0$. Note one must use the input admittance, since the derivation for it also assumes only one voltage source in the loop at $\phi_1 = 0$.

7.5 The Single Capacitor Loop

The case of a single capacitor in the periphery of the loop is a particularly interesting one and provides a good example of the expressions and methodology. We know by (5.4) that a loop, if small enough, acts more inductively than capacitively and therefore, if a capacitor is added to the periphery, we expect intuitively that at some point in the spectrum a resonance of significance is likely to occur. The questions are: Where? What is the quality of that resonance? What value of capacitor might optimise it? and How might we tune it?

7.5.1 Constant Value Capacitor

As an arbitrary beginning, introduce a capacitor in the periphery of value, $C = \varepsilon_0 b l_\varepsilon$, as given by the definitions in (4.6). For this capacitor $X_C = \xi/(k_b l_\varepsilon)$ and l_ε is a constant. The capacitive reactance is therefore a parabolic function of k_b, as needed to represent a capacitor over frequency. Therefore, a pleasant arbitrary value to begin with is $l_\varepsilon = 1.0$. For all loops, $M = 12$.

We now move the capacitor around the periphery beginning at $\phi = 0$ in stages, calculating the admittance as we go using (7.27). Each stage places the capacitor an angle of $\Delta\phi = \pi/6$ from the previous until the admittance for seven positions on the loop have been calculated, ending at $\phi = \pi$. There is no need to go further due to symmetry; more than that would be repetitious.

Figure 7.3 shows the results of the computation. A very full, high quality resonance does appear in the sub-wavelength region (the region where $k_b < 0.5$) at roughly $k_b = 0.34$. Careful calculations show that the resonance for a PEC, $\Omega = 12$ loop occurs very near 0.3384, depending on code implementation. Figure 7.4 shows the current configurations around each loop and Fig. 7.5 gives the radiation patterns.

It would be useful to try to compute the Q of the sub-wavelength resonance, and this can be arranged when the capacitor appears at $\phi = 0$, since in that case $q = 1$ and Z_k^1 and Y_1^p each have only one member; the counters $k = 1$ and $p = 1$ only. We have from (7.12), (7.15) and (7.27)

$$Y_{in} = \sum_{n=-\infty}^{\infty} y_{n1}\left(1 - Z_1^1[F^{-1}]_1^1 Y_1^1\right)$$

$$F_1^1 = 1 + Y_1^1 Z_1^1 \quad \text{and} \quad Y_1^1 = \sum_{n=-\infty}^{\infty} \frac{1}{Z'_n} = Y_{cl} \tag{7.33}$$

Since the load $Z_1^1 = Z_1 = -jX_C$, (7.15) yields

$$F_1^1 = 1 + Z_1 Y_{cl} \qquad (F^{-1})_1^1 = \frac{1}{1 + Z_1 Y_{cl}}$$

$$\text{and} \quad F_1^1 = Z_1^1[F^{-1}]_1^1 Y_1^1 = \frac{Z_1 Y_{cl}}{1 + Z_1 Y_{cl}} \tag{7.34}$$

Therefore

$$Y_{in} = \sum_{n=-\infty}^{\infty} \frac{1}{Z'_n}\left[1 - \frac{Z_1 Y_{cl}}{1 + Z_1 Y_{cl}}\right] = \left[\frac{1}{1 + Z_1 Y_{cl}}\right] \sum_{n=-\infty}^{\infty} \frac{1}{Z'_n} = \frac{Y_{cl}}{1 + Z_1 Y_{cl}} \tag{7.35}$$

Or

$$Z_{in} = Z_{cl}\left(1 + \frac{Z_1}{Z_{cl}}\right) = Z_{cl} - jX_C \tag{7.36}$$

The capacitor is in series with the input impedance of the closed loop, which seems reasonable given that the capacitor is in series with the input terminals of the loop at $\phi_1 = 0$. The circuit elements for a PEC or lossy thin-wire loop with a capacitor at $\phi = 0$ are:

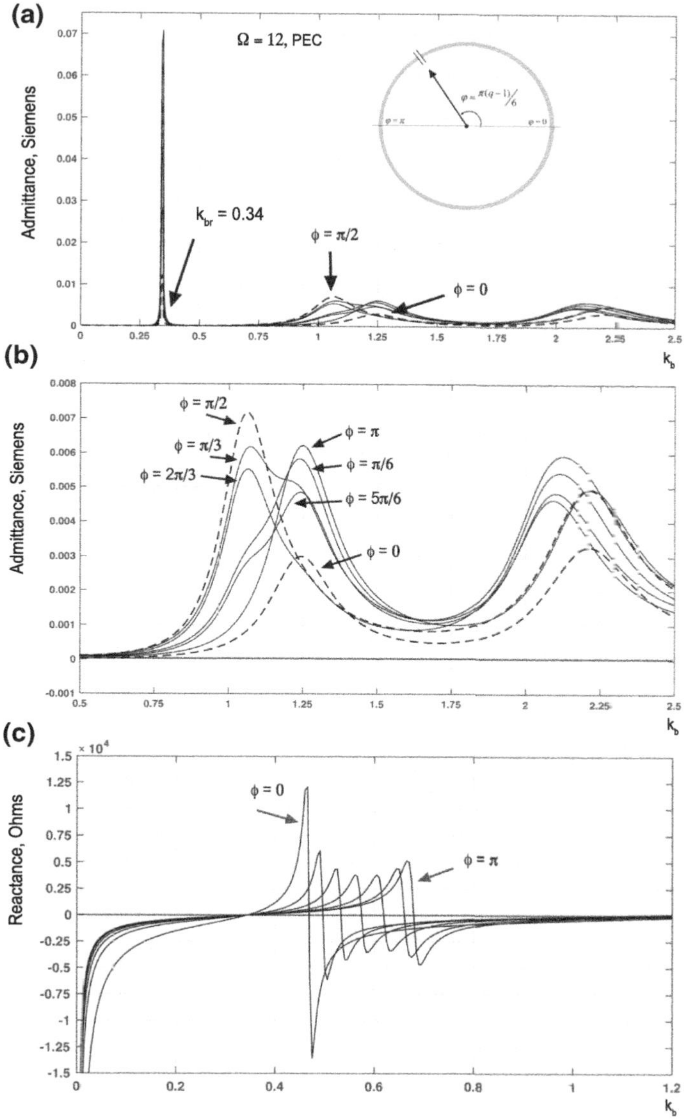

Fig. 7.3 A single capacitor of value, $l_\varepsilon = 1.0$, placed in the periphery of a thin-wire PEC loop, moved by an angle $\pi/6$ through seven positions (q = 1–7) from $\phi = 0$ to $\phi = \pi$. The sub-wavelength resonance they cause, shown in (**a**), remains at $k_{br} \approx 0.34$ and its quality factor is constant. The conductance peaks vary for each capacitor, with the highest occurring for the capacitor at $\phi = \pi$ at 342 mS. The wavelength resonances, shown in (**b**), blue shift with placement near $\phi = 0$ and $\phi = \pi$. An expansion of bandwidth for these small resonances can also occur, as seen in (**b**) for $\pi/3$ and $5\pi/6$. The anti-resonance moves significantly, as shown in (**c**)

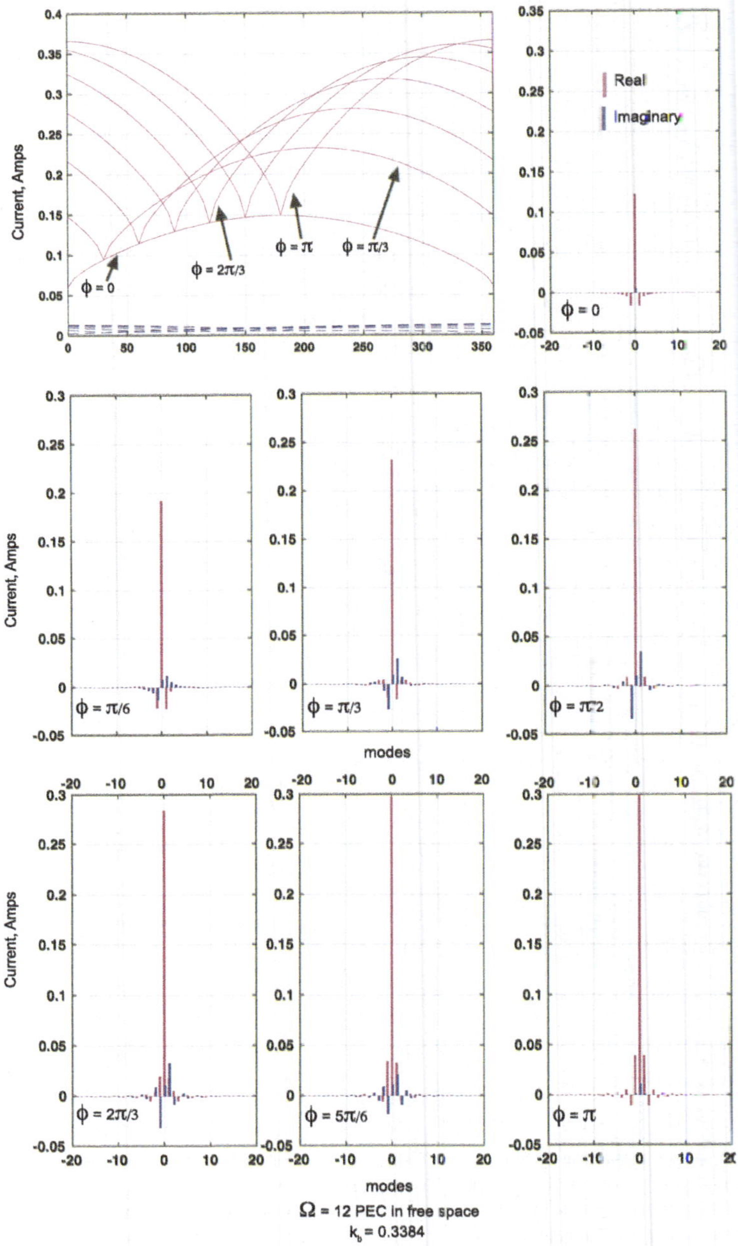

Fig. 7.4 The current and current coefficients for the loops with capacitive loading as described in Fig. 7.3. Only the real parts of the current are large enough to show; the imaginary parts are near 0. Note that current maximum on the circumference grows by 3 times as the capacitor moves toward the 180 degree position. The imaginary components are negligible, since this is resonance. The coefficients are symmetric only in the case of placements at $\phi = 0$ and $\phi = \pi$

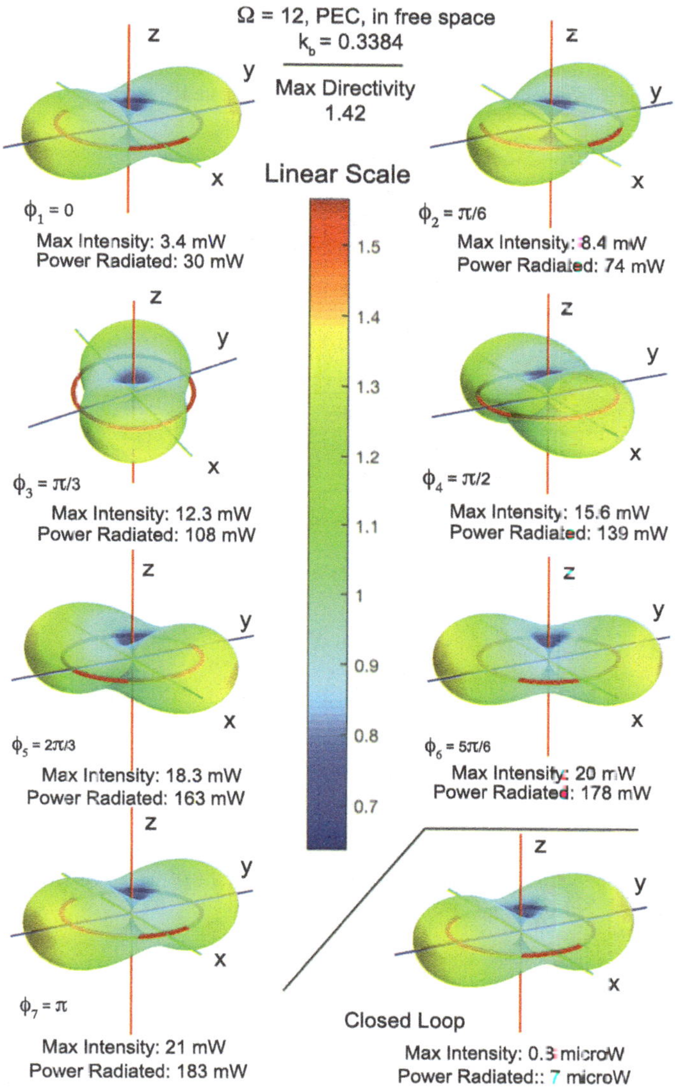

Fig. 7.5 Radiation patterns for the capacitively loaded loops of the preceding figures. Notice that the pattern rotates counterclockwise around the loop as the capacitor moves in that same direction. The directivity is nearly the same as the closed loop at the same frequency in all cases, but the intensity of radiation and total power radiated is substantially higher due to the high quality factor

Table 7.1 The computational results due to (7.27) for seven $M = 12$ capacitively loaded $\Omega = 12$, PEC loops, keeping the capacitor value at $l_\varepsilon = 1.0$. The second table shows the circuit elements calculated using (7.36) and (7.37), and a substantial quality factor in the hundreds for this resonance. The quality factors of the remaining six loops cannot be calculated

ϕ (of cap)	k_{br}	X_q ohms	R_{in} ohms	X_{in} ohms
0	0.3384	-j1114	16.9	0
$\pi/6$	0.3384	-j1114	6.9	0
$\pi/3$	0.3384	-j1114	4.7	0
$\pi/2$	0.3384	-j1114	3.7	0
$2\pi/3$	0.3384	-j1114	3.2	0
$5\pi/6$	0.3384	-j1114	2.9	0
π	0.3384	-j1114	2.8	0

	Resonance									
Ω	M at $\phi = 0$	k_{br} at $X_{in} = 0$	L μH	X_L Ohms	C_{cl} pF	$X_{C,cl}$ Ohms	C_1 pF	$X_{C,1}$ Ohms	C pf	Q
12	1	0.3384	11	2344	3.8	-j1232	4.2	-j1114	2.0	139

$$X_{in} = X_{L,cl} - (X_{C,cl} + X_{C,1}) \quad \text{therefore,}$$

$$X_{C,cl} + X_{C,1} = \frac{\xi_0}{k_{br}}\left(\frac{1}{l_{\varepsilon,cl}} + \frac{1}{l_{\varepsilon,1}}\right) \quad \text{and} \quad l_\varepsilon = \frac{l_{\varepsilon,cl} l_{\varepsilon,1}}{l_{\varepsilon,cl} + l_{\varepsilon,1}}$$

$$R_{in} = R_{cl}, \quad L_{in} = \mu_0 b l_{\mu,cl}, \quad C_{in} = \varepsilon_0 b l_\varepsilon \quad \text{and} \quad Q = \frac{\sqrt{L_{in}/C_{in}}}{R_{in}} \tag{7.37}$$

The calculated values using a load capacitor of $l_{\varepsilon_1} = 1.0$ appears in Table 7.1. Moving the capacitor to $\phi_7 = \pi$, provides the same results as an $M = 2$ loaded loop with the capacitor at $q = 2$. In this case,

$$\tilde{Z} = \begin{bmatrix} 0 & 0 \\ 0 & Z_2 \end{bmatrix} \quad \tilde{Y} = \begin{bmatrix} Y_{cl} & -Y_{cl} \\ -Y_{cl} & Y_{cl} \end{bmatrix} \quad \tilde{f} = \begin{bmatrix} 1 & -Y_{cl}Z_2 \\ 0 & 1 + Y_{cl}Z_2 \end{bmatrix} \tag{7.38}$$

yielding

$$\tilde{F} = \frac{\begin{bmatrix} 0 & 0 \\ -Y_{cl}Z_2 & Y_{cl}Z_2 \end{bmatrix}}{1 + Y_{cl}Z_2} \qquad Z_{in} = \frac{Z_{cl}^2}{Z_{cl} + Z_2} \tag{7.39}$$

which cannot easily be separated into the inductive and capacitive parts. The same difficulties remain for the other positions q.

Note that L and C are functions of the diameter of the loop, but neither R nor $\sqrt{L/C}$ are, and therefore Q is not. The L and C given in the table were calculated using a loop radius, $b = 3$ m.

Several interesting points are evident:

First, the resonant frequency in the sub-wavelength region is the same for all angular positions of the capacitor, $k_{br} = 0.3384$ or roughly 0.34, perhaps not surprisingly. These numbers will vary slightly depending on code implementation. The quality factor, Q, for the loop when the capacitor is placed at $\phi = 0$ is about 143. This may be true of all the remaining positions, but it cannot be calculated using the proposed methods. The bandwidth would therefore be $\Delta k_b = k_{br}/Q = 0.0024$.
The radiation resistance, R_{in}, changes, decreasing from 16.4 ohms at $\phi = 0$ by about 6 times to 2.7 ohms at $\phi = \pi$, which implies that conductance and radiation must increase as the capacitor approaches $\phi = \pi$. If the quality factor is constant, the loop's $\sqrt{L/C}$ would also fall, since R falls.

Second, an expansion of the spectrum around the resonances in the wavelength region show that some of the capacitors (those near 0 and π) cause a blue shift in the primary resonance ($k_{br} \approx 1.0$) of about 25%. This is a rather large shift. The second resonance (near $k_b \approx 2.0$) does not shift as much. Moreover, capacitors at $\phi = \pi/3$ and $5\pi/6$ broaden the bandwidth considerably.

Third, an expansion of the reactance curves around the anti-resonance shows that it blue shifts significantly from $k_b = 0.46$ to $k_b = 0.68$. This is a 50% increase in frequency as the capacitor moves around the loop.

Fourth, Fig. 7.4 shows how the current along the loop circumference changes as the capacitor is moved. Only the real part is shown, since the imaginary part is quite small, but it too shows a sharp change in the current structure at the capacitor in each case. Therefore the radiation patterns will have some structure to them as Fig. 7.5 indicates. There the closed loop pattern at $k_b = 0.34$ is compared with the patterns associated with the capacitors.

Basically, the directivity is the intensity divided by the total power radiated. We can see that the values for the closed loop are in the micro-watt region, while those for the capacitively loaded loop are thousands of times larger. The best loading for intensity is a capacitor at $\phi = \pi$. This behaviour can only be due to the very high resonance. Since these are PEC loops, there is no loss.

The driving source is located at $\phi = 0$, on the positive x axis. Hence the pattern shoots out perpendicular, to the right and left, of the position of the capacitor (the capacitor forms a minimum node in the pattern where it is located and at 180 degrees from its location). Finally, there is no radiation in the Z-direction normal to the plane of the loop, in all cases.

7.5.2 Tuning by Varying the Capacitance Value, l_ε

Suppose now we place the capacitor at $\phi = \pi$ and vary the parameter l_ε. This will change the capacitor value and doing so will generate a number of sub-wavelength resonances, all with different quality factors. Figure 7.6 shows the results,

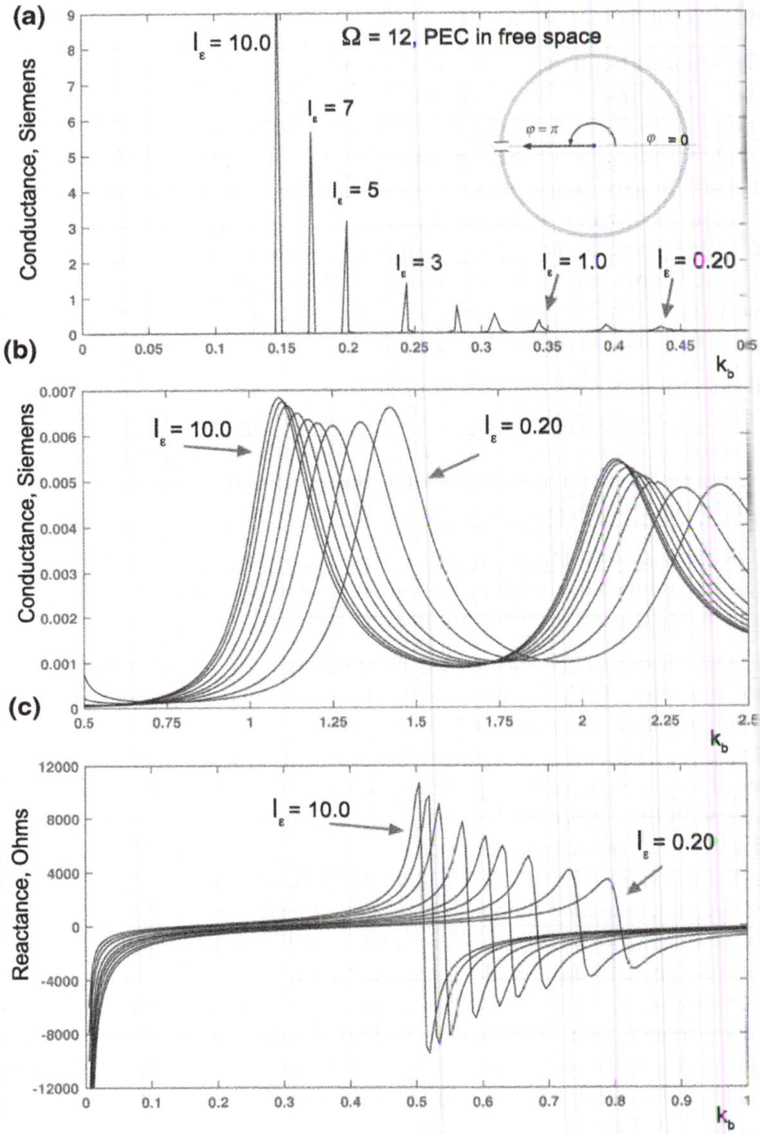

Fig. 7.6 The effect of placing various single capacitors of value, $l_\varepsilon = \{0.2, 0.5, 1.0, 1.5, 2.0, 3.0, 5.0, 7.0, 10.0\}$ at $\phi = \pi$. **a** Conductance in the sub-wavelength region. **b** Conductance in the wavelength region. **c** The induced anti-resonances, which blue shift as the reactance of the capacitor is decreased

Table 7.2 The computational results due to (7.27) varying the capacitor value l_ε, placed at $\phi = \pi$ in the $\Omega = 12$, PEC loop. The graphical results are in Fig. 7.6. The values will vary slightly depending on code implementation. The k_{br} values are given to four places because the resonance is so large that a slight bit off can miss it entirely. $X_2 = \xi_0/(k_{br} l_\varepsilon)$. Since the capacitor is not at $\phi = 0$, the Q cannot be calculated

l_ε (of cap)	k_{br}	X_2 ohms	R_{in} ohms	X_{in} ohms
0.2	0.4257	885	7.0	0
0.5	0.3863	975	4.7	0
1.0	0.3384	1121	2.8	0
1.5	0.3043	1239	1.8	0
2.0	0.2786	1353	1.3	0
3.0	0.2420	1558	0.8	0
5.0	0.1980	1904	0.4	0
7.0	0.1716	2198	0.3	0
10.0	0.1464	2576	0.2	0

using $l_\varepsilon = \{0.2, 0.5, 1.0, 1.5, 2, 5, 7, 10\}$. Table 7.2 gives computational details and Fig. 7.7 show the radiation patterns. The currents have the same shape as that for the capacitor at $\phi = \pi$ in Fig. 7.4.

First, the sub-wavelength region shows very high-Q resonances. Unfortunately, there is no way to calculate the Q, as noted in Sect. 7.5.1; it must be measured from a plot of the expression. The values of the resonances will vary slightly depending upon code implementation, and high resolution along k_b is required.

Second, the resonances blue shift as the capacitor value gets smaller. The capacitor then allows an easy way of tuning the loop. Note, that at a driving wavelength of 3 m, the capacitor value for $l_\varepsilon = 0.2$ is 5.3 pF, and at $l_\varepsilon = 10$ the value is 265 pF. At 3 mm in the GHz region, the values are a thousand times smaller.

Third, the anti-resonances also blue shift significantly.

7.6 The Effects of Multiple Capacitors

Multiple capacitors spread around the periphery of the loop also have the interesting effect of moving the resonance. Form six loops where each loop gains one more capacitor than the previous; that is, $M = 1, 2, 3, 4, 5, 6$. Place capacitors at all q positions of the corresponding loop with value $l_\varepsilon = 1.5$. In other words, form 6 different loops with the number of capacitors changing from $q = 1$ through 6, each capacitor having the same value. The input admittance results in Fig. 7.8, displaying resonances that move out of the sub-wavelength region and moving toward the wavelength region as M approaches 6. Figure 7.9 shows the current distribution and the Fourier current coefficients for these loops.

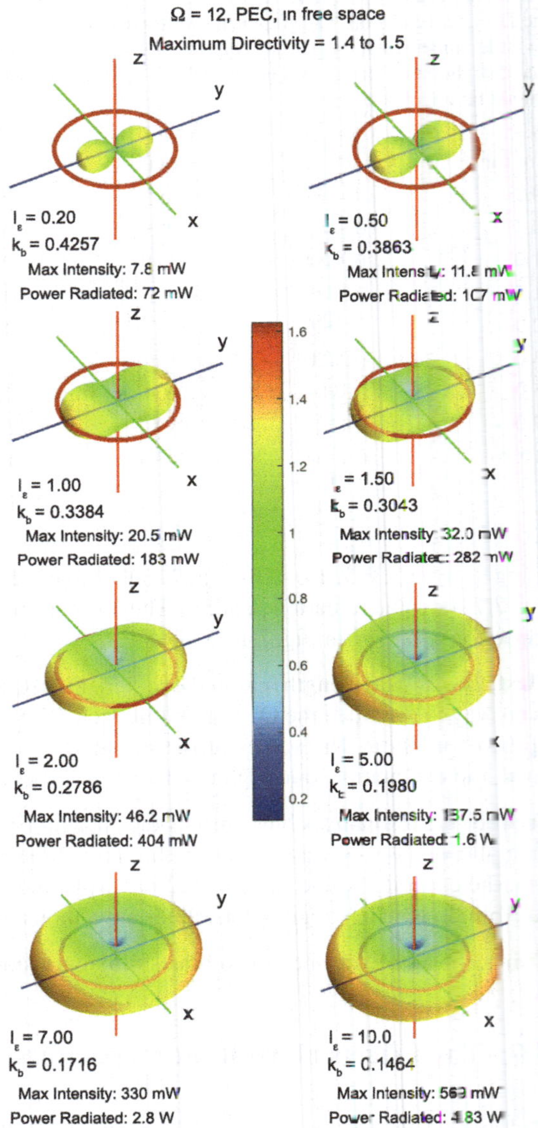

Fig. 7.7 The radiation patterns of loops with eight of the capacitors in Fig. 7.6 placed at $\phi = \pi$. The scale is in decibels. As the higher intensity loops are approached, the pattern becomes more circular in the XY plane

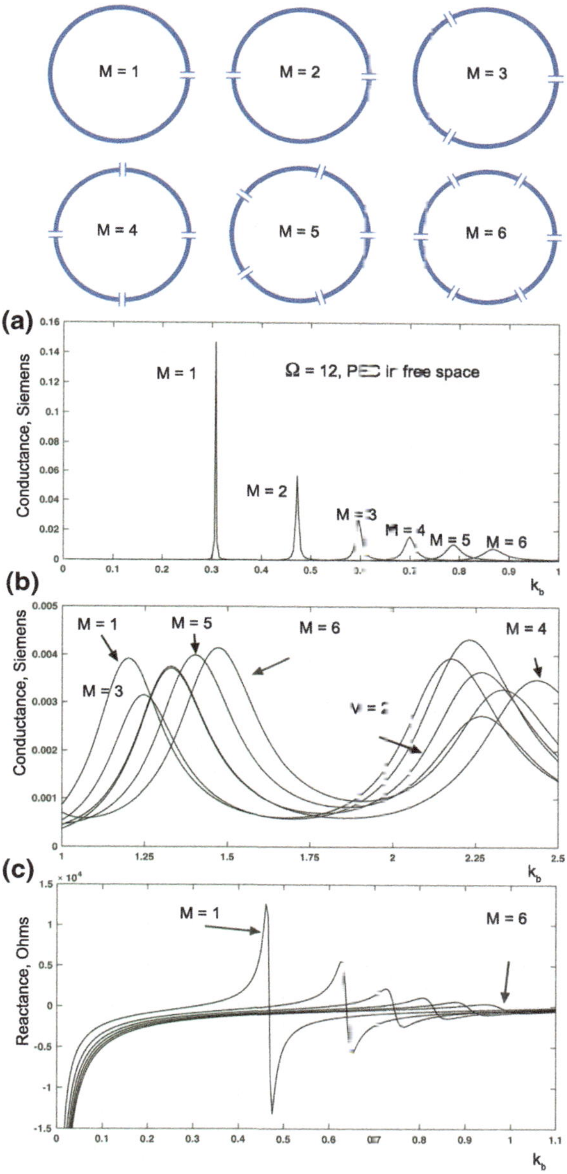

Fig. 7.8 The effect created in the 6 loops shown where all capacitors have the same value $l_\varepsilon = 1.5$. This corresponds to computational results in Table 7.3. The loop is energized at $\phi = 0$ with a delta function voltage. In **a** and **b**, the resonances created. In **c**, the anti-resonances of the same loops. Note how both the resonances and the anti-resonances weaken and shift toward higher frequencies as the number of capacitors increases

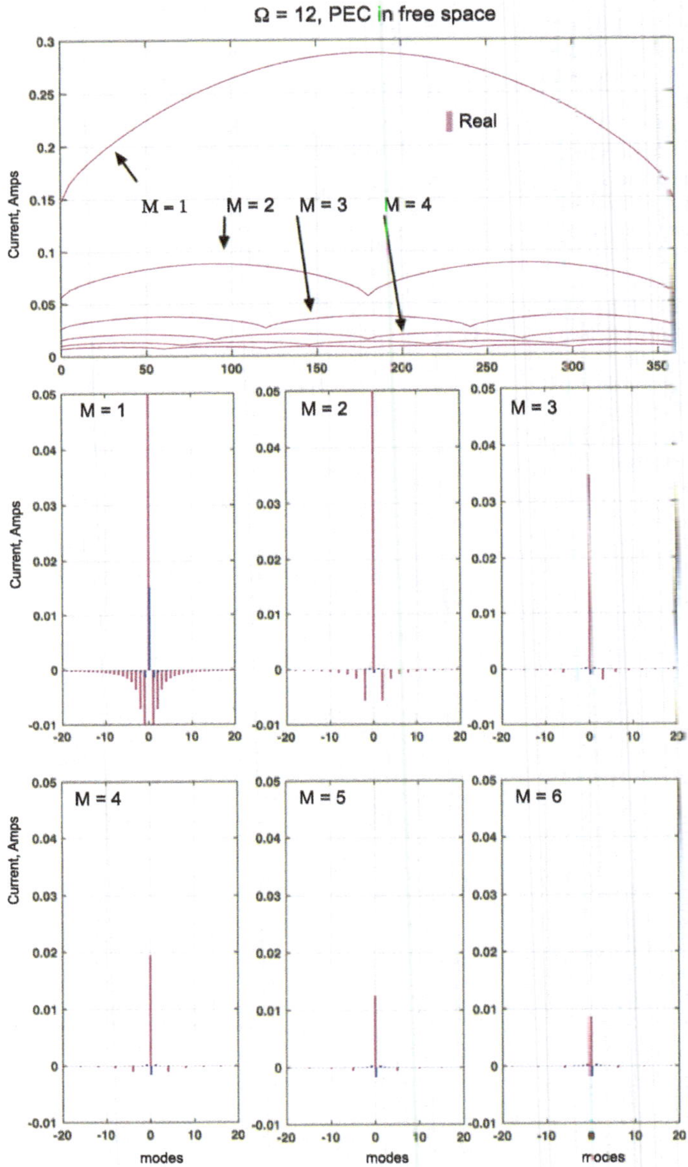

Fig. 7.9 The effect of loading M = 6 different loops with 1 through 6 capacitors all with value $l_\varepsilon = 1.5$. Currents around each loop are at the top. Note that the single capacitor at $\phi = 0$ shows the largest current. The imaginary parts are quite small as the current coefficients show. The currents diminish as the capacitors are added. The radiated power, however, increases due to the k_b^2 term in (7.32)

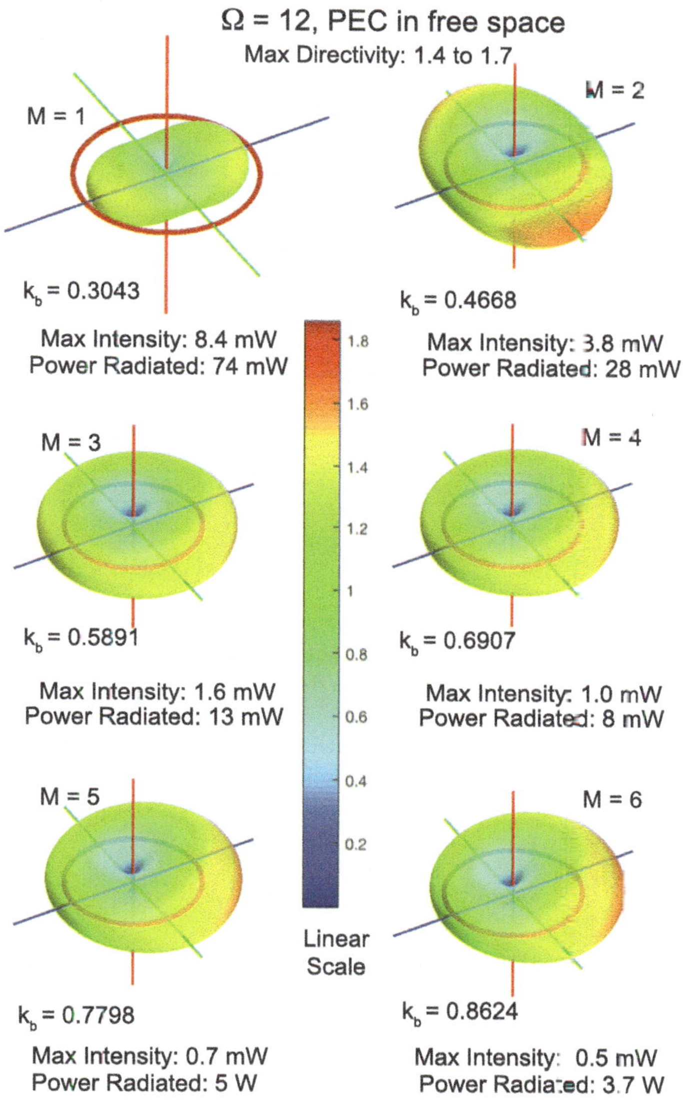

Fig. 7.10 The radiation patterns of the six loops shown in Fig. 7.8 with capacitor value $l_\varepsilon = 1.5$. As more capacitors are added, the resonance blue-shifts and the power radiated decreases. The pattern becomes more donut like, but one direction being stronger than the others. The driving source is at $\phi = 0$ on the x axis

Table 7.3 The computational results due to (7.27) for $M = 1$ through 6 capacitively loaded $\Omega = 12$, PEC loops, keeping the capacitor value at $l_\varepsilon = 1.5$. The second table shows the circuit elements calculated using (7.36) and (7.37), and a substantial quality factor in the hundreds for this resonance. The quality factors of the remaining five loops cannot be calculated

M (of cap)	k_{br}	X (of cap) ohms	R_{in} ohms	X_{in} ohms
1	0.3043	826	7	0
2	0.4668	538	18	0
3	0.5891	427	37	0
4	0.6907	364	63	0
5	0.7798	322	96	0
6	0.8624	291	136	0

	Resonance									
Ω	M at $\phi = 0$	k_{br} at $X_{in} = 0$	L μH	X_L Ohms	C_{cl} pF	$X_{C,cl}$ Ohms	C_1 pF	$X_{C,1}$ Ohms	C pf	Q
12	1	0.3043	7.4	1416	8.85	591	6.3	826	3.7	208

Table 7.3 gives the resonances to four decimals for these loops, although the exact values vary depending on code implementation. Note that the quality factor of the single capacitor placed at $\phi = 0$ is again in the hundreds.

Figure 7.10 shows the radiation patterns for all loops described in Table 7.3. Note that the main load of the radiation pattern swivels from the X axis to the Y axis as M goes from 1 to 2, and then become more circular in the XY plane, but less intense.

References

1. K. Iizuka, IEEE Trans. Antennas Propag. **13**(1), 7 (1965)
2. R.F. Harrington, Radio Sci. **1**(3), 347 (1966)
3. L.G. McCoy, QST **LII**(3), 17 (1968)
4. K. Patterson, Electronics (1967)
5. J.B. McKinley II, Two pictures of closed circuit resonant loop antennas (ccrl) (1968). Personal papers
6. J. Nagar, R.J. Chaky, M.F. Pantoja, A.F. McKinley, D.H. Werner, Optimization of far-field radiation from impedance-loaded nanoloops accelerated by an exact analytical formulation (2013)

Chapter 8
Thick PEC Rings

8.1 Thick Rings in Use in the Short-Wavelength Regions

The models developed in previous chapters have all relied on the thin-wire approximation (see Sect. 2.5.3 for a description). This made a solution of the governing equations simple enough to solve in the 1950s and 1960s. In this chapter we want to take a look at these coupled differential equations again without that restriction and see if these can be separated enough to draw some conclusions on nano-scaled ring behaviour. We focus now on nano-scaled rings, because they cannot easily be made thin. All nanoscaled rings in the literature invariably have Ω on the order of 7 or 8. Figure 4 found in [1] and Fig. 1 found in [2] show typical rings. Gapped rings will be covered in a later chapter.

Cylindrical dipole antennas have been studied over many years for the purposes of finding closed form solutions for their radiated electric fields and hence their radiation patterns. Not all of these studies focused on thin wires, but most did. Those that did often used "reduced" forms for R in the integral kernel: $\mathtt{exp}(-jkR)/R$, instead of the exact form. The exact form for the cylindrical tube is in spherical coordinates and appears as $R^2 = (z - z')^2 + \rho - 2\rho b\mathtt{cos}(\phi - \phi')$. Werner [3], in 1993, using an expansion suggested by Wang [4], managed an exact solution for even thick wires, but with uniform current along the length of the dipole. A few years later, he [5] managed a full closed solution with algorithms using exact series representations of the kernel and efficient numerical procedures for their calculation. He assumed Fourier current coefficients of the form

$$I_n(z') = A_n + B_n\mathtt{sin}[\beta(z - z')] + C_n\mathtt{cos}[\beta(z - z')] \tag{8.1}$$

It would be nice to be able to extend these considerations to thick loops, but the fact that precious little has appeared in the journals suggest that the task is not

A. McKinley, *The Analytical Foundations of Loop Antennas and Nano-Scaled Rings*, Signals and Communication Technology,
https://doi.org/10.1007/978-981-13-5893-7_8

trivial.[1] The reason is the distance function, R, in the kernel, given by Eq. (2.55). Moreover, our interest is not in the near or far field radiation patterns of the ring, but in the impedance function when driven. For that we need to solve for the Fourier coefficients of the current.

8.2 The Governing Equations

The general governing equations for the current on the loop, derived from the linear Maxwell equations, are given in Eq. (2.120). Section 2.3.3, in a discussion of skin depth, suggests that $\Omega = 8$ metal wires are fully penetrated by impinging $\vec{E}$ fields at frequencies above 145 THz, and thinner wires by roughly 48 THz. There are two regions, then, of concern: (1) the region lower than 150 GHz, where metals are perfectly conducting (PEC), and (2) the region between 150 GHz and the 48 THz, where metals suffer loss from a surface impedance. The theory for metal wires with the thin-wire approximation does not adequately serve for thick wires, because the circumference of the wire becomes a substantial current pathway. In other words, $\vec{J}_\psi$ exists, leading quite possibly to helical currents.

8.2.1 *Perfectly Conducting, Thick Ring Equations Applicable at Low Frequency to 150 GHz*

Examining the lossless region then, we may assume that no field appears within the wire and therefore no current flows there. The appropriate renditions of the governing equations are given in Eqs. (2.122) and (2.123). Repeating:

$$
\begin{aligned}
E_\sigma &= 0 \\
E_\phi &= \frac{-j\xi}{k(b+a\cos\psi)}\Bigg[\frac{\partial}{\partial\phi}\int_{S'}\frac{e^{-jkR}}{4\pi R}\left(a\frac{\partial J_{\phi'}}{\partial\phi'}+\frac{\partial(b+a\cos\psi')J_{\psi'}}{\partial\psi'}\right)d\phi' d\psi' \\
&\quad + k^2 a(b+a\cos\psi)\int_{S'}\frac{e^{-jkR}}{4\pi R}(b+a\cos\psi')\times \\
&\quad \Big(J_{\phi'}\cos(\phi-\phi')+J_{\psi'}\sin\psi'\sin(\phi-\phi')\Big)d\phi' d\psi'\Bigg]
\end{aligned}
$$

[1]One interesting paper by Scharstein and Wilson [6] appeared in 2005 looking at conduction in a thick ring given electrostatic excitation. Bibby [7] numerically calculated currents in illuminated thick loops, which we will see below in a later section.

$$E_\psi = \frac{-j\xi}{k}\left[\frac{\partial}{\partial\psi}\int_{S'}\frac{e^{-jkR}}{4\pi R}\left(\frac{\partial J_{\phi'}}{\partial\phi'}+\frac{1}{a}\frac{\partial(b+a\cos\psi')J_{\psi'}}{\partial\psi'}\right)a d\phi' d\psi' \right.$$
$$+k^2\int_{S'}\frac{e^{-jkR}}{4\pi R}a(b+a\cos\psi')\times\Big(-J_{\phi'}\sin\psi\sin(\phi-\phi')$$
$$\left. + J_{\psi'}\left(\sin\psi'\sin\psi\cos(\phi-\phi')+\cos\psi'\cos\psi\right)\Big)d\phi' d\psi'\right] \quad (8.2)$$

The distance R can no longer be the thin-wire variation of Eq. (2.45); we must use the more expansive, surface-to-surface form, Eq. (2.37), or one of its approximations.

$$R^2 = 4b\left[b^+ + \frac{a}{b/a}(\cos\psi'\cos\psi)\right]\sin^2\frac{(\phi-\phi')}{2}$$
$$+4a^2\sin^2\frac{(\psi-\psi')}{2} \quad (8.3)$$

The two governing equations given just above are coupled differential-integral equations in J_ϕ and J_ψ. We now seek solutions. To simplify notation we take the definitions

$$b' \equiv b + a\cos\psi, \quad b^+ \equiv b' + a\cos\psi' \quad \text{and} \quad 2\gamma = \phi - \phi' \quad (8.4)$$

Reducing the Integrals for $\mathbf{E}_\phi$.

We shall solve the internal integrals first and substitute them into the full integral later, since many of the internal integrals have common characteristics.

(1) The first integral of E_ϕ from Eq. (8.2) is

$$E_{\phi 1} = \frac{-j\xi}{2}\int_{-\pi}^{\pi}\left[\frac{a}{k(b+a\cos\psi)}\frac{\partial J_{\phi'}}{\partial\phi'}\frac{\partial}{\partial\phi}\left(\frac{1}{2\pi}\int_{-\pi}^{\pi}\frac{e^{-jkR}}{R}d\psi'\right)\right]d\phi' \quad (8.5)$$

We shall use approximation Eq. (2.41) for R, repeated here:

$$R^2 = 4bb^+\sin^2\gamma \quad (8.6)$$

The integration of the internal integral is accomplished using Leibniz's rule with constant integral limits. We may restate the required integral as

$$\frac{1}{2\pi}\int_{-\pi}^{\pi}\frac{e^{-jkR}}{R}d\psi' = A\frac{1}{2\pi}\int_{-\pi}^{\pi}\frac{\partial e^{-jkR}}{\partial\psi}d\psi' = A\frac{\partial}{\partial\psi}\left[\frac{1}{2\pi}\int_{-\pi}^{\pi}e^{-jkR}d\psi'\right] \quad (8.7)$$

where A is to be determined. To find A

$$\frac{\partial e^{-jkR}}{\partial\psi} = -jke^{-jkR}\frac{\partial R}{\partial\psi} = j2kba\sin^2\gamma\sin\psi\frac{e^{-jkR}}{R} \quad (8.8)$$

Therefore,

$$A = \frac{-j}{2kba\sin^2\gamma\sin\psi} \quad \text{giving}$$

$$\frac{1}{2\pi}\int_{-\pi}^{\pi}\frac{e^{-jkR}}{R}d\psi' = \frac{-j}{2kba\sin\psi\sin^2\gamma}\frac{\partial}{\partial\psi}\left(\frac{1}{2\pi}\int_{-\pi}^{\pi}e^{-jkR}d\psi'\right) \tag{8.9}$$

The integral on the right can now be solved separately. The root in R can be expanded, since the last two terms with the cosines are quite a bit smaller than b, even for $\Omega = 6$:

$$\begin{aligned}
R^2 &= 4bb^+\sin^2\gamma \\
R &= 2\,|\sin\gamma|\sqrt{bb^+} = 2\,|\sin\gamma|\sqrt{b(b'+a\cos\psi')} \\
&\approx 2\sqrt{b}\,|\sin\gamma|\left[\sqrt{b'}+\frac{a\cos\psi'}{2\sqrt{b'}}\right] \\
&= 2\sqrt{bb'}\,|\sin\gamma| + a\sqrt{b/b'}\,|\sin\gamma|\cos\psi' \\
&= z + z'\cos\psi'
\end{aligned} \tag{8.10}$$

There are two terms to R, which then separates the exponential into two parts:

$$\frac{1}{2\pi}\int_{-\pi}^{\pi}e^{-jkR}d\psi' = e^{-jkz}\left(\frac{1}{2\pi}\int_{-\pi}^{\pi}e^{-jkz'\cos\psi'}d\psi'\right) = \mathbf{J}_0(kz')e^{-jkz} \tag{8.11}$$

where $\mathbf{J}_0$ is a Bessel of the first kind. The desired integral solution then for (1) is

$$\frac{1}{2\pi}\int_{-\pi}^{\pi}\frac{e^{-jkR}}{R}d\psi' = \frac{-j\mathbf{J}_0(kz')}{2kba\sin\psi\sin^2\gamma}\frac{\partial}{\partial\psi}e^{-jkz} = \frac{\mathbf{J}_0(kz')e^{-jkz}}{z} \tag{8.12}$$

We now differentiate with respect to ϕ. We have

$$\begin{aligned}
\frac{\partial}{\partial\phi}\left[\frac{1}{2\pi}\int_{-\pi}^{\pi}\frac{e^{-jkR}}{R}d\psi'\right] &= \frac{\partial}{\partial\phi}\left[\frac{\mathbf{J}_0(kz')e^{-jkz}}{z}\right] \\
&= -\frac{bk_b'\sin(2\gamma)e^{-jkz}}{2z^2}\left[\frac{\mathbf{J}_1(kz')}{(b'/a)}+\frac{\mathbf{J}_0(kz')}{kz}+j2\mathbf{J}_0(kz')\right]
\end{aligned} \tag{8.13}$$

The first full integral of E_ϕ is

$$\begin{aligned}
E_{\phi 1} &= \frac{-j\xi}{2}\int_{-\pi}^{\pi}\left[\frac{a}{kb'}\frac{\partial J_{\phi'}}{\partial\phi'}\frac{\partial}{\partial\phi}\left(\frac{1}{2\pi}\int_{-\pi}^{\pi}\frac{e^{-jkR}}{R}d\psi'\right)\right]d\phi' \\
&= \frac{j\xi ba}{4}\int_{-\pi}^{\pi}\frac{\sin(2\gamma)e^{-jkz}}{z^2}\left[\frac{\mathbf{J}_1(kz')}{(b'/a)}+\frac{\mathbf{J}_0(kz')}{kz}+j2\mathbf{J}_0(kz')\right]\frac{\partial J_{\phi'}}{\partial\phi'}d\phi'
\end{aligned} \tag{8.14}$$

where γ, z and z' are all functions of ϕ'.
(2) The second full integral of E_ϕ from Eq. (8.2) is

$$E_{\phi 2} = \frac{-j\xi}{2}\int_{-\pi}^{\pi}\left[kaJ_{\phi'}\cos(\phi-\phi')\left(\frac{1}{2\pi}\int_{-\pi}^{\pi}\frac{e^{-jkR}}{R}(b+a\cos\psi')d\psi'\right)\right]d\phi' \tag{8.15}$$

Following the procedure used in (1) above, the internal integral may then be restated in the following way:

$$\frac{1}{2\pi}\int_{-\pi}^{\pi}\frac{e^{-jkR}}{R}(b+a\cos\psi')d\psi' = \frac{-j}{2kba\sin\psi\sin^2\gamma}\frac{\partial}{\partial\psi}\left(\frac{1}{2\pi}\int_{-\pi}^{\pi}e^{-jkR}(b+a\cos\psi')d\psi'\right) \tag{8.16}$$

The integral on the right side can be solved by using the same approximation for R as in (1), namely, Eq. (8.10), and separating the exponential as above.

$$\begin{aligned}\frac{1}{2\pi}\int_{-\pi}^{\pi}e^{-jkR}(b+a\cos\psi')d\psi' &= e^{-jkz}\left(\frac{b}{2\pi}\int_{-\pi}^{\pi}e^{-jkz'\cos\psi'}d\psi' + \frac{a}{2\pi}\int_{-\pi}^{\pi}e^{-jkz'\cos\psi'}\cos\psi' d\psi'\right)\\ &= e^{-jkz}\left(b\mathbf{J}_0(kz') + ja\mathbf{J}_1(kz')\right)\end{aligned} \tag{8.17}$$

where $\mathbf{J}_0$ and $\mathbf{J}_1$ are Bessels of the first kind. The parameters z and z' are functions of ψ. Taking the derivative with respect of ψ, we have for the integral of (8.16)

$$\begin{aligned}&\frac{1}{2\pi}\int_{-\pi}^{\pi}\frac{e^{-jkR}}{R}(b+a\cos\psi')d\psi'\\ &= \frac{je^{-jkz}}{z}\left[\left(a+\frac{b}{b'/a}\right)\mathbf{J}_1(kz') + j\left(\left(b+\frac{a}{8(b'/a)}\right)\mathbf{J}_0(kz') - \frac{a}{8(b'/a)}\mathbf{J}_2(kz')\right)\right]\end{aligned} \tag{8.18}$$

This may now be substituted into Eq. (8.15) to give

$$\begin{aligned}E_{\phi 2} &= \frac{-j\xi}{2}\int_{-\pi}^{\pi}\left[kaJ_{\phi'}\cos(\phi-\phi')\left(\frac{1}{2\pi}\int_{-\pi}^{\pi}\frac{e^{-jkR}}{R}(b+a\cos\psi')d\psi'\right)\right]d\phi'\\ &= \frac{\xi ka}{2}\int_{-\pi}^{\pi}\frac{e^{-jkz}}{z}\left[\left(a+\frac{b}{b'/a}\right)\mathbf{J}_1(kz') + j\left(\left(b+\frac{a}{8(b'/a)}\right)\mathbf{J}_0(kz') - \frac{a}{8(b'/a)}\mathbf{J}_2(kz')\right)\right]\\ &\quad\times J_{\phi'}\cos(\phi-\phi')d\phi'\end{aligned} \tag{8.19}$$

(3) The third full integral for E_ϕ from Eq. (8.2) is

$$E_{\phi 3} = \frac{-j\xi}{2}\int_{-\pi}^{\pi}\left[\frac{1}{k(b+a\cos\psi)}\frac{\partial(b+a\cos\psi')J_{\psi'}}{\partial\psi'}\frac{\partial}{\partial\phi}\left(\frac{1}{2\pi}\int_{-\pi}^{\pi}\frac{e^{-jkR}}{R}d\varphi'\right)\right]d\psi' \tag{8.20}$$

The approach here is the same as that taken by Hallen; see Eq. (3.30). For the internal integral, we use

$$\frac{\partial}{\partial\phi}\left[\frac{1}{2\pi}\int_{-\pi}^{\pi}\frac{e^{-jkR}}{R}d\phi'\right] = \frac{\partial}{\partial\phi}\left[\frac{1}{2\pi}\int_{-\pi}^{\pi}\frac{e^{-jkR}-1}{R}d\phi' + \frac{1}{2\pi}\int_{-\pi}^{\pi}\frac{1}{R}d\phi'\right] \tag{8.21}$$

For the first integral on the right we use $2\gamma = \phi - \phi'$ without affecting the integral variable or limits, and we use as before

$$R^2 = 4bb^+\sin^2\gamma \tag{3.22}$$

Then,

$$\begin{aligned}\frac{1}{2\pi}\int_{-\pi}^{\pi}\frac{e^{-jkR}-1}{R}d\phi' &= -\frac{j}{2\pi}\int_{-\pi}^{\pi}\left[\int_0^k e^{-jxR}dx\right]d\phi' \\ &= -\frac{j}{2\pi}\int_{-\pi}^{\pi}\left[\int_0^k e^{-j2x\sqrt{bb^+}|\sin\gamma|}dx\right]d\phi'\end{aligned} \tag{8.23}$$

Taking the derivative with respect to ϕ under the integral yields

$$\frac{\partial}{\partial\phi}\left(\frac{1}{2\pi}\int_{-\pi}^{\pi}\frac{e^{-jkR}-1}{R}d\phi'\right) = -\frac{j}{2\pi}\int_{-\pi}^{\pi}\left[\int_0^k\frac{\partial}{\partial\phi}e^{-j2x\sqrt{bb^+}|\sin\gamma|}dx\right]d\phi' \tag{8.24}$$

Executing the derivative

$$\frac{\partial}{\partial\phi}e^{-j2x\sqrt{bb^+}|\sin\gamma|} = -j\frac{x\sqrt{bb^+}\sin\gamma\cos\gamma}{|\sin\gamma|}e^{-j2x\sqrt{bb^+}|\sin\gamma|}$$

gives

$$\begin{aligned}&= -\frac{\sqrt{bb^+}}{2\pi}\int_{-\pi}^{\pi}\left[\int_0^k\frac{x\sin\gamma\cos\gamma}{|\sin\gamma|}e^{-j2x\sqrt{bb^+}|\sin\gamma|}dx\right]d\phi' \\ &= -\sqrt{bb^+}\int_0^k x\left[\frac{1}{2\pi}\int_{-\pi}^{\pi}\frac{\sin\gamma\cos\gamma}{|\sin\gamma|}e^{-j2x\sqrt{bb^+}|\sin\gamma|}d\phi'\right]dx\end{aligned} \tag{8.25}$$

The internal integral may be solved separately. Using the change of variables

$$2\gamma = \phi - \phi' \text{ then } d\phi' = -2d\gamma \tag{8.26}$$

and the limits flip, so that the internal integral above becomes

$$\begin{aligned}&\frac{1}{2\pi}\int_{-\pi}^{\pi}\frac{\mathtt{sin}\gamma\mathtt{cos}\gamma}{|\mathtt{sin}\gamma|}e^{-j2x\sqrt{bb^-}|\mathtt{sin}\gamma|}d\phi'\\&=\frac{1}{\pi}\int_{-\pi/2}^{\pi/2}\left[\mathtt{cos}(2x\sqrt{bb^+}\,|\mathtt{sin}\gamma|)-j\mathtt{sin}(2x\sqrt{bb^+}\,|\mathtt{sin}\gamma|)\right]\frac{\mathtt{sin}\gamma\mathtt{cos}\gamma}{|\mathtt{sin}\gamma|}d\gamma\end{aligned} \tag{8.27}$$

The kernel is an interesting one. We have, from Eqs. (9.1.42) and (9.1.43) in [8]

$$\begin{aligned}&\mathtt{cos}(z\,|\mathtt{sin}\gamma|)=\mathtt{cos}(z\mathtt{sin}\gamma)=\mathbf{J}_0(z)+2\sum_{m=1}^{\infty}\mathbf{J}_{2m}(z)\mathtt{cos}(2m\gamma)\\&\mathtt{sin}(z\,|\mathtt{sin}\gamma|)=2(\mathtt{sgn}\gamma)\sum_{m=0}^{\infty}\mathbf{J}_{2m+1}(z)\mathtt{sin}((2m+1)\gamma)\quad \text{if}-\pi<\theta<\pi\end{aligned} \tag{8.28}$$

These lead to the following integrals with $z^+ = 2x\sqrt{bb^+}$

$$\begin{aligned}=\mathbf{J}_0(z^+)&\left[\frac{1}{\pi}\int_{-\pi/2}^{\pi/2}\frac{\mathtt{sin}(2\gamma)}{2\,|\mathtt{sin}\gamma|}d\gamma\right]+\sum_{m=1}^{\infty}\mathbf{J}_{2m}(z^+)\left[\frac{1}{\pi}\int_{-\pi/2}^{\pi/2}\frac{\mathtt{sin}(2\gamma)\,\mathtt{cos}(2m\gamma)}{|\mathtt{sin}\gamma|}d\gamma\right]\\&-j\sum_{m=0}^{\infty}\mathbf{J}_{2m+1}(z^+)\left[\frac{1}{\pi}\int_{-\pi/2}^{\pi/2}\mathtt{sgn}(\gamma)\frac{\mathtt{sin}(2\gamma)\,\mathtt{sin}((2m+1)\gamma)}{|\mathtt{sin}\gamma|}\right]d\gamma\end{aligned} \tag{8.29}$$

The kernels of the first and second terms have a discontinuity at $\gamma = 0$ due to the absolute value; this turns them into odd functions. Thus their integrals over the limits are 0. The kernel of the third term is an odd function and therefore its integral is also 0. Therefore the first integral on the right side of Eq. (8.21)

$$\frac{\partial}{\partial\phi}\left(\frac{1}{2\pi}\int_{-\pi}^{\pi}\frac{e^{-jkR}-1}{R}d\phi'\right)=0 \tag{8.30}$$

With the same R, the second integral on the right of the breakout in Eq. (8.21) has a pole at $\theta = \{0, \pi\}$. Using the more accurate version for R, namely,

$$R^2 = 4bb^+\mathtt{sin}^2\gamma + 4a^2 \tag{8.31}$$

we have

$$\frac{\partial}{\partial\phi}\left[\frac{1}{2\pi}\int_{-\pi}^{\pi}\frac{1}{R}d\phi'\right]=\left[\frac{1}{2\pi}\int_{-\pi}^{\pi}\frac{\partial}{\partial\phi}\frac{1}{\sqrt{4bb^{+}\mathtt{sin}^2(\phi-\phi')/2+4a^2}}d\phi'\right]$$
$$=-\frac{1}{\pi}\int_{-\pi/2}^{\pi/2}\frac{bb^{+}\mathtt{sin}2\gamma}{4(bb^{+}\mathtt{sin}^2\gamma+a^2)^{3/2}}d\gamma$$
$$=0 \tag{8.32}$$

The kernel is an odd function around zero and periodic over the range $-\pi/2$ to $\pi/2$ and therefore the integral begets zero, no matter the value of ϕ. The internal integral of the full integral in Eq. (8.20) is 0 and therefore the full integral is zero.

$$E_{\phi 3}=\frac{j\xi}{2}\int_{-\pi}^{\pi}\left[\frac{1}{k(b+a\mathtt{cos}\psi)}\frac{\partial(b+a\mathtt{cos}\psi')J_{\psi'}}{\partial\psi'}\frac{\partial}{\partial\phi}\left(\frac{1}{2\pi}\int_{-\pi}^{\pi}\frac{e^{-jkR}}{R}d\phi'\right)\right]d\psi'$$
$$=\boxed{0}$$

(4) The fourth full integral for E_ϕ from Eq. (8.2) is

$$E_{\phi 4}=\frac{-j\xi}{2}\int_{-\pi}^{\pi}\left[ka(b+a\mathtt{cos}\psi')J_{\psi'}\mathtt{sin}\psi'\left(\frac{1}{2\pi}\int_{-\pi}^{\pi}\frac{e^{-jkR}}{R}\mathtt{sin}(\phi-\phi')d\phi'\right)\right]d\psi' \tag{8.33}$$

Using Eq. (8.22) for R, we have for the internal integral:

$$=\frac{1}{2\pi}\int_{-\pi}^{\pi}\frac{e^{-j2k\sqrt{bb^{+}}|\mathtt{sin}((\phi-\phi')/2)|}\mathtt{sin}(\phi-\phi')}{2\sqrt{bb^{+}}\,|\mathtt{sin}((\phi-\phi')/2)|}d\phi'$$
$$=\frac{1}{2\pi}\int_{-\pi/2}^{\pi/2}\frac{e^{-j2k\sqrt{bb^{+}}|\mathtt{sin}(\gamma)|}\mathtt{sin}(2\gamma)}{2\sqrt{bb^{+}}\,|\mathtt{sin}(\gamma)|}d\gamma \tag{8.34}$$

The real part and imaginary parts of the kernel are both odd functions around ϕ and periodic over the range $-\pi/2$ to $\pi/2$ periodic in π. A discontinuity in the real part occurs at $\phi'=\phi$, but a proper integration through the discontinuity yields 0. The imaginary part has no discontinuity and is simple, hence it also integrates to 0. The desired integral for (4) therefore is $\boxed{0}$.

Reduced Integrals for E_ϕ

The ϕ component of the Electric Field is now reduced to

$$
\begin{aligned}
E_\phi = &\frac{j\xi ba}{4}\int_{-\pi}^{\pi}\frac{\sin(2\gamma)e^{-jkz}}{z^2}\left[\frac{\mathbf{J}_1(kz')}{(b'/a)}+\frac{\mathbf{J}_0(kz')}{kz}+j2\mathbf{J}_0(kz')\right]\frac{\partial J_{\phi'}}{\partial\phi'}d\phi' \\
&+\frac{\xi ka}{2}\int_{-\pi}^{\pi}\frac{e^{-jkz}}{z}\left[\left(a+\frac{b}{b'/a}\right)\mathbf{J}_1(kz')+j\left(\left(b+\frac{a}{8(b'/a)}\right)\mathbf{J}_0(kz')-\frac{a}{8(b'/a)}\mathbf{J}_2(kz')\right)\right] \\
&\times J_{\phi'}\cos(\phi-\phi')d\phi'
\end{aligned}
\tag{8.35}
$$

Reducing the Integrals for E_ψ

The integrals for the ψ component of the E field of the thick loop are given in Eq. (8.2).

$$
\begin{aligned}
E_\psi = &\frac{-j\xi}{k}\left[\frac{\partial}{\partial\psi}\int_{S'}\frac{e^{-jkR}}{4\pi R}\left(\frac{\partial J_{\phi'}}{\partial\phi'}+\frac{1}{a}\frac{\partial(b+a\cos\psi')J_{\psi'}}{\partial\psi'}\right)d\phi' d\psi'\right. \\
&+k^2\int_{S'}\frac{e^{-jkR}}{4\pi R}a(b+a\cos\psi')\times\Big(-J_{\phi'}\sin\psi\sin(\phi-\phi') \\
&\left.+J_{\psi'}\left(\sin\psi'\sin\psi\cos(\phi-\phi')+\cos\psi'\cos\psi\right)\Big)d\phi' d\psi'\right]
\end{aligned}
\tag{8.36}
$$

(1) The first full integral of E_ψ is

$$
E_{\psi 1} = \frac{-j\xi}{2}\int_{-\pi}^{\pi}\left[\frac{1}{k}\frac{\partial J_{\phi'}}{\partial\phi'}\frac{\partial}{\partial\psi}\left(\frac{1}{2\pi}\int_{-\pi}^{\pi}\frac{e^{-jkR}}{R}d\psi'\right)\right]d\phi'
\tag{8.37}
$$

The internal integral here is the same as that in Eq. (8.7). The results are in Eq. (8.12).

$$
\frac{1}{2\pi}\int_{-\pi}^{\pi}\frac{e^{-jkR}}{R}d\psi' = \frac{\mathbf{J}_0(kz')e^{-jkz}}{z}
\tag{8.38}
$$

where $z = 2\sqrt{bb'}|\sin\gamma|$ and $z' = a\sqrt{b/b'}|\sin\gamma|$. The parameter b' is $b + a\cos\psi$, a function of ψ, and therefore z and z' are also functions of ψ. Differentiating with respect to ψ, we have

$$
\frac{\partial}{\partial\psi}\left(\frac{\mathbf{J}_0(kz')e^{-jkz}}{z}\right) = \frac{z'}{z}\sin\psi e^{-jkz}\left[\frac{\mathbf{J}_0(kz')}{z}-\frac{z'}{z}k\mathbf{J}_1(kz')+jk\mathbf{J}_0(kz')\right]
$$

The first full integral for E_ψ from Eq. (8.37) therefore is

$$
E_{\psi 1} = -\frac{-j\xi}{2}\sin\psi\int_{-\pi}^{\pi}\left[e^{-jkz}\frac{z'}{z}\left[\frac{\mathbf{J}_0(kz')}{kz}-\frac{z'}{z}\mathbf{J}_1(kz')+j\mathbf{J}_0(kz')\right]\frac{\partial J_{\phi'}}{\partial\phi'}\right]d\phi'
\tag{8.39}
$$

(2) The second integral for E_ψ is

$$E_{\psi 2} = \frac{j\xi}{2}\int_{-\pi}^{\pi}\left[kaJ_{\phi'}\sin\psi\sin(\phi-\phi')\left(\frac{1}{2\pi}\int_{-\pi}^{\pi}\frac{e^{-jkR}}{R}(b+a\cos\psi')d\psi'\right)\right]d\phi' \tag{8.40}$$

The internal integral appears in $E_{\phi 2}$. The solution is given in Eq. (8.18). Therefore, we have for the full integral

$$E_{\psi 2} = \frac{j\xi}{2}\int_{-\pi}^{\pi}\left[kaJ_{\phi'}\sin\psi\sin(\phi-\phi')\left(\frac{je^{-jkz}}{z}\left[\left(a+\frac{b}{b'/a}\right)\mathbf{J}_1(kz')\right.\right.\right.$$
$$\left.\left.\left.+j\left(\left(b+\frac{a}{8(b'/a)}\right)\mathbf{J}_0(kz')-\frac{a}{8(b'/a)}\mathbf{J}_2(kz')\right)\right]\right)\right]d\phi' \tag{8.41}$$

(3) The third integral of E_ψ is

$$E_{\psi 3} = -\frac{j\xi}{2}\int_{-\pi}^{\pi}\left[\frac{1}{ka}\frac{\partial(b+a\cos\psi')J_{\psi'}}{\partial\psi'}\frac{\partial}{\partial\psi}\left(\frac{1}{2\pi}\int_{-\pi}^{\pi}\frac{e^{-jkR}}{R}d\phi'\right)\right]d\psi' \tag{8.42}$$

The internal integral is the same as that found in $E_{\phi 3}$. Using the same procedure, we break it out into two terms, as shown in Eq. (8.21).

$$\frac{\partial}{\partial\psi}\left[\frac{1}{2\pi}\int_{-\pi}^{\pi}\frac{e^{-jkR}}{R}d\phi'\right] = \frac{\partial}{\partial\psi}\left[\frac{1}{2\pi}\int_{-\pi}^{\pi}\frac{e^{-jkR}-1}{R}d\phi' + \frac{1}{2\pi}\int_{-\pi}^{\pi}\frac{1}{R}d\phi'\right] \tag{8.43}$$

Taking the first term as shown in Eq. (8.23)

$$\frac{1}{2\pi}\int_{-\pi}^{\pi}\frac{e^{-jkR}-1}{R}d\phi' = -\frac{j}{2\pi}\int_{-\pi}^{\pi}\left[\int_0^k e^{-j2x\sqrt{bb^+}|\sin\gamma|}dx\right]d\phi' \tag{8.44}$$

where $b^+ = b' + a\cos\psi'$. Taking the derivative with respect to ψ,

$$\frac{\partial}{\partial\psi}\left[\frac{1}{2\pi}\int_{-\pi}^{\pi}\frac{e^{-jkR}-1}{R}d\phi'\right] =$$
$$\frac{1}{2\pi}\int_{-\pi}^{\pi}\left[\int_0^k(x\,|\sin\gamma|)\frac{ab\sin\psi}{\sqrt{bb'}}e^{-j2x\sqrt{bb^+}|\sin\gamma|}dx\right]d\phi'$$
$$= \frac{ab\sin\psi}{\sqrt{bb'}}\int_0^k x\left[\frac{1}{2\pi}\int_{-\pi}^{\pi}|\sin\gamma|\,e^{-j2x\sqrt{bb^+}|\sin\gamma|}d\phi'\right]dx \tag{8.45}$$

The internal integral may be solved separately. Again using the change of variables, $2\gamma = \phi - \phi'$,

$$= \frac{1}{\pi}\int_{-\pi/2}^{\pi/2}\left[\cos(2x\sqrt{bb^+}\,|\sin\gamma|) - j\sin(2x\sqrt{bb^+}\,|\sin\gamma|)\right]|\sin\gamma|\,d\gamma \tag{8.46}$$

Using Eq. (8.28), these lead to the integral

$$\begin{aligned} = \frac{1}{\pi}\int_{-\pi/2}^{\pi/2}\Bigg[&\left(\mathbf{J}_0(z^+)\,|\sin\gamma| + 2\sum_{m=1}^{\infty}\mathbf{J}_{2m}(z^+)\cos(2m\gamma)\,|\sin\gamma|\right) \\ &- j(\mathrm{sgn}\gamma)\left(2\sum_{m=0}^{\infty}\mathbf{J}_{2m+1}(z^+)\sin((2m+1)\gamma)\,|\sin\gamma|\right)\Bigg]d\gamma \end{aligned} \tag{8.47}$$

Here $z^+ = 2x\sqrt{bb^+}$. The Bessels all come out of the integral since they are not functions of ϕ'. Orthogonality does not work here first, due to the absolute value of the $|\sin\gamma|$ and second, due to the difference in the argument of the function and the variable of integration. We have for the first term

$$2\int_{-\pi/2}^{\pi/2}|\sin\gamma|\,d\gamma = \boxed{2} \tag{8.48}$$

For the second term,

$$2\int_{-\pi/2}^{\pi/2}\cos(2m\gamma)\,|\sin\gamma|\,d\gamma = c_m \equiv \boxed{\frac{-4}{4m^2-1}} \tag{8.49}$$

For the third term, we have

$$2\int_{-\pi/2}^{\pi/2}(\mathrm{sgn}\gamma)\sin((2m+1)\gamma)|\sin\gamma|d\gamma = \begin{cases}0 \text{ if } m \neq 0\\ \pi \text{ if } m = 0\end{cases} \tag{8.50}$$

The result of the integral Eq. (8.47) is

$$\frac{1}{\pi}\left[2\mathbf{J}_0(z^+) + \sum_{m=1}^{\infty}c_m\mathbf{J}_{2m}(z^+) - j\pi\mathbf{J}_1(z^+)\right] \tag{8.51}$$

This must now be integrated as directed by Eq. (8.45):

$$\frac{1}{\pi}\int_0^k x\left[2\mathbf{J}_0(z^+) + \sum_{m=1}^{\infty}c_m\mathbf{J}_{2m}(z^+) - j\pi\mathbf{J}_1(z^+)\right]dx \tag{8.52}$$

The second term raises the problem of where to limit the summation. We note that the largest the argument can become is $z^+_{max} \approx 2k_b$, or roughly 5. As m increases, the Bessel and c_m both decrease rapidly over the range $0 < k_b < 5$, such that only the Bessels that will matter are $2m = 0$, 1, and 2. But the summation begins at $m = 1$, therefore the only term that matters in the summation is $\mathbf{J}_2(z^+)$. The integral therefore, to solve is:

$$\begin{aligned}
&\frac{1}{\pi}\int_0^k x\left[2\mathbf{J}_0(2x\sqrt{bb^+}) + c_2\mathbf{J}_2(2x\sqrt{bb^+}) - j\pi\mathbf{J}_1(2x\sqrt{bb^+})\right]dx \\
&= \frac{k}{\pi\sqrt{bb^+}}\mathbf{J}_1(2k\sqrt{bb^+}) - \frac{c_2}{2\pi bb^+}\left[\mathbf{J}_0(2k\sqrt{bb^+}) + k\sqrt{bb^+}\mathbf{J}_1(2k\sqrt{bb^+}) - 1\right] \\
&\quad - j\frac{k\pi}{4\sqrt{bb^+}}\left[\mathbf{H}_0(2k\sqrt{bb^+})\mathbf{J}_1(2k\sqrt{bb^+}) - \mathbf{H}_1(2k\sqrt{bb^+})\mathbf{J}_0(2k\sqrt{bb^+})\right]
\end{aligned} \tag{8.53}$$

After rearranging a bit, the end result of the first integral on the right of Eq. (8.43) above with the derivative taken inside is

$$\begin{aligned}
&\frac{\partial}{\partial\psi}\left[\frac{1}{2\pi}\int_{-\pi}^{\pi}\frac{e^{-jkR}-1}{R}d\phi'\right] \\
&= \frac{ka\mathtt{sin}\psi}{2\pi\sqrt{b'b^+}}\left[\frac{1}{k\sqrt{bb^+}}\left(\frac{\mathbf{J}_0(2k\sqrt{bb^+})}{4} - 1\right) + \frac{9}{4}\mathbf{J}_1(2k\sqrt{bb^+})\right. \\
&\quad \left. + j\frac{\pi^2}{2}\left[\mathbf{H}_0(2k\sqrt{bb^+})\mathbf{J}_1(2k\sqrt{bb^+}) - \mathbf{H}_1(2k\sqrt{bb^+})\mathbf{J}_0(2k\sqrt{bb^+})\right]\right]
\end{aligned} \tag{8.54}$$

The second term of Eq. (8.43) has a pole at $\gamma = \{0, \pi\}$ if we use the same R. Using, then, the more accurate

$$R^2 = 4bb^+\mathtt{sin}^2\gamma + 4a^2 \tag{8.55}$$

We have,

$$\begin{aligned}
\frac{\partial}{\partial\psi}\left[\frac{1}{2\pi}\int_{-\pi}^{\pi}\frac{1}{R}d\phi'\right] &= \frac{1}{4a\pi}\frac{\partial}{\partial\psi}\left[\int_{-\pi}^{\pi}\frac{1}{\sqrt{(bb^+/a^2)\mathtt{sin}^2\gamma + 1}}d\phi'\right] \\
&= \frac{1}{2\pi a}\frac{\partial}{\partial\psi}\left[\int_{-\pi/2}^{\pi/2}\frac{1}{\sqrt{(bb^+/a^2)\mathtt{sin}^2\gamma + 1}}d\gamma\right]
\end{aligned} \tag{8.56}$$

This is an Incomplete Elliptic integral of the First Kind, $\mathbb{F}$.

$$\begin{aligned}
&= \frac{1}{2\pi a}\frac{\partial}{\partial\psi}\left[\mathbb{F}\left(\gamma\left| -\frac{bb^+}{a^2}\right)\right._{-\pi/2}^{\pi/2}\right] = \frac{1}{\pi a}\frac{\partial}{\partial\psi}\left[\mathbb{K}\left(-\frac{bb^+}{a^2}\right)\right] \\
&= \frac{\mathtt{sin}\psi}{\pi b^+}\left[\frac{\mathbb{E}(-\frac{bb^+}{a^2}) - (\frac{bb^+}{a^2}+1)\mathbb{K}(-\frac{bb^+}{a^2})}{(b/a)\left(\frac{bb^+}{a^2}+1\right)}\right]
\end{aligned} \tag{8.57}$$

The solution to Eq. (8.43) is Eq. (8.54) plus Eq. (8.57). Substituting into Eq. (8.42) yields

$$
\begin{aligned}
E_{\psi 3} &= -\frac{j\xi}{2}\int_{-\pi}^{\pi}\left[\frac{1}{ka}\frac{\partial(b+a\cos\psi')J_{\psi'}}{\partial\psi'}\frac{\partial}{\partial\psi}\left(\frac{1}{2\pi}\int_{-\pi}^{\pi}\frac{e^{-jkR}}{R}d\phi'\right)\right]d\psi' \\
&= -\frac{j\xi\sin\psi}{4\pi\sqrt{b'b^{+}}}\int_{-\pi}^{\pi}\left[\frac{1}{k\sqrt{bb^{+}}}\left(\frac{\mathbf{J}_0(2k\sqrt{bb^{+}})}{4}-1\right)+\frac{9}{4}\mathbf{J}_1(2k\sqrt{bb^{+}})\right. \\
&\quad + j\frac{\pi^2}{2}\left[\mathbf{H}_0(2k\sqrt{bb^{+}})\mathbf{J}_1(2k\sqrt{bb^{+}})-\mathbf{H}_1(2k\sqrt{bb^{+}})\mathbf{J}_0(2k\sqrt{bb^{+}})\right] \\
&\quad \left.\times\frac{\partial(b+a\cos\psi')J_{\psi'}}{\partial\psi'}\right]d\psi'
\end{aligned}
\tag{8.58}
$$

(4) The fourth term of the ψ component of E_ψ is

$$
E_{\psi 4} = -\frac{j\xi}{2}\int_{-\pi}^{\pi}\left[ka(b+a\cos\psi')J_{\psi'}\sin\psi'\sin\psi\left(\frac{1}{2\pi}\int_{-\pi}^{\pi}\frac{e^{-jkR}}{R}\cos(\phi-\phi')d\phi'\right)\right]d\psi'
\tag{8.59}
$$

The internal integral is quite like the fourth integral of the E_ϕ component, but with $\sin(\phi-\phi')$ replaced with $\cos(\phi-\phi')$ (see (8.33). Using Hallen's approach (see Eq. (3.30),

$$
\begin{aligned}
&\frac{1}{2\pi}\int_{-\pi}^{\pi}\frac{e^{-jkR}}{R}\cos(\phi-\phi')d\phi' \\
&= \frac{1}{2\pi}\int_{-\pi}^{\pi}\frac{e^{-jkR}-1}{R}\cos(\phi-\phi')d\phi'+\frac{1}{2\pi}\int_{-\pi}^{\pi}\frac{\cos(\phi-\phi')}{R}d\phi'
\end{aligned}
\tag{8.60}
$$

Now using Eq. (8.23), and the short version for R, Eq. (8.22), the first integral on the right, with $2\gamma=\phi-\phi'$, becomes

$$
\begin{aligned}
\frac{1}{\pi}\int_{-\pi/2}^{\pi/2}\frac{e^{-jkR}-1}{R}\cos 2\gamma d\gamma &= -\frac{j}{\pi}\int_{-\pi/2}^{\pi/2}\left[\int_0^k e^{-jxR}dx\right]\cos 2\gamma d\gamma \\
&= -j\int_0^k\left[\frac{1}{\pi}\int_{-\pi/2}^{\pi/2}e^{-j2x\sqrt{bb^{+}}|\sin\gamma|}\cos 2\gamma d\gamma\right]dx
\end{aligned}
\tag{8.61}
$$

Using the equations in Eq. (8.28), the internal integral leads to

$$
\begin{aligned}
&= \frac{1}{\pi}\int_{-\pi/2}^{\pi/2}\left[\cos(2x\sqrt{bb^{+}}|\sin\gamma|)-j\sin(2x\sqrt{bb^{+}}|\sin\gamma|)\right]\cos 2\gamma d\gamma \\
&= \frac{1}{\pi}\int_{-\pi/2}^{\pi/2}\left[\left(\mathbf{J}_0(z^{+})\cos 2\gamma+2\sum_{m=1}^{\infty}\mathbf{J}_{2m}(z^{+})\cos(2m\gamma)\cos 2\gamma\right)\right. \\
&\quad \left.-j(\mathrm{sgn}\gamma)\left(2\sum_{m=0}^{\infty}\mathbf{J}_{2m+1}(z^{+})\sin((2m+1)\gamma)\cos 2\gamma\right)\right]d\gamma
\end{aligned}
\tag{8.62}
$$

Again $z^+ = 2x\sqrt{bb^+}$. The Bessels all come out of the integral since they are not functions of γ. Orthogonality works here in this case, and therefore

$$\int_{-\pi/2}^{\pi/2} \mathtt{cos}2\gamma d\gamma = 0$$

$$\int_{-\pi/2}^{\pi/2} \mathtt{cos}(m2\gamma)\mathtt{cos}2\gamma d\gamma = \begin{cases} 0 \text{ if } m \neq 1 \\ \pi \text{ if } m = 1 \end{cases}$$

$$\int_{-\pi/2}^{\pi/2} (\mathtt{sgn}\gamma)\mathtt{sin}((2m+1)\gamma)\mathtt{cos}2\gamma d\gamma$$

$$\equiv \frac{\mathtt{sgn}(\gamma)((1-2m)\mathtt{cos}((2m+3)\gamma) - (2m+3)\mathtt{cos}(\gamma - 2m\gamma) + 4m + 2) + 4m - 2}{8m^2 + 8m - 6}\Bigg|_{-\pi/2}^{\pi/2}$$

$$= d_m = \frac{2m+1}{m^2 + m - 3/4} \tag{8.63}$$

The solution to the first integral of the breakout given by Eq. (8.60) is

$$\frac{1}{\pi}\int_{-\pi/2}^{\pi/2} \frac{e^{-jkR} - 1}{R}\mathtt{cos}2\gamma d\gamma = -\frac{j}{\pi}\int_0^k \left[2\pi \mathbf{J}_2(z^+) - j2\sum_{m=0}^{\infty} d_m \mathbf{J}_{2m+1}(z^+)\right]$$

$$= -\int_0^k \left[\frac{1}{\pi}\sum_{m=0}^{\infty} d_m \mathbf{J}_{2m+1}(z^+) + j2\mathbf{J}_2(z^+)\right] dx \tag{8.64}$$

Using the same logic regarding the maximum of z^+ as given in Eq. (8.52), we may limit the summation to $m = 0$. This gives $d_m = -4/3$ and

$$= \int_0^k \left[\frac{4}{3\pi}\mathbf{J}_1(z^+) - j2\mathbf{J}_2(z^+)\right] dx$$

$$= \frac{2}{3}\left(\frac{1 - \mathbf{J}_0(k\sqrt{bb^+})}{2\sqrt{bb^+}}\right) - j\left[\frac{\left(2\pi k\sqrt{bb^+}\mathbf{H}_0(2\sqrt{bb^+}) - 4\right)\mathbf{J}_1(2\sqrt{bb^+})}{2\sqrt{bb^+}}\right.$$

$$\left. + \left(2k - \pi k\mathbf{H}_1(2\sqrt{bb^+})\right)\mathbf{J}_0(\sqrt{bb^+})\right] \tag{8.65}$$

Here $\mathbf{H}$ is the Struve function.

The second integral of the breakout given by Eq. (8.60)

$$\frac{1}{\pi}\int_{-\pi/2}^{\pi/2} \frac{\mathtt{cos}2\gamma}{R} d\gamma \tag{8.66}$$

introduces a pole at $\phi' = \phi$ if we use the shortened form for R. Therefore, using the more accurate form, Eq. (8.31), we have for the internal integral

$$= \frac{1}{a\pi}\int_{-\pi/2}^{\pi/2} \frac{\cos 2\gamma}{\sqrt{\frac{bb^+}{a^2}\sin^2\gamma + 1}} d\gamma$$

$$= \frac{1}{a\pi}\left[\frac{(\frac{bb^+}{a^2}+2)\mathbb{F}(\gamma|-\frac{bb^+}{a^2}) - 2\mathbb{E}(\gamma|-\frac{bb^+}{a^2})}{bb^+/a^2}\right]_{-\pi/2}^{\pi/2}$$

$$= \frac{2}{a\pi}\left[\frac{(\frac{bb^+}{a^2}+2)\mathbb{K}(-\frac{bb^+}{a^2}) - 2\mathbb{E}(-\frac{bb^+}{a^2})}{bb^+/a^2}\right] \quad (8.67)$$

$\mathbb{K}(-z^2)$ is the Complete Elliptical Integral of the First Kind with parameter $m = k^2$ and $\mathbb{E}(-z^2)$ is the Complete Elliptical Integral of the Second Kind with parameter $m = z^2$.

The solution to Eq. (8.60) is the sum of Eqs. (8.65) and (8.67): The final solution to the integral for the fourth component of E_ψ is found by substituting this into Eq. (8.59).

$$E_{\psi 4} = -\frac{j\xi}{2}\int_{-\pi}^{\pi}\left[ka(b + a\cos\psi')J_{\psi'}\sin\psi'\sin\psi \times \left(\frac{2}{3}\left(\frac{1-\mathbf{J}_0(k\sqrt{bb^+})}{2\sqrt{bb^+}}\right)\right.\right.$$
$$- j\left[\frac{\left(2\pi k\sqrt{bb^+}\mathbf{H}_0(2\sqrt{bb^+}) - 4\right)\mathbf{J}_1(2\sqrt{bb^+})}{2\sqrt{bb^+}}\right.$$
$$\left. + \left(2k - \pi k\mathbf{H}_1(2\sqrt{bb^+})\right)\mathbf{J}_0(\sqrt{bb^+})\right]$$
$$\left.\left. + \frac{2}{a\pi}\left[\frac{(\frac{bb^+}{a^2}+2)\mathbb{K}(-\frac{bb^+}{a^2}) - 2\mathbb{E}(-\frac{bb^+}{a^2})}{bb^+/a^2}\right]\right)\right]d\psi' \quad (8.68)$$

(5) The fifth full integral for the ψ component of the electric field is

$$-\frac{j\xi}{2}\int_{-\pi}^{\pi}\left[ka(b + a\cos\psi')J_{\psi'}\cos\psi'\cos\psi\left(\frac{1}{2\pi}\int_{-\pi}^{\pi}\frac{e^{-jkR}}{R}d\phi'\right)\right]d\psi' \quad (8.69)$$

The internal integral may be expanded into two portions:

$$\frac{1}{2\pi}\int_{-\pi}^{\pi}\frac{e^{-jkR}}{R}d\phi' = \frac{1}{2\pi}\int_{-\pi}^{\pi}\frac{e^{-jkR}-1}{R}d\phi' + \frac{1}{2\pi}\int_{-\pi}^{\pi}\frac{1}{R}d\phi' \quad (8.70)$$

The first is expanded as before; see, for example, Eq. (8.23):

$$\frac{1}{2\pi}\int_{-\pi}^{\pi}\frac{e^{-jkR}-1}{R}d\phi' = -\frac{j}{2\pi}\int_{-\pi}^{\pi}\left[\int_0^k e^{-jxR}dx\right]d\phi'$$
$$= -j\int_0^k\left[\frac{1}{2\pi}\int_{-\pi}^{\pi}e^{-j2x\sqrt{bb^+}|\sin\gamma|}d\phi'\right]dx \quad (8.71)$$

The internal integral is expanded again using Eq. (8.26) and the identities, Eq. (8.28), substituted.

$$
\begin{aligned}
&= \frac{1}{\pi}\int_{-\pi/2}^{\pi/2}\left[\cos(2x\sqrt{bb^+}\,|\sin\gamma|) - j\sin(2x\sqrt{bb^+}\,|\sin\gamma|)\right]d\gamma \\
&= \frac{1}{\pi}\int_{-\pi/2}^{\pi/2}\left[\left(\mathbf{J}_0(z^+) + 2\sum_{m=1}^{\infty}\mathbf{J}_{2m}(z^+)\cos(2m\gamma)\right)\right. \\
&\quad \left. - j(\operatorname{sgn}\gamma)\left(2\sum_{m=0}^{\infty}\mathbf{J}_{2m+1}(z^+)\sin((2m+1)\gamma)\right)\right]d\gamma
\end{aligned} \tag{8.72}
$$

The Bessels are not functions of γ and may be withdrawn from the integral. The integration of the first term is 0 and the second has a value only for $m = 0$; however, the summation begins at $m = 1$. Therefore, the second term is also 0. The integration for the third term is

$$
= -\frac{j}{\pi}\sum_{m=0}^{\infty}\frac{4}{2m+1}\mathbf{J}_{2m+1}(z^+) \tag{8.73}
$$

The first integral in the breakout is

$$
\frac{1}{\pi}\int_{-\pi/2}^{\pi/2}\frac{e^{-jkR}-1}{R}d\gamma = -\frac{1}{\pi}\sum_{m=0}^{\infty}\frac{4}{2m+1}\int_0^k \mathbf{J}_{2m+1}(z^+)dx \tag{8.74}
$$

The second term of the breakout exhibits a pole at $\phi' = \phi$ if we continue to use the short form for R. Therefore using the more accurate form, Eq. (8.31), we have

$$
\begin{aligned}
\frac{1}{\pi}\int_{-\pi/2}^{\pi/2}\frac{1}{R}d\gamma &= \frac{1}{\pi a}\int_{-\pi/2}^{\pi/2}\frac{1}{\sqrt{(bb^+/a^2)\sin(\gamma)+1}}d\gamma \\
&= \frac{1}{\pi a}\mathbb{F}\left(\gamma\middle| -\frac{bb^+}{a^2}\right)\Bigg|_{-\pi/2}^{\pi/2} = \frac{1}{\pi a}\mathbb{K}\left(-\frac{bb^+}{a^2}\right)
\end{aligned} \tag{8.75}
$$

where $\mathbb{K}$ is a Complete Elliptic Integral of the First Kind. The full integral for the fifth component of E_ψ is therefore

$$
\begin{aligned}
E_{\psi 5} &= -\frac{j\xi}{2}\int_{-\pi}^{\pi}\left[ka(b+a\cos\psi')J_{\psi'}\cos\psi'\cos\psi\left(\frac{1}{2\pi}\int_{-\pi}^{\pi}\frac{e^{-jkR}}{R}d\phi'\right)\right]d\psi' \\
&= -\frac{j\xi}{2}\int_{-\pi}^{\pi}\left[ka(b+a\cos\psi')J_{\psi'}\cos\psi'\cos\psi\right. \\
&\quad \left.\times\left(-\frac{1}{\pi}\sum_{m=0}^{\infty}\frac{4}{2m+1}\int_0^k \mathbf{J}_{2m+1}(z^+)dx + \frac{1}{\pi a}\mathbb{K}\left(-\frac{bb^+}{a^2}\right)\right)\right]d\psi'
\end{aligned} \tag{8.76}
$$

Reduced Integrals for E_ψ

The ψ component of the Electric Field is now reduced to

$$
\begin{aligned}
E_\psi = &-\frac{-j\xi}{2}\sin\psi\int_{-\pi}^{\pi}\left[e^{-jkz}\frac{z'}{z}\left[\frac{\mathbf{J}_0(kz')}{kz}-\frac{z'}{z}\mathbf{J}_1(kz')+j\mathbf{J}_0(kz')\right]\frac{\partial J_{\phi'}}{\partial\phi'}\right]d\phi'\\
&+\frac{j\xi}{2}\int_{-\pi}^{\pi}\left[kaJ_{\phi'}\sin\psi\sin(\phi-\phi')\left(\frac{je^{-jkz}}{z}\left[\left(a+\frac{b}{b'/a}\right)\mathbf{J}_1(kz')\right.\right.\right.\\
&\left.\left.\left.+j\left(\left(b+\frac{a}{8(b'/a)}\right)\mathbf{J}_0(kz')-\frac{a}{8(b'/a)}\mathbf{J}_2(kz')\right)\right]\right)\right]d\phi'\\
&-\frac{j\xi\sin\psi}{4\pi\sqrt{b'b^+}}\int_{-\pi}^{\pi}\left[\frac{1}{k\sqrt{bb^+}}\left(\frac{\mathbf{J}_0(2k\sqrt{bb^+})}{4}-1\right)+\frac{9}{4}\mathbf{J}_1(2k\sqrt{bb^+})\right.\\
&\left.+j\frac{\pi^2}{2}\left[\mathbf{H}_0(2k\sqrt{bb^+})\mathbf{J}_1(2k\sqrt{bb^+})-\mathbf{H}_1(2k\sqrt{bb^+})\mathbf{J}_0(2k\sqrt{bb^+})\right]\frac{\partial(b+a\cos\psi')J_{\psi'}}{\partial\psi'}\right]d\psi'\\
&-\frac{j\xi}{2}\int_{-\pi}^{\pi}\left[ka(b+a\cos\psi')J_{\psi'}\sin\psi'\sin\psi\times\left(\frac{2}{3}\left(\frac{1-\mathbf{J}_0(k\sqrt{bb^+})}{2\sqrt{bb^+}}\right.\right.\right.\\
&-j\left[\frac{\left(2\pi k\sqrt{bb^+}\mathbf{H}_0(2\sqrt{bb^+})-4\right)\mathbf{J}_1(2\sqrt{bb^+})}{2\sqrt{bb^+}}+\left(2k-\pi k\mathbf{H}_1(2\sqrt{bb^+})\right)\mathbf{J}_0(\sqrt{bb^+})\right]\\
&\left.\left.\left.+\frac{2}{a\pi}\left[\frac{(\frac{bb^+}{a^2}+2)\mathbb{K}(-\frac{bb^+}{a^2})-2\mathbb{E}(-\frac{bb^+}{a^2})}{bb^+/a^2}\right]\right)\right)\right]d\psi'\\
&-\frac{j\xi}{2}\int_{-\pi}^{\pi}\left[ka(b+a\cos\psi')J_{\psi'}\cos\psi'\cos\psi\right.\\
&\left.\times\left(-\frac{1}{\pi}\sum_{m=0}^{\infty}\frac{4}{2m+1}\int_0^k\mathbf{J}_{2m+1}(z^+)dx+\frac{1}{\pi a}\mathbb{K}\left(-\frac{bb^+}{a^2}\right)\right)\right]d\psi'
\end{aligned}
\tag{8.77}
$$

8.2.2 *Solutions to the Perfectly Conducting, Thick Ring Equations Applicable to 150 GHz*

These two sets of integral equations, Eqs. (8.72) and (8.77) have not appeared in the literature before this writing and have not yet been solved. Indeed, little work has focused on the current distribution in thick rings. One interesting paper, however, appeared in 2013 [7] that uses an illuminating TEM wave as the source on rings with thicknesses $\Omega \in \{5.06, 6.45, 7.83,$ and $9.22\}$ [7]. These are not round numbers due to the authors' use of b/a as the measure of thickness, rather than Ω. Their coordinate system is shown in Fig. 8.1. The illuminating wave approaches the ring from the $-\hat{z}$ direction broadside with the $\vec{H}$ field in the $\hat{y}$ direction only, as shown in Fig. 8.2.

Figures 8.3a, b show the currents, J_θ and J_ϕ, calculated using the authors' specialized "Normally Integrated Magnetic Field Integral Equation" formulation [9]. They provide therefore numerical results of the current magnitudes over $b = \lambda/2\pi$. The

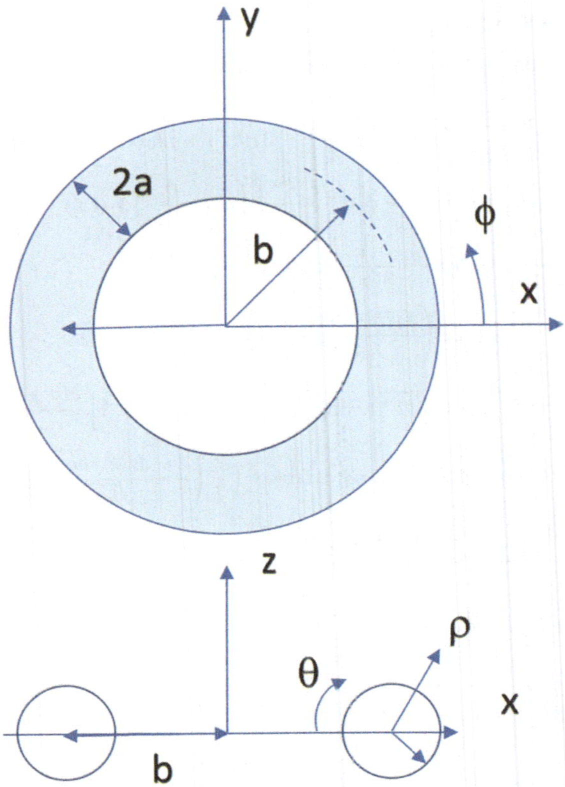

Fig. 8.1 The coordinates matched to Bibby et al.

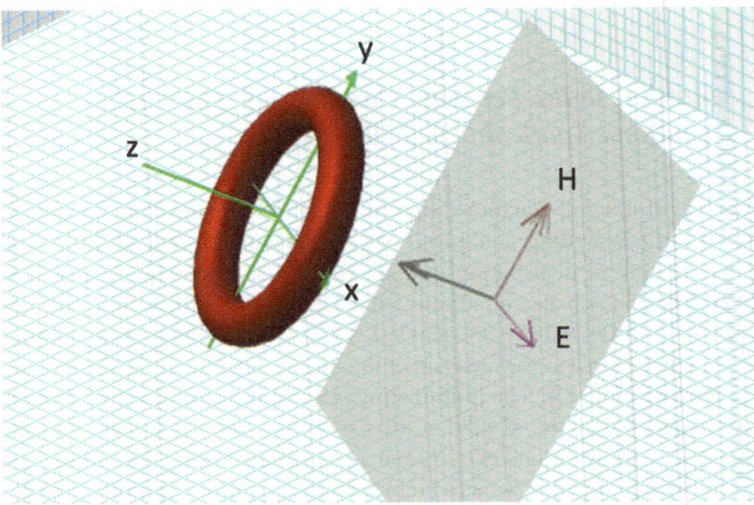

Fig. 8.2 The plane wave in [7] illuminating the ring

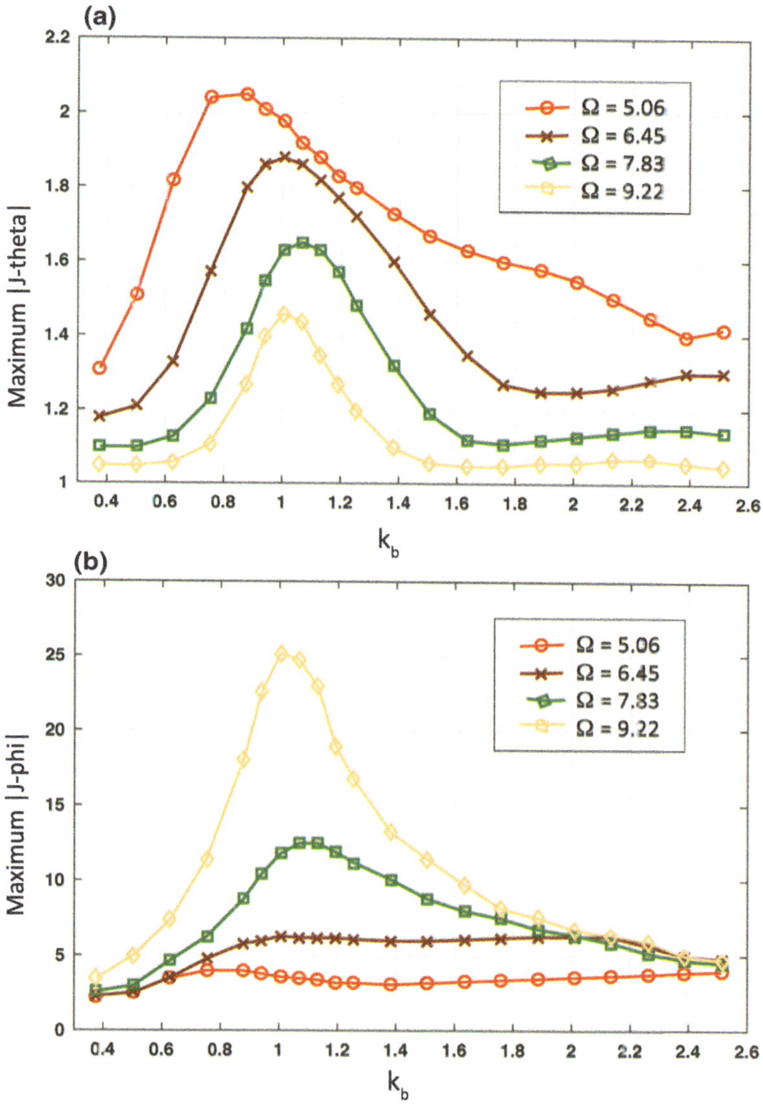

Fig. 8.3 Current densities on the illuminated thick ring: **a** J_θ **b** J_ϕ. The units are not specified in [7]

figures reproduce their data using our standard k_b on the horizontal axis. The main resonances of both currents for the two thinnest wires are about the same as for the thin-wire equivalents, 1.088 and 1.15 (see Table 4.1). These thick rings do not show a second resonance. The bandwidths are considerably wider than the thin wire loops as we might expect, leading to much diminished Qs.

References

1. A.L. Koh, D.W. McComb, S.A. Maier, H. Low, J.K.W. Yang, Microelectronics and nanometer structures. J. Vac. Sci. Technol. B **28**(6), C6O45 (2010). https://doi.org/10.1116/1.3501351
2. A.W. Clark, J.M. Cooper, Small **7**(1), 119 (2011). https://doi.org/10.1002/smll.201001433
3. D. Werner, IEEE Trans. Antennas Propag. **41**(8), 1009 (1993). https://doi.org/10.1109/8.241641
4. W.X. Wang, IEEE Trans. Antennas Propag. **39**(4), 434 (1991)
5. D. Werner, IEEE Trans. Antennas Propag. **46**(3), 373 (1998). https://doi.org/10.1109/8.662656
6. R.W. Scharstein, H.B. Wilson, Electromagnetics **25**(1), 1 (2005). https://doi.org/10.1080/02726340590522102
7. M.M. Bibby, C.M. Coldwell, A.F. Peterson, IEEE Trans. Antennas Propag. **61**(7), 3658 (2013). https://doi.org/10.1109/TAP.2013.2258316
8. I. Stegun, M. Abramowitz, *Handbook of Mathematical Functions With Formulas, Graphs and Mathematical Tables*, vol. 55, Applied Mathematics Series (US. Governent Printing Office, WDC, 1964)
9. M.M. Bibby, C.M. Coldwell, A.F. Peterson, IEEE Trans. Antennas Propag. **55**(9), 2530 (2007). https://doi.org/10.1109/TAP.2007.904136

Part III
Coupled Rings in One, Two and Three Dimensions

Abstract Small rings set out in the form of an array of some sort, couple their fields; that is, the electromagnetic field created either by their driving sources or by an illuminating plane wave causes secondary currents and secondary fields with which the first fields mutually interact. This is, of course, a well known phenomenon, since it is the basis for Faraday's and Lenz's Laws. The phenomenon was not used much in the RF region for propagation with large loop antennas, except perhaps in Yagi-Uda type arrangements and in a number of other limited scenarios. However, in 1999, a rather novel 3D configuration of small rings was suggested as a mechanism for engaging the illuminating magnetic field, rather than the electric field, and this engagement was shown to cause a negative permeability averaged over the 3D structure. The rings themselves were eventually called "meta-atoms", after the structure itself, which was called a "meta-material". This part first covers coupled, PEC, thinwire loop antennas and rings, workable up to 150 GHz. The first meta-material ever described and the first meta-material structure built were, in fact, configured for 10 GHz.

Chapter 9
Meta-atoms & Rings as Large-Scale Atoms

9.1 Introduction

Meta-atoms come in varying shapes and sizes. They each work in concert with others of similar nature to produce a macro-effect, just as atoms working together with other atoms in a molecule yield a macro-effect. In the case of the first meta-material, that effect was a negative permeability and a negative dielectivity leading to a negative index of refraction, a phenomenon that does not exist in nature. In later developments, 2D meta-surfaces were created to focus, deflect and polarise incoming plane waves. Figure 9.1 in an article by Powell [1] shows a few of these various types of meta-atoms. They can be spiral rings, split-rings, V-shaped, spherical, thick horseshoe like and even twisted crosses.

Powell's article suggests a general approach for discovering the response of an arbitrarily shaped meta-atom to an arbitrary incident plane wave illumination over a wide bandwidth. The key is to find the modal current distribution first by solving Maxwell's equations using volume integral equations (as is presented in Sect. 2.4), and then writing a simple matrix of impedance functions that represents these modes as oscillator circuit elements (similar to Chap. 4, but using a more Physics approach). Use of that matrix in an eigenvalue problem leads to the eigenvalue impedance matrix with elements up to any number of modes. That eigenvalue matrix gives a very good approximation of the characteristic modal behaviour of the meta-atom. It can be interpreted as the sum of equivalent responses of each mode. We shall return to its results for split-rings later in the chapter. The point here is that a general method exists for understanding the scattering behaviour of meta-atoms of any shape in terms of its resonant modes.

The focus in this chapter is on PEC rings, because we have no analytical material in the previous chapters of this volume on dielectric rings. Studies of the behaviour of dielectric rings is, at the moment, on the cutting edge. An analytical look at such rings requires solutions to all three of the governing equations to yield current distributions within the core of the ring material, and these have not been solved; the main reasons

A. McKinley, *The Analytical Foundations of Loop Antennas and Nano-Scaled Rings*, Signals and Communication Technology,
https://doi.org/10.1007/978-981-13-5893-7_9

being that it requires a clearer understanding of the effects of dielectric material characteristics on charge movement and it requires an understanding of how charge moves in a thick ring on the order of $\Omega = 6, 7$ and 8.

9.2 Plane Wave Illumination of Meta-atom Rings

The plane wave illumination of a closed loop or closed ring of any thickness is discussed in some detail in Sect. 2.6.3. The phenomenon yields an $\vec{E}$ field at a point Q on the ring given by Eqs. (2.148) and (2.150), repeated here:

$$\begin{aligned}\vec{E}_Q &= E_0 e^{-j\vec{k}\cdot\vec{R}}\left[(-\cos\gamma\sin\phi_p - \sin\gamma\cos\theta_p\cos\phi_p)\,\hat{x}\right.\\ &\quad \left.+ (\cos\gamma\cos\phi_p - \sin\gamma\cos\theta_p\sin\phi_p)\,\hat{y} + \sin\gamma\sin\theta_p\,\hat{z}\right]\\ &= E_0 e^{-j\vec{k}\cdot\vec{R}}\\ &\quad \times \left[(\cos\gamma\cos\psi_q\sin\alpha - \sin\gamma\cos\theta_p\cos\psi_q\cos\alpha + \sin\gamma\sin\theta_p\sin\psi_q)\,\hat{\sigma}\right.\\ &\quad + (\cos\gamma\cos\alpha + \sin\gamma\cos\theta_p\sin\alpha)\,\hat{\phi}\\ &\quad \left.- (\cos\gamma\sin\psi_q\sin\alpha - \sin\gamma\cos\theta_p\sin\psi_q\cos\alpha - \sin\gamma\sin\theta_p\cos\psi_q)\,\hat{\psi}\right]\end{aligned}$$

The various angles are defined in Fig. 2.12, and $\alpha = \phi_q - \phi_p$. Two orientations are discussed in that section. The first is a broadside illumination with $\theta_p = 0$ and the second a side illumination with $\theta_p = \pi/2$. We repeat the illumination orientations here, but with gapped rings.

Some confusion exists in the literature between the electric and magnetic resonances of gapped, or split, rings. The early literature tended to call the source of any resonances in these split-rings "magnetic" resonances, due to the penetration of the $\vec{H}$ through the plane of the loop. However, these resonances will be additionally stimulated by the $\vec{E}$ field when it appears across the gap in s-polarisation at the same time (see Fig. 9.3a). Also see [2–5]. In addition, a rather interesting $k_b = 1.0$ resonance exists, which has been confused with the standard harmonic resonances. This resonance, discussed in Sect. 2.6.3, has not been understood until now.

Orientation 1 $\theta_p = 0$: In this orientation, shown in Fig. 9.1, ϕ_p can be 0 or $\pi/2$ without changing the outcome, due to symmetry. The $\vec{E}$ field can be oriented in one of two directions, with either (a) $\gamma = 0$ or (b) $\gamma = \pi/2$.

(a) $\gamma = 0$: Substituting $\theta_p = 0$, $\phi_p = 0$ and $\gamma = 0$, reduces Eq. (2.150) to

$$\begin{aligned}\vec{E}_Q &= E_0 e^{-jkR}\,\hat{y}\\ &= E_0 e^{-jkR}\left[\cos\psi_q\sin\phi_q\hat{\sigma} + \cos\phi_q\hat{\phi} - \sin\psi_q\sin\phi_q\hat{\psi}\right]\end{aligned} \tag{9.1}$$

Both fields lie in the plane of the ring on impact. The electric field appears only along the $\hat{y}$ axis on the periphery of the ring, no matter the wavelength, λ, of the incoming

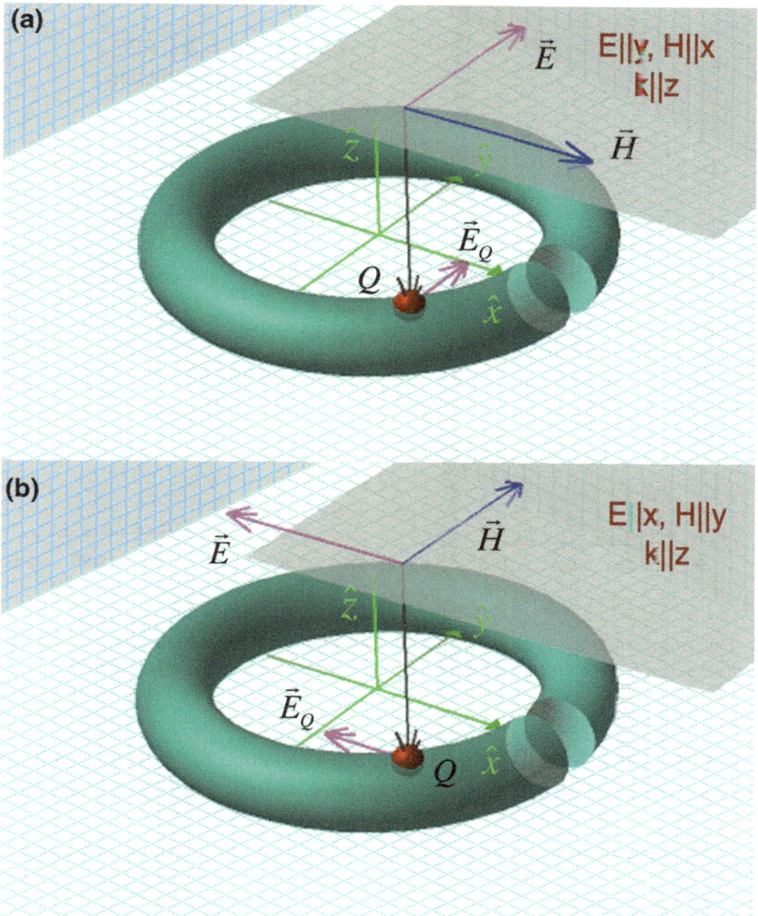

Fig. 9.1 Wire rings illuminated broadside: $\theta_p = 0$, $\phi_p = 0$. In **a**, $\gamma = 0$ puts the $\vec{E}$ field across the gap (s-polarisation). In **b**, $\gamma = \pi/2$ puts it perpendicular to the gap (p-polarisation). Configuration **a** generates the harmonic resonances and the sub-wavelength resonance. Configuration **b** stimulates the $k_b = 1.0$ resonance due to the wave pulsing through the ring

wave. The simulations, shown in Fig. 9.2 show the sub-wavelength resonance and the standard harmonics, blue-shifted due to the added capacitance of the gap. These resonances are explained by the s-polarisation across the gap of the incoming electric field. Since the angle ϕ_p is measured from the x-axis, the cosine in the $\hat{\phi}$ component of $\vec{E}$ displays a maximum there and the electric field is therefore at maximum density across the gap for any given illuminating wavelength. In other words, the gap acts like a delta function generator and produces these resonances. The magnetic field has no effect on ring currents, since it must penetrate the plane of the ring to induce current. **Case (a) generates the sub-wavelength and the standard harmonic resonances.**

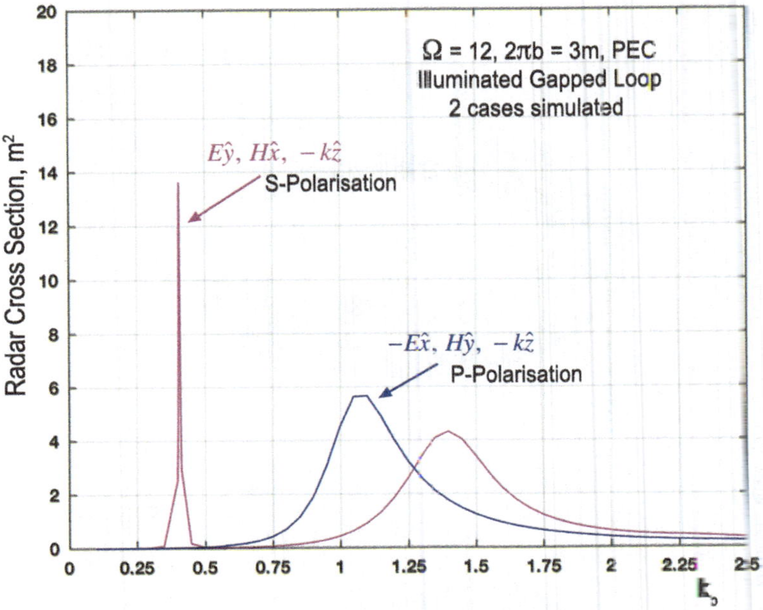

Fig. 9.2 The ring resonances generated by the two plane wave configurations in Fig. 9.1. The standard resonances match with **a** and the broad $k_b = 1.0$ resonance matches with **b**

As the loop thickens, the resonances blue shift because the current tends to flow on the inside surface with circumference $2\pi(b-a)$ instead of $2\pi b$ in the thin-wire case.

(b) $\gamma = \pi/2$: Substituting the angles properly reduces the main equations to

$$\begin{aligned}\vec{E}_Q &= -E_0 e^{-jkR}\hat{x} \\ &= E_0 e^{-jkR}\left[-\cos\psi_q \cos\phi_q \hat{\sigma} + \sin\phi_q \hat{\phi} + \sin\psi_q \cos\phi_q \hat{\psi}\right]\end{aligned} \tag{9.2}$$

As in (a), the two fields impact in the plane of the ring. The $\vec{E}$ field this time is constant along $-\hat{x}$, independent of the illuminating wavelength. In this case, however, the field is minimum at $\phi = 0$ and does not stimulate the harmonic resonances, nor the sub-wavelength resonance. Moreover, the magnetic field does not induce currents. The simulations in Fig. 9.2 show the broad $k_b = 1.0$ resonance discovered in Sect. 2.6.3. This is a resonance due to pulsing by the illuminating wave, independent of the gap and the standard harmonic resonances. **Case (b) will generate only the broad** $k_b = 1.0$ **resonance.**

Orientation 2 In the second orientation where $\theta_p = \pi/2$, there are two cases: $\phi_p = 0$ and $\phi_p = \pi/2$. The first is shown in Fig. 9.3 and the second in Fig. 9.5

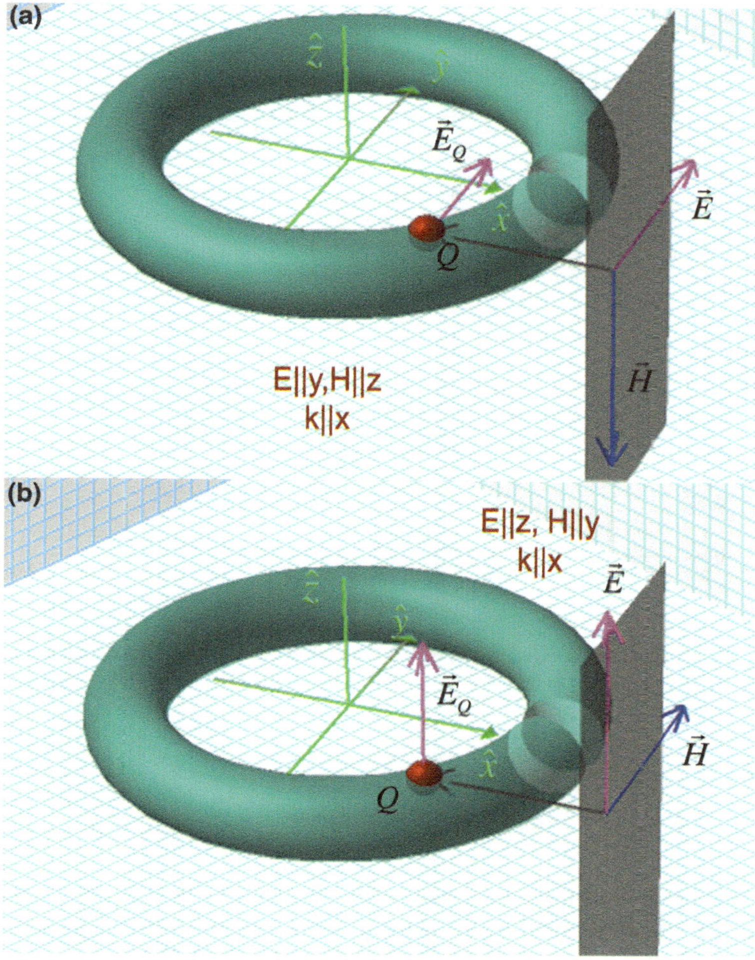

Fig. 9.3 Rings illuminated from the side, as in Fig. 9.5, but from the adjacent side. Two configurations of the plane wave are again shown, **a** the electric field is in the plane in s-polarisation across the gap and the magnetic field penetrates the plane of the ring; **b** when the electric field is perpendicular to the plane of the ring and the magnetic field is in the plane. Resonances are shown in Fig. 9.4

Case 1, Fig. 9.3: $\phi_p = 0$, orients the plane wave in the yz plane. There are two orientations of the $\vec{E}$ field: one with $\gamma = 0$ and the other with $\gamma = \pi/2$.
(a) $\gamma = 0$: Substituting the angles into the main equation yields

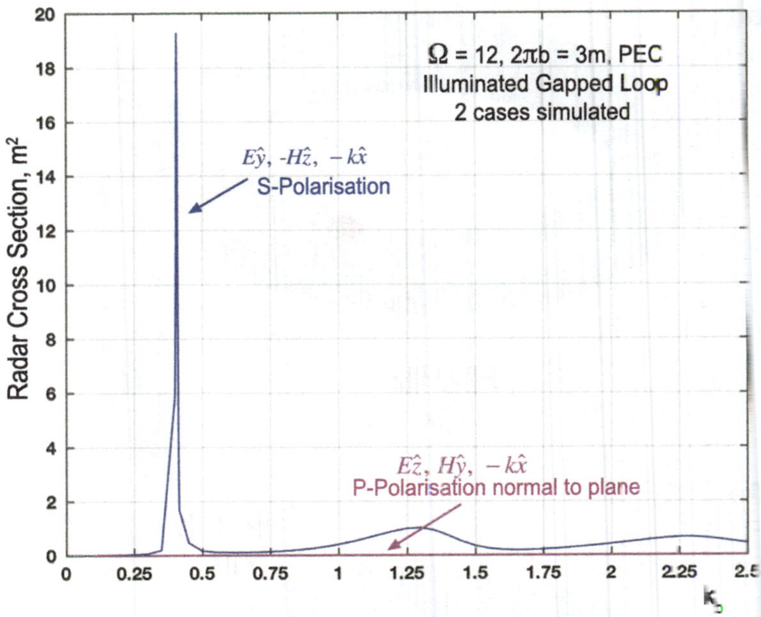

Fig. 9.4 The resonances associated with Fig. 9.3. The sub-wavelength and standard harmonics are stimulated by configuration (**a**) in that figure, while configuration (**b**) stimulates no resonances

$$\begin{aligned}\vec{E}_Q &= E_0 e^{-jkR}\hat{y}\\ &= E_0 e^{-jkR}\left[\cos\psi_q\sin\phi_q\hat{\sigma} + \cos\phi_q\hat{\phi} - \sin\psi_q\sin\phi_q\hat{\psi}\right]\end{aligned} \tag{9.3}$$

The $\vec{E}$ field lies in the plane of the ring, while the magnetic field penetrates the ring along $\hat{z}$. The simulations in Fig. 9.4 show a very strong sub-wavelength resonance and the standard harmonics. These are explained by the S-Polarisation of the $\vec{E}$ across the gap, where the $\hat{\phi}$ component cosine is maximum; the field acts like a delta-function generator. Moreover, the induction stimulated by the penetration of the $\vec{H}$ field through the plane of the ring adds additional stimulus to these resonances. The combination imparts more energy into the ring than either one of the fields acting alone. **This is a particularly effective illumination because all resonances of the loop are stimulated by both the magnetic field and the electric field.**
(**b**) $\gamma = \pi/2$: Substituting yields

$$\begin{aligned}\vec{E}_Q &= E_0 e^{-jkR}\hat{z}\\ &= E_0 e^{-jkR}\left[\sin\psi_q\hat{\sigma} + \cos\psi_q\hat{\psi}\right]\end{aligned} \tag{9.4}$$

The simulations indicate that none of the resonances are stimulated. The magnetic field does not penetrate the loop and therefore does not stimulate currents and the electric field is perpendicular to the plane of the loop and therefore has no effect on the gap and does not pulse the loop. **Therefore, neither field stimulates resonances in this configuration.**

Case 2, Fig. 9.5: $\phi_p = \pi/2$, orients the plane wave in the xz plane. There are two orientations of the $\vec{E}$ field: one with $\gamma = 0$ and the other with $\gamma = \pi/2$.
(a) $\gamma = 0$: Substituting the angles reduces the main equation to

$$\begin{aligned}\vec{E}_Q &= -E_0 e^{-jkR}\hat{x} \\ &= E_0 e^{-jkR}\left[-\cos\psi_q \cos\phi_q \hat{\sigma} + \sin\phi_q \hat{\phi} + \sin\psi_q \cos\varphi_q \hat{\psi}\right]\end{aligned} \tag{9.5}$$

The simulations in Fig. 9.6 indicate the sub-wavelength resonance and the standard harmonic resonances. These appear because of the stimulation by the magnetic field penetrating the plane of the loop. The electric field does not participate since it approaches from the side in the plane in p-polarisation to the gap. Nor does the electric field stimulate the $k_b = 1.0$ resonance. **This configuration generates the sub-wavelength and standard harmonic resonances.**
(b) $\gamma = \pi/2$: Substituting the angles reduces the main equation to

$$\begin{aligned}\vec{E}_Q &= E_0 e^{-jkR}\hat{z} \\ &= E_0 e^{-jkR}\left[\sin\psi_q \hat{\sigma} + \cos\psi_q \hat{\psi}\right]\end{aligned} \tag{9.6}$$

The simulations in Fig. 9.6 indicate that no resonances are stimulated and this is because the electric field is in p-polarisation with the gap and the magnetic field does not induce currents. The electric field also does not pulse the ring to create the $k_b = 1.0$ resonance. **This configuration does not generate any of the ring's resonances.**

9.3 Gap Capacitance

The gap in a nano-scaled split-ring is often thought of as a capacitor set in the periphery that modifies the zero-order inductance of the loop and provides by doing so the basis for the sub-wavelength resonance. Certainly, placing a lumped capacitor in the periphery does generate a resonance in the sub-wavelength region (see Sect. 7.5). And illuminating a split-ring also stimulates a sub-wavelength resonance, as noted in the previous section. This correspondence suggests we can work backwards from the resonance, knowing the ring's inductance and capacitance (see Sect. 4.2.1), to the capacitance of the gap.

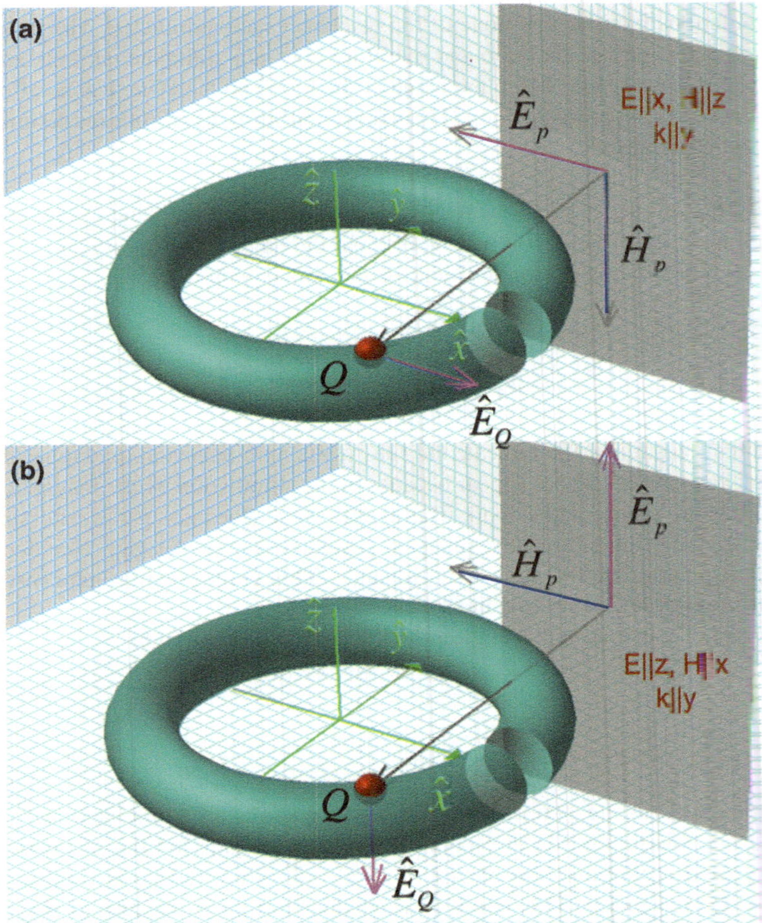

Fig. 9.5 Rings illuminated from the side in two configurations of the plane wave. In **a** the electric field is in the plane of the ring and perpendicular to the gap. The magnetic field penetrates the plane of the ring. In **b** the two field are flipped 90°. Resonances are shown in Fig. 9.6

We first look at the standard models for gap capacitance, then simulate split-rings of various sizes including various width gaps. Results of post-processing the gap capacitance are compared with the standard gap models to see if any of them approach the correct values. Such characterisation leads us to an intuitive understanding of how gaps affect ring behaviour.

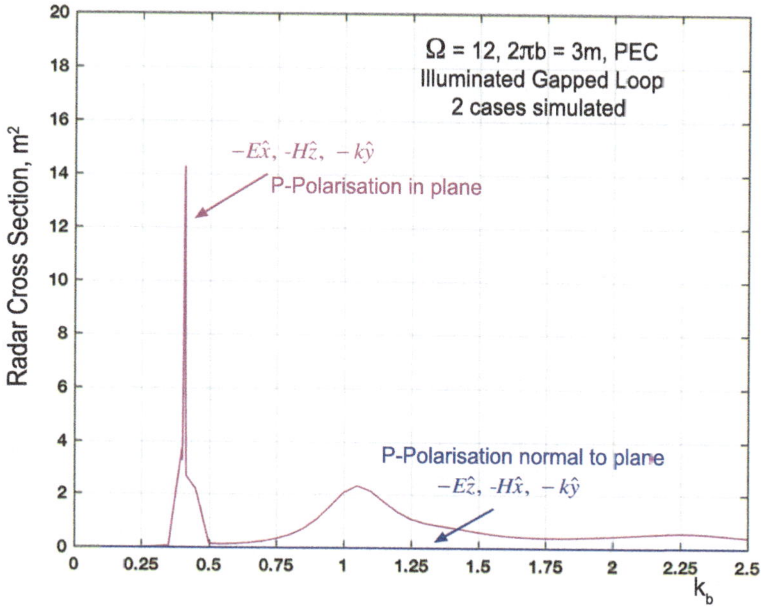

Fig. 9.6 The resonances associated with the plane waves in Fig. 9.5. Configuration (**a**) generates the sub-wavelength and standard resonances. (**b**) does not stimulate any of the ring's resonances

9.3.1 *Standard Models*

For a cylindrical or square ring in the cross-section, the gap capacitance is often calculated as if it were a flat plate capacitor. It is assumed that this sort of behaviour holds, even as high as optical frequencies. The flat-plate model has not been seriously challenged in the literature and remains to be done.

In this section, therefore, we will indicate an effect of adding a gap that does not seem to have been noticed and then suggest a method for determining the effective capacitance of a given gap. Three flat plate capacitance models have been suggested:

- **The Standard Flat-Plate Model**: two flat plates with area, A, are separated by gap width, g, of air [6, 7]. The capacitive value is $C_{gap} = \varepsilon_0 A/g$. The relation between l_ε and the gap ratio, g/a, can be derived if a round wire is assumed:

$$C_{gap} = C_{fp} = \varepsilon b l_\varepsilon = \varepsilon \frac{\pi a^2}{g} = \varepsilon b \frac{\pi}{(b/a)(g/a)}$$

or

$$l_{\varepsilon,fp} = \frac{\pi}{(b/a)(g/a)}. \tag{9.7}$$

- **Fringe Effect Addition**: the electric field is assumed to extend outside the gap somewhat, thus affecting the overall capacitance of the gap. The literature has used $C_{fringe} = \varepsilon_0(h + w + g)$ to represent the effect in rectangular cross-sectional rings [8, 9]. For the round wire, we convert to half the circular circumference, because $h + w$ is half the circumference of the rectangle, plus the gap, g, and use $C_{frng} = \varepsilon(\pi a + g)$. The relationship becomes:

$$C_{gap} = \varepsilon b \left(\frac{\pi}{(b/a)(g/a)} + \frac{\pi}{b/a} + \frac{g/a}{b/a} \right)$$

or

$$l_{\varepsilon,frng} = l_{\varepsilon,fp} \left(1 + (g/a) + \frac{(g/a)^2}{\pi} \right). \tag{9.8}$$

- **Surface Capacitance Addition**: the charge on the surface of the ring is assumed to add capacitance. This may be no more than an attempt to include the capacitance of the closed ring, but it is hard to tell from the description [8]. The total capacitance, then, is the flat-plate plus the fringe plus the surface capacitance. The extra capacitance becomes $C_{surf} = (2/\pi)\varepsilon(h + w)ln(4b/g)$. Again using πa for $h + w$, we have

$$C_{gap} = C_{fp} + C_{frng} + \varepsilon b \left(\frac{2}{b/a} \mathrm{ln} \frac{4(b/a)}{g/a} \right)$$

or

$$l_{\varepsilon,surf} = l_{\varepsilon,fp} \left(1 + (g/a) + \frac{(g/a)^2}{\pi} \right) + \frac{2}{b/a} \mathrm{ln} \frac{4(b/a)}{g/a} \tag{9.9}$$

The literature of split-rings often calculates approximately the ring's resonance, and most authors realise that to do so the ring must be small compared with wavelength. One typical line of demarkation is the sub-wavelength zone, defined as a loop with circumference less than 0.50λ. Moreover, most authors use the low frequency approximation for the inductance of the ring, even in the GHz range, namely, $L = \mu b[\mathrm{ln}(8b/a) - 2]$, with $\mu = \mu_0$ of free space. Section 5.4 compares the exact inductance of a small, closed thin ring with this approximation and its associated Fig. 5.4 shows that a more proper demarkation is 0.10λ.

Consequently, if the above gap capacitances are incorrect and the approximate inductance is used for a ring larger in circumference than 0.10λ, the resulting resonance may err considerably. This error can be rectified somewhat by using the more accurate equations for the total L of the closed loop given by Eq. (4.14), even if only the first mode is included. But the closed loop also has capacitance, not in the zero mode, but in all the other modes, given by the same equation. This capacitance is in series with any added gap capacitance and hence reduces the overall capacitance of the ring, increasing the resonant frequency.

A better approach is to use the full analytical expressions for the thin loop, found in Sect. 4.2, even if the studied ring is thick. They would provide a closer approximation to the correct resonance until the analytical expressions for the thick ring are developed.

As an example, add a capacitor, X_q to a thin ring in series with the driving delta-function generator at $\phi = 0$ (see Sect. 7.5 for full details). Since it is located at the driving input, the input impedance becomes

$$\begin{aligned} Z &= Z_{CL} + Z_q = R_{CL} + iX_{CL} - iX_q \\ &= R + iX_L - iX_C - iX_q \\ &= R + i(X_L - (X_C + X_q)) \end{aligned} \tag{9.10}$$

where the subscript "CL" (for "closed loop") was dropped on the R, X_L, and X_C. The capacitive reactances of the closed loop and of the lumped capacitor add: $X_C + X_q$. The values for L and C of the closed loop are known using the equations noted just above. The value of C_q is known, since it is the value of the added capacitor. Using $\omega = k_b c/b$ and

$$L = \mu_0 b l_\mu, \quad C = \varepsilon_0 b l_\varepsilon \quad \text{and} \quad C_q = \varepsilon_0 b l_{\varepsilon q} \tag{9.11}$$

the associated reactances are:

$$\begin{aligned} X_L = b_x k_b l_\mu, \quad X_C = \frac{b_x}{k_b l_\varepsilon} \quad &\text{and} \quad X_q = \frac{b_x}{k_b l_{\varepsilon q}} \\ X_C + X_q = \frac{1}{\omega}\left(\frac{1}{C} + \frac{1}{C_q}\right) &= \frac{1}{\omega}\frac{1}{C'} = \frac{b_x}{k_b l'_\varepsilon}. \end{aligned} \tag{9.12}$$

The sub-wavelength resonance occurs where the reactance goes to zero: that is, where $X_L = (X_C + X_q)$. Solving this using (9.12), the resonance is:

$$k_{br} = \frac{1}{\sqrt{l_\mu l'_\varepsilon}} \tag{9.13}$$

The effect of the added capacitance on the resonances of the loop can be visualised, as in Fig. 9.7, which shows how the total reactance of an $\Omega = 10$, $2\pi b = 300\,\mu$m, gold loop changes when series capacitors are added. A small added capacitive reactance, X_q, introduces a resonance in the lower portion of the sub-wavelength region, indicated by the ring reactance crossing zero, which then blue-shifts as the reactance increases negatively.

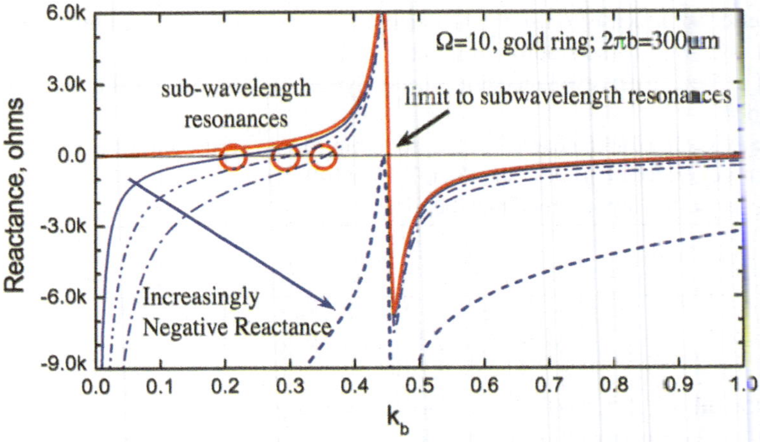

Fig. 9.7 Adding a capacitor in series with a loop adds negative reactance to the loop. This causes a sub-wavelength resonance to occur where the curve crosses zero. The resonances blue shift (move to the right) as the magnitude of the reactance increases until a limit is reached and the resonances disappear (dashed curve)

9.3.2 Testing the Models

We wish now to develop a method for determining whether the gap follows any of these models. To do so, we rely on the observation that a broadside illumination of a ring with a single gap generates a large resonance in the sub-wavelength region, just as a ring with a single capacitor driven by a delta function generator generates a large resonance in the same region (see Sect. 7.5.1). The gap can be varied, which varies the resonance (see Sect. 7.5.2).

We now make the assertion that the gap behaves as if it were a capacitor that generates the resonance. A difficulty arises because gapped loops are generally illuminated rather than driven by a generator. We therefore need to choose a form of illumination that generates the resonances when a gap is present. This would be that given by an S-polarised broadside illuminating wave, as shown in Fig. 9.2 or by a side illuminating wave as shown in Fig. 9.4 or Fig. 9.6. The effect of the broadside illumination of the first figure was discussed in Sect. 9.2, Orientation 1, Eq. (9.1). In Sect. 3.5, the convergence of the Storer solution was discussed and the Fourier coefficients of the illuminated field were calculated using a modified version of Eq. (3.84). The results are that this broadside wave yields the same current coefficients as a driven loop:

$$I_n = j\frac{2b}{\xi a_n}E_{Q,n} = \frac{V_0}{Z_n} \qquad (9.14)$$

Table 9.1 Gap values, g/a, leading to sub-wavelength resonances, k_{br}, next to capacitance values required to generate those resonances calculated from theory. These are then compared with the capacitance values calculated from the three models. $l_{\varepsilon g}$ is due to a lumped capacitor in a driven ring placed at $\phi = \pi$

$\Omega = 12$, PEC, $b/a = 64.2$						$\Omega = 8$, PEC and gold, $b/a = 8.7$					
g/a	k_{br}	$l_{\varepsilon g}$	$l_{\varepsilon,fp}$	$l_{\varepsilon,frng}$	$l_{\varepsilon,surf}$	g/a	k_{br}	$l_{\varepsilon g}$	$l_{\varepsilon,fp}$	$l_{\varepsilon,frng}$	$l_{\varepsilon,surf}$
0.10	0.377	0.59	0.489	0.540	0.784	0.10	N/A	N/A	3.62	4.00	5.34
0.15	0.397	0.40	0.326	0.377	0.609	0.15	0.300	2.36	2.41	2.79	4.04
0.20	0.409	0.32	0.245	0.297	0.520	0.20	0.322	1.83	1.81	2.19	3.38
0.25	0.417	0.26	0.196	0.249	0.465	0.25	0.338	1.43	1.45	1.84	2.97
0.30	0.422	0.23	0.163	0.217	0.427	0.30	0.351	1.15	1.21	1.60	2.70
0.40	0.430	0.17	0.122	0.177	0.379	0.40	0.370	0.80	0.90	1.31	2.34
0.50	0.435	0.14	0.098	0.155	0.349	0.50	0.383	0.60	0.72	1.14	2.12
0.75	0.442	0.09	0.065	0.126	0.308	0.75	0.405	0.29	0.48	0.93	1.81
1.00	0.446	0.07	0.049	0.113	0.286	1.00	0.420	0.10	0.36	0.84	1.65
2.00	0.456	0.01	0.024	0.105	0.256	2.00	N/A	N/A	0.18	0.77	1.43
Anti-resonance $k_b = 0.458$						Anti-resonance $k_b = 0.429$					

Therefore we ought to be able to place a capacitor in the loop anywhere and illuminate it and the gapped ring; the resonance should appear in each case and be identical, as long as the gap acts like a capacitor. The gap capacitance should be in series with the closed loop capacitance and therefore the capacitance of the ring becomes

$$C_T = \frac{C_{CL} C_g}{C_{CL} + C_g} \tag{9.15}$$

where T refers to the ring's total capacitance, C_{CL} refers to the closed ring's capacitance without a gap and C_g is the capacitance of the gap.

The examination will be done with PEC rings of two thicknesses, $\Omega = 12$ and 8 and with an $\Omega = 8$ gold ring with circumference $\lambda = 0.3$ mm (about 1 THz). A gap in a wire as thin as $\Omega = 12$ really makes no sense; most gaps are placed in wires as thick as $\Omega = 8$ and thicker, and even though $\Omega = 8$ is on the edge of accuracy for the thin-wire approximation, we shall assume that the theory is sufficiently accurate for this test. The gaps will vary in width between $g = 0.1a$ and $2.0a$; corresponding sub-wavelength resonances found by the numerical engines, in this case CST [10], will be noted. The goal is to calculate the capacitances needed to produce these resonances using the theory, given in Sect. 7.5.1 and to see which of the three models most accurately represents the values calculated by theory. As noted in Fig. 7.3, the capacitor may be placed anywhere in the periphery and the loop driven at $\phi = 0$. We will place it at $\phi = \pi$. The value is varied until the resulting resonance matches the numerical resonance.

Table 9.1 shows the resulting matched capacitors vs gap ratios and compares them with the values calculated using each of the models described in the previous section.

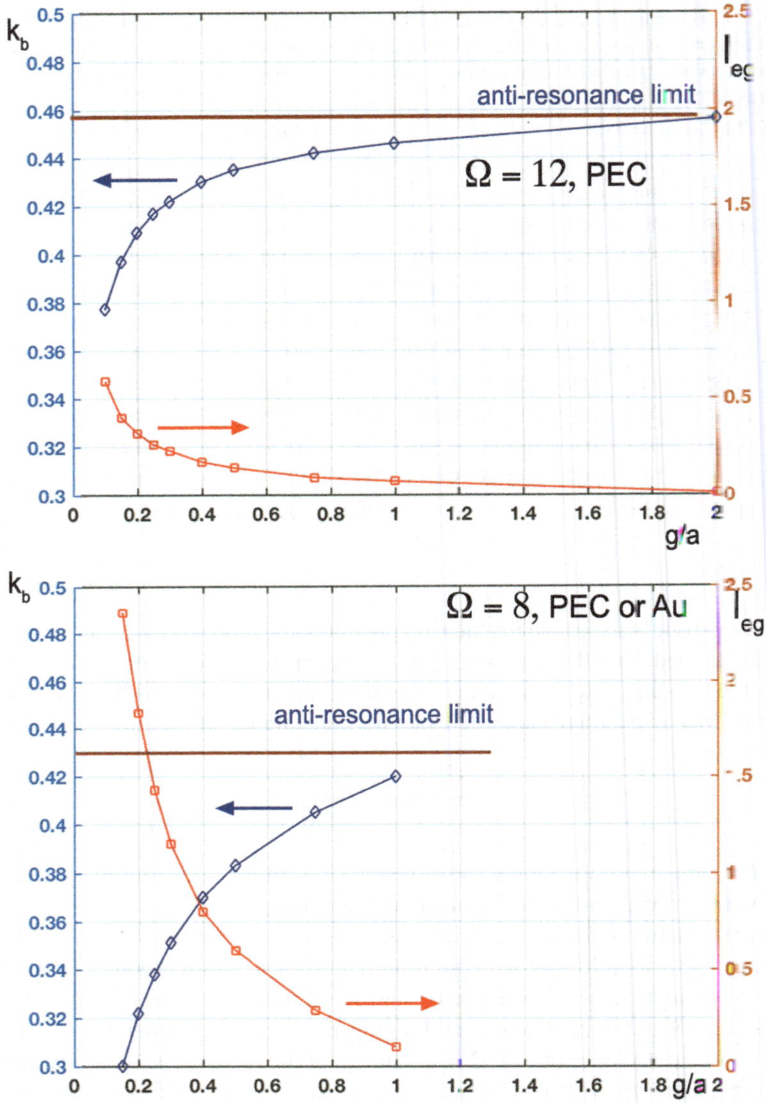

Fig. 9.8 The resonances due to the gaps correlated with the those of lumped capacitors placed at the gap, as calculated in Sect. 7.5.1. Lines guide the eye, suggesting exponential relationships

Figure 9.8 shows the resonances created by the various gaps and the corresponding calculated capacitors that produce those resonances. In the case of the thicker wire, it is readily apparent that gaps can produce resonances no lower than $k_b = 0.30$. Gaps

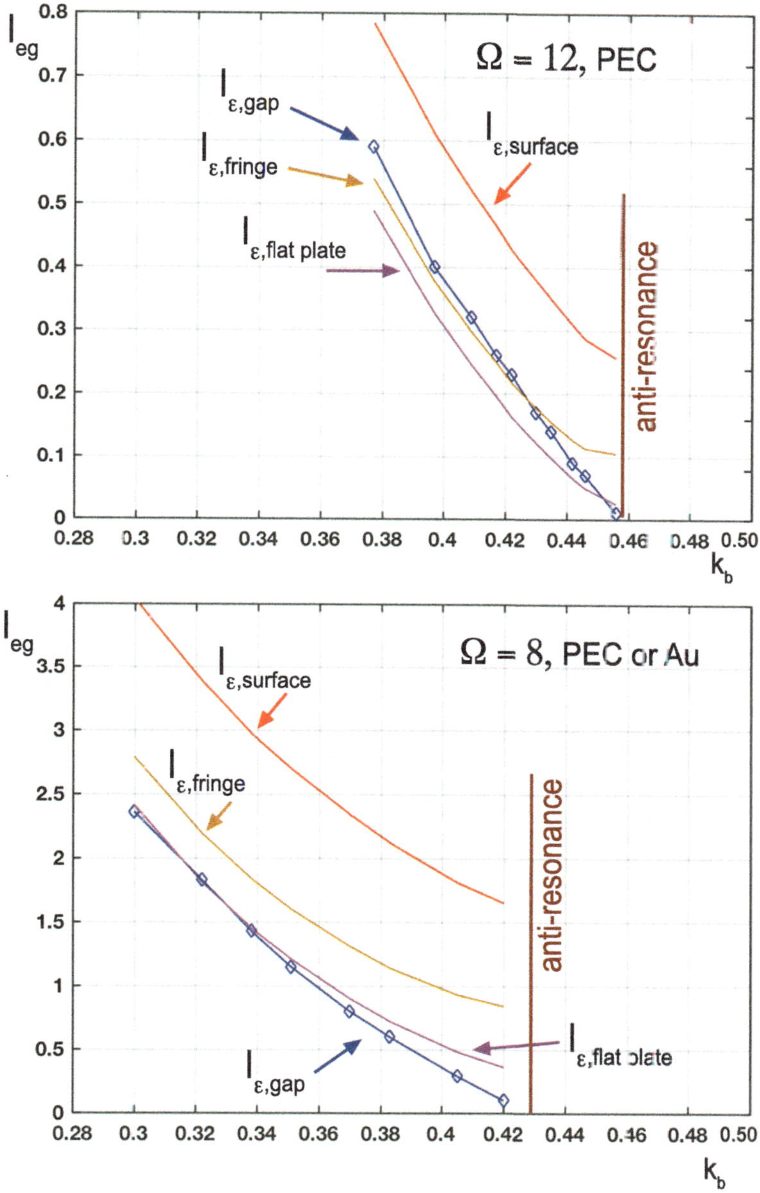

Fig. 9.9 A comparison of the capacitor models found in the literature with theoretically calculated capacitance. The model marked "fringe" performs best for thin loops, while the flat-plate model is adequate for thick loops

wider than $1.2a$ run into the anti-resonance limit. The boundaries for thinner wires are slightly different.

Figure 9.9 shows how each of the capacitor models compare with the theoretically calculated capacitors. For thick wires, the flat plate capacitance appears to be a completely adequate model except near the anti-resonance limit where it can be off by as much as 4 times. For thin wires, the fringe model becomes important. In neither case does the surface model reflect correct capacitance.

It does need to be noted that these results apply up to wavelengths on the order of 2 μm for thick gold wires.

Lines guide the eye, suggesting exponential relationships.

References

1. D.A. Powell, Phys. Rev. B **90**, 075108 (2014). https://doi.org/10.1103/PhysRevB.90.075108
2. N. Katsarakis, T. Koschny, M. Kafesaki, E.N. Economou, C.M. Soukoulis, Appl. Phys. Lett. **84**(15), 2943 (2004). https://doi.org/10.1063/1.1695439
3. C. Enkrich, M. Wegener, S. Linden, S. Burger, L. Zschiedrich, F. Schmidt, J.F. Zhou, T. Koschny, C. Soukoulis, Phys. Rev. Lett. **95**, 20 (2005). https://doi.org/10.1103/physrevlett.95.203901
4. C. Rockstuhl, F. Lederer, C. Etrich, T. Zentgraf, J. Kuhl, H. Giessen, Opt. Express 14, 8827 (2006)
5. C. Rockstuhl, T. Zentgraf, E. Pshenay-Severin, J. Petschulat, A. Chipouline, J. Kuhl, T. Pertsch, H. Giessen, F. Lederer, Opt. Express **15**(14), 8871 (2007). https://doi.org/10.1364/OE.15.008871
6. S. Tretyakov, Metamaterials **1**(1), 40 (2007). https://doi.org/10.1016/j.metmat.2007.02.004
7. D.J. Shelton, D.W. Peters, M.B. Sinclair, I. Brener, L.K. Warne, L.I. Basilio, K.R. Coffey, G.D. Boreman, Opt. Express **18**(2), 1085 (2010). https://doi.org/10.1364/OE.18.001085
8. O. Sydoruk, E. Tatartschuk, E. Shamonina, L. Solymar, J. Appl. Phys. **105**, 4 (2009)
9. V. Delgado, O. Sydoruk, E. Tatartschuk, R. Marques, M. Freire, L. Jelinek, Metamaterials **3**(2), 57 (2009). https://doi.org/10.1016/j.metmat.2009.03.001
10. C.S.T. AG. Microwave studio. Darmstadt, Germany, www.cst.com (2012)

Chapter 10
Coupled Loops and Rings

10.1 Coupled Rings as Meta-atoms

One of the most important applications of coupled loops occurs with Yagi-Uda type configurations at low frequencies and with double split-rings used in meta-materials at higher frequencies. Rings and loops under these conditions couple tightly when they are close together and "loosely", so to speak, when they are distant from each other. Indeed, once the loops are far enough away, any radiation from one of the loops reaches the other as a plane wave, and then the methods of Sects. 2.6.3 and 9.2 will apply.

The far-field radiation patterns from the single loop, with and without gaps or capacitors, are found in Sects. 5.2, 7.4 and 6.5. At the high GHz, THz, and optical frequencies losses of the material apply, and this circumstance is covered in Chap. 6. Near-field radiation patterns at low frequencies are given in Sect. 5.2.1, following [1] and expanded to high frequencies in Sect. 6.5 following [2]. These are used here, but we introduce some coupling complications that are attended to in [3].

Analytical expressions for closely coupled loops have been developed using a quasi-static approximation under special conditions. Integral-based equations also exist for identical coaxial loops with extensions to arbitrary arrays of loops (see the papers cited above for the literature review).

The contribution in these papers is a method of solution for the current in a target loop due to the currents in the source loop, stimulated by a delta-function generator. The result is exact analytical solutions for the mutual coupling between lossy rings of arbitrary radii, location and orientation. When the rings are close together, the resulting expressions for the current in the coupled ring involve complicated integrals over spherical Hankel functions of the second kind. The numerical integrations of these expressions require sophisticated routines to arrive at accurate solutions. Asymptotic approximations of several of the functions, however, yields tractable analytical expressions that lead to a more intuitive understanding of the physics involved in the coupling.

A. McKinley, *The Analytical Foundations of Loop Antennas and Nano-Scaled Rings*, Signals and Communication Technology,
https://doi.org/10.1007/978-981-13-5893-7_10

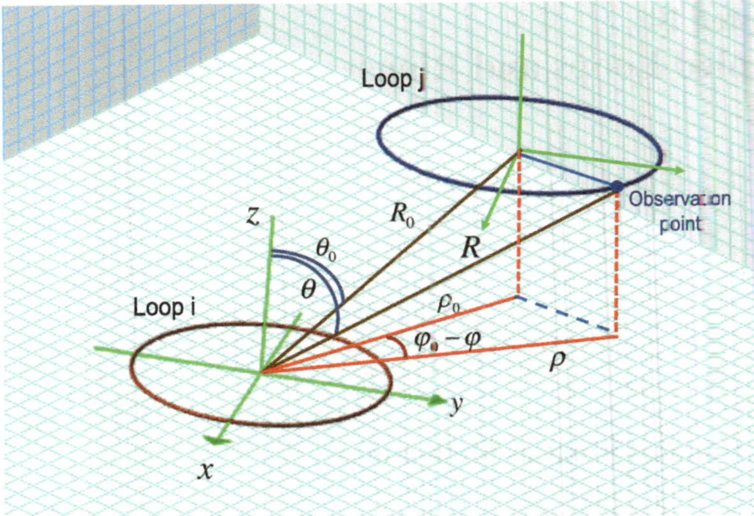

Fig. 10.1 The geometry of the observed point on loop j with respect to the source loop i

10.2 Near-Field Coupling of Thin Rings

Following [3], the source loop is designated by the subscript i and the target loop is designated by the subscript j. Both loops are "quasi-closed", meaning that in the source loop, an infinitesimal gap occurs at $\phi = 0$ (for the delta-function generator) and, in the target gap, an infinitesimal gap occurs at $\phi' = 0$ (for the terminals needed to measure the impedance of the loop). Moreover, neither loop has any loads. The analysis could be done with loads, were it necessary; however, the resulting equations, of course, would be that much more complicated. The coordinate system for one loop with respect to the other loop is shown in Fig. 10.1.

The authors assume that the current distribution in the source loop is given by Fourier series with symmetric coefficients, as we have done in Chap. 3. However, they assimilate the 2 into Z_n.

$$I(\phi) = \sum_{-\infty}^{\infty} I_n e^{jn\phi} = \sum_{0}^{\infty} \frac{V_0}{Z_n} e^{jn\phi} = \sum_{0}^{\infty} Y_n e^{jn\phi} \tag{10.1}$$

Z_n has the surface impedance with loss inserted (see Chap. 6). The radiated wave generates similar coefficients in the target loop; namely,

$$I_n' = Y_n' V_0 = \frac{V_0}{Z_n'} \tag{10.2}$$

The tangential component of the illuminating $\vec{E}$ field stimulates a voltage, V_{ij}, induced in the target loop by the source loop:

$$V_{ij} = b_j \int_0^{2\pi} \vec{E}_i \cdot \hat{t} d\phi' \tag{10.3}$$

where $\hat{t}$ is the tangential unit vector and ϕ' is the angle from the $\hat{x}'$ axis. Since $\{\hat{x}, \hat{y}, \hat{z}\}$ and $\{\hat{x}', \hat{y}', \hat{z}'\}$ are the same,

$$\hat{t} = -\texttt{sin}\phi'\hat{x} + \texttt{cos}\phi'\hat{y} \tag{10.4}$$

The Fourier coefficients of the current in the target loop are therefore:

$$I_j(\phi') = V_{ij}\left[\sum_{p=0}^{\infty} Y'_{pj}\texttt{cos}(p(\phi' - \phi))\right] \tag{10.5}$$

where the admittance coefficients are summed over the modes, p. One can write the self-admittance of the source loop:

$$Y_{ii} = \sum_{p=0}^{\infty} Y'_{p,i} \tag{10.6}$$

while the mutual admittance of the two loops is found using

$$Y_{ji} = \frac{I_j(\phi = 0)}{V_i} \tag{10.7}$$

where V_i is the voltage of the generator on the source loop. See [4–6] for more on self and mutual inductance in arrays of coaxially stacked loops.

The current at the input terminals of loop j is given by

$$I_j = \sum_{j=1}^{N} Y_{ji} V_i \tag{10.8}$$

The voltage stimulating loop j due to radiation from loop i requires the computation of (10.3), where E_i is the radiated field. The components of the radiated near-field are given in [1, see Eqs. (46)–(48)]. The resulting currents involve integrals of spherical Bessel functions and are not fully analytical. These are the "'pseudo-analytical" solutions for the current.

To make them more tractable, a number of approximations are employed, allowing a more analytical solution. The authors give the solution for the current for the general case [3, see Eq. (24)], where the rings are placed anywhere and two more solutions where the rings are co-planar [3, see Eq. (25)], and where stacked [3, see Eq. (27)].

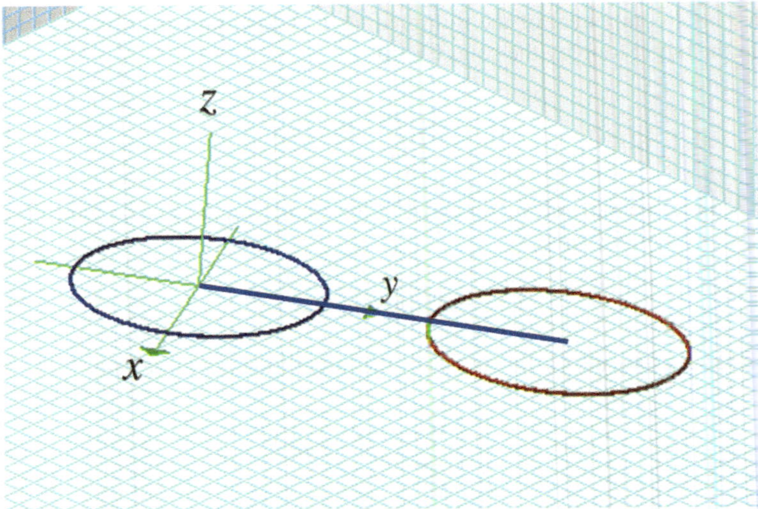

Fig. 10.2 Two coupled rings in the same plane

The authors checked calculations of the pseudo-analytical and of the analytical solutions for the current in a number of two loop cases against simulations in FEKO (a full-wave integral equation solver). PEC loops were simulated and experimentally tested at 2 GHz. Gold loops were simulated at 600 nm and 3000 nm. Agreement wth calculations of the current was strong in all case. The authors did not pursue calculations of the radiation patterns.

Results in the planar case: Figure 10.2 shows two identical thin ($\Omega = 12$) PEC loops laid out in the $\{xy\}$ plane (in other words at $\theta = 90°$, $\phi = 90°$) with centres separated along the $\hat{y}$ axis by $14b$, $7b$ and $4b$ where b is the loop radius. A delta-generator drives the source loop; two resonances in the current are generated in the target loop, one just above $k_b = 1.0$, the second just above $k_b = 2.0$. Results differ depending upon how far apart the loops are located and whether the radial component of the radiation field is included in the calculation.

When all three components of the radiation field are applied to the target loop, the two resonances are distinct with both peaks becoming more and more distinct as the loops move closer together. Nothing unexpected there.

1. What is unexpected, however, is that the $k_b = 2$ peak is more distinct than the $k_b = 1.0$ resonance when far apart. As the loops come closer to each other the peak at $k_b = 1.0$ becomes more distinct faster than does the $k_b = 2.0$ peak until it surpasses the latter in height, somewhere between the $7b$ and the $4b$ positions.
2. Also unexpected is that the $k_b = 1$ peak completely disappears in the calculations if the radial component, E_r, is not included in the formulation. It is of course there in the simulations and in the results. In other words, that resonance appears to be due solely to the radial component of the field.

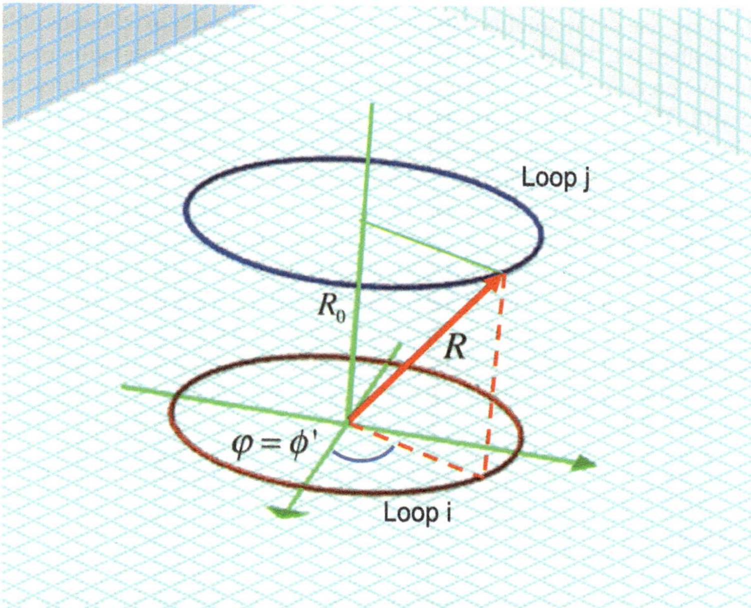

Fig. 10.3 The geometry of two coupled rings situated along the same axis

3. The gold nano-ring simulations were compared with calculations in which the gold loss was included. As expected from discussions in Chap. 6, the main resonance red-shifts and saturates due to loss. That resonance occurred at $k_b \approx 0.09$ for the 600 nm ring. The 3000 nm ring was not simulated in the planar case.

Results of the stacked Yagi-Uda case: In this case (see Fig. 10.3), the PEC loops show both resonances and both grow stronger as the loops approach each other. The distances tested were the same as before: $14b$, $7b$, and $4b$ centre to centre. The first resonance is quite a bit stronger than the second, on the order of 4 to 8 times, depending on the separation distance.

10.3 Optimization Procedure for the Design of Coupled Nano-Loop Antennas

Nagar et al. [7] present a formal inverse design procedure for Yagi-Uda type loop antennas based on the optimisation of multiple objectives. One of the examples given was a multiple director nano-loop Yagi-Uda. Optimisations, as usually implemented, rely on single parameter optimisations; multi-objective optimisations are much trickier and complicated. These authors use an evolutionary, multiple-objective algorithm (MOEA) that generates a set of "Pareto-optimal" solutions. "Pareto-optimality" implies that every one of the solutions in the set is optimal to those in other possible sets. This set is called the "Pareto front". The algorithm evolves a population of designs over time by "breeding" successful population members and removing those that are unsuccessful. The particular algorithm they use is one called "BORG" [8].

The analytical theory suggests that a peak in the broadside directivity appears at $k_b = 1.16$ (340 THz) for a gold nano-loop of circumference 1.0 μm. Using this frequency, the authors sought to maximise the directivity and the height of the array. The results produced an array of length 0.815λ with a reflector, driver and four directors and directivity maximum of 14.1 dB. The input parameters were loop and wire radii and distance between the loops. A delta-function voltage generator placed at $\phi = 0$ drives the array. The runs finish quite quickly due to the series representations of the integrals involved. A second optimisation sought to maximise gain and minimise height. The result was a director and a slightly larger reflector, total length 0.165λ with a gain of 7.2 dB.

References

1. D. Werner, IEEE Trans. Antennas Propag. **44**(2), 157 (1996). https://doi.org/10.1109/8.481642
2. M.F. Pantoja, J. Nagar, B. Lu, T. Yue, D.H. Werner, in 2016 10th European Conference on Antennas and Propagation (EuCAP) (2016), pp. 1–4. https://doi.org/10.1109/EuCAP.2016.7481565
3. J. Nagar, M. Pantajo, IEEE Trans. Antennas Propag. **65**(12), 6889–6903 (2017). https://doi.org/10.1109/TAP.2017.2754411
4. K. Iizuka, R. King, J. Harrison, IEEE Trans. Antennas Propag. **14**(4), 440 (1966). https://doi.org/10.1109/TAP.1966.1138711
5. S. Ito, N. Inagaki, T. Sekiguchi, IEEE Trans. Antennas Propag. **19**(4), 469 (1971). https://doi.org/10.1109/TAP.1971.1139954
6. A. Shoamanesh, L. Shafai, IEEE Trans. Antennas Propag. **27**(5), 711 (1979). https://doi.org/10.1109/TAP.1979.1142168
7. J. Nagar, S.D. Campbell, Q. Ren, J.A. Easum, R.P. Jenkins, D.H. Werner, IEEE J. Multiscale Multiphysics Comput. Tech. **2**, 147 (2017). https://doi.org/10.1109/JMMCT.2017.2734810
8. D. Hadka, P. Reed, Evol. Comput. **21**(2), 231 (2013)

Appendix A
Bessel Functions

Abstract The Bessel functions are used extensively throughout this volume. Their definitions and some identities are provided here for convenience and are referenced from text.

A.1 Bessel of the First Kind

This can be defined in several different ways. Abramovitz and Stegun [1, 9.1.8] give

$$\begin{aligned}\mathbf{J}_n(z) &= \frac{1}{\pi}\int_0^{\pi} \cos(n\theta - z\sin\theta)d\theta \\ &= \frac{j^{-n}}{\pi}\int_0^{\pi} \mathrm{e}^{jz\cos\theta}\cos n\theta d\theta \end{aligned} \tag{A.1}$$

Here n refers to an integer and z to a complex number. MathWorld [2] gives, in addition

$$\mathbf{J}_n(z) = \frac{j^{-n}}{2\pi}\int_0^{2\pi} \mathrm{e}^{j(n\theta+z\cos\theta)}d\theta = \frac{j^{-n}}{2\pi}\int_{-\pi}^{\pi} \mathrm{e}^{j(n\theta+z\cos\theta)}d\theta \tag{A.2}$$

The rightmost form and the following forms may be checked for validity using a mathematical tool, such as MatLab [3] or The Graphing Calculator [4].

$$\mathbf{J}_n(z) = \frac{j^{-n}}{2\pi}\int_0^{2\pi} \mathrm{e}^{-j(n\theta-z\cos\theta)}d\theta \tag{A.3}$$

$$\mathbf{J}_n(z) = \frac{j^{n}}{2\pi}\int_0^{2\pi} \mathrm{e}^{j(n\theta-z\cos\theta)}d\theta \tag{A.4}$$

A. McKinley, *The Analytical Foundations of Loop Antennas and Nano-Scaled Rings*, Signals and Communication Technology,
https://doi.org/10.1007/978-981-13-5893-7

$$\mathbf{J}_n(z) = \frac{j^n}{2\pi}\int_0^{2\pi} \mathrm{e}^{-j(n\theta+z\cos\theta)}d\theta \tag{A.5}$$

$$= \frac{j^n}{2\pi}\int_{-\pi}^{\pi} \mathrm{e}^{-jz\cos\theta}\cos n\theta d\theta \tag{A.6}$$

$$= \frac{j^{-n}}{2\pi}\int_{-\pi}^{\pi} \mathrm{e}^{jz\cos\theta}\cos n\theta d\theta \tag{A.7}$$

In addition,

$$\frac{2n}{z}\mathbf{J}_n(z) = \mathbf{J}_{n+1}(z) + \mathbf{J}_{n-1}(z) \quad \text{and} \quad \mathbf{J}_n'(z) = \frac{n}{z}\mathbf{J}_n(z) - \mathbf{J}_{n+1}(z) \tag{A.8}$$

A.2 Modified Bessel of the First and Second Kind

Abramovitz and Stegun [1, 9.6.16] give
First Kind

$$\mathbf{I}_0(z) = \frac{1}{\pi}\int_0^{\pi} e^{\pm z\cos\theta}d\theta = \frac{1}{\pi}\int_0^{\pi} \cosh(z\cos\theta)d\theta \tag{A.9}$$

$$\mathbf{I}_n(z) = \frac{1}{\pi}\int_0^{\pi} e^{z\cos\theta}\cos(n\theta)d\theta \tag{A.10}$$

Second Kind

$$\mathbf{K}_0(z) = \int_0^{\infty} \mathrm{e}^{\pm jz\sinh v}dv = \int_0^{\infty} \cos(z\sinh v)dv \quad z > 0 \tag{A.11}$$

A.3 Lommel-Weber Function

Abramovitz and Stegun [1, 12.3.3] give

$$\mathbf{\Omega}_n(z) = \frac{1}{\pi}\int_0^{\pi} \sin(n\theta - z\sin\theta)d\theta \tag{A.12}$$

A.4 Combinations

$$\frac{1}{\pi}\int_0^{\pi} e^{\pm j(n\theta - z\mathtt{sin}\theta)})d\theta = \frac{1}{\pi}\int_0^{\pi} \mathtt{cos}(n\theta - z\mathtt{sin}\theta) \pm j\frac{1}{\pi}\int_0^{\pi} \mathtt{sin}(n\theta - z\mathtt{sin}\theta)$$
$$= \mathbf{J}_n(z) \pm j\boldsymbol{\Omega}_n(z) \tag{A.13}$$

References

1. I. Stegun, M. Abramowitz, *Handbook of Mathematical Functions with Formulas, Graphs and Mathematical Tables*. Applied Mathematics Series 55 (US. Governnent Printing Office, WDC, 1964)
2. E.W. Weisstein, From MathWorld–A Wolfram Web Resource (2017), http://mathworld.wolfram.com/BesselFunctionoftheFirstKind.html
3. Mathworks, Natick, MA 01760-2098, *MATLAB R2016a* (2016)
4. P. Tech, *The Graphing Calculator*. 2710 Martin Luther King Jr., Way, Berkeley, CA 94703 (2018)

The manufacturer's authorised representative in the EU is Springer Nature Customer Service Centre GmbH, Europaplatz 3, 69115 Heidelberg, Germany. If you have any concerns regarding our products, please contact ProductSafety@springernature.com

Printed and bound by CPI Group (UK) Ltd, Croydon, CR0 4YY

23/07/2026

02175034-0004

The Beverly Hills Housewife

ΟΔΟΣ

JAMES CAHILL

The Beverly Hills Housewife

Hockney's Californian Muse
and the World Beyond the Pool

With 68 illustrations, 33 in colour

T&H

FRONTISPIECE: Betty Freeman, Corfu, August 1981, photographer unknown

First published in the United Kingdom in 2026 by
Thames & Hudson Ltd, 6–24 Britannia Street, London WC1X 9JD

First published in the United States of America in 2026 by
Thames & Hudson Inc., 500 Fifth Avenue, New York, New York 10110

Edited and designed by Andrew Brown

EU Authorized Representative: Interart S.A.R.L.
19 rue Charles Auray, 93500 Pantin, Paris, France
productsafety@thameshudson.co.uk
interart.fr

A CIP catalogue record for this book is available from the British Library

Library of Congress Control Number 2026943731

ISBN 978-0-500-02881-0
01

Printed and bound in Spain by Estellaprint

MIX
Paper | Supporting responsible forestry
FSC® C009279

Contents

1

Flight to Los Angeles

'... and now it was the hour when in all the houses all around the pretty women were putting on perfume and enameled bracelets and kissing the pretty children goodnight, the hour of apparent grace and promised music ...'

JOAN DIDION, *PLAY IT AS IT LAYS*, 1970[1]

In the garden

The woman in pink looks into the middle distance, beyond the bounds of the picture. Is she aware of being watched? Does she know that she is someone's subject? There is something arresting about the way she holds herself, there on the garden terrace – solemn yet serene at the centre of a self-contained world. It is a pristine sort of place, with its minimalist portico and parade of plate-glass windows, all straight lines and smooth surfaces. The light is uniform. The near absence of shadows suggests noontime, and yet the woman's elegant pink dress points to an hour later in the afternoon. Behind her, beyond the glass, the interiors of the house are a cluster of rectangles: a strip of orange carpet, the black oblong of an inner room, the bright aperture of a window.

She is outside, and yet the house itself seems to have turned inside out. The garden has become a room. A recliner covered in zebra skin lies to one side of the terrace. The woman might be about to lie down, although her poised bearing makes it seem unlikely: she is ready for her close-up, not a nap. On a wall that juts from the rear facade of the house, dividing the portico, is the stuffed head of an antelope. It is the kind of thing one would expect to find in a hunting lodge, not a hard-edged modernist villa.

The horned creature, docile in death, faces the woman – their two heads are practically aligned. She ignores it.

The objects around her seem like proxy bodies: the recliner, the pygmy palm tree that rises from the grass in the foreground, the stack of boulders on the right of the picture. Each thing is a silent partner of the woman in pink, herself apparently static and silent, as though lost in thought.

She is the focus of the picture, its gravitational centre: a pillar of pink against the grey of the house and the patio. But she is also, as far as the picture is concerned, nameless. A pink blank. The title, *Beverly Hills Housewife*, assigns her a place – imparts an identity – but in doing so, it displaces her, burying her inside her role. Where her gaze is directed, or what she is thinking as she stands there in the sunlight, whether she has just stepped outside or is about to turn and go indoors, are impossible to tell. Her impermeability instils the ordinary scene, a lifestyle-magazine vision of bourgeois affluence, with tension.

The title was her own suggestion. Not *A Portrait of Betty Freeman* but *Beverly Hills Housewife*. There is a joke in this, a sense of 'Now I must play my part.' Seeming to explain the picture, the title only underlines the inexplicability of the subject. Perhaps she did not want to be discovered – only seen.

Beverly Hills Housewife was the first portrait of an identifiable subject that David Hockney began in Los Angeles. Until 1966, his Californian subjects had been not only nameless but unreal: beautiful boys translated from the pages of magazines, a green-frocked art collector he had dreamt up. And yet, more than any portrait he made, *Beverly Hills Housewife* is concerned with the concealments of portraiture, its feints and deflections. The sealed planes of the house, the bars of light on the windows, restate those deflections in particularized form. Each slice of bouncing sunlight seems to repeat and amplify the central ambiguity.

The opaqueness of the picture is interrupted, even so, by pockets of transparency. The reflections themselves might be clues as to what we cannot see – rebounding perhaps from the surface of a pool somewhere in the garden. The opened glass door reveals the house's interior.

David Hockney, *Beverly Hills Housewife*, 1966–7
Acrylic on two canvases, 182.9 × 365.8 cm (72 × 144 in.)

That front window, visible in the background, discloses a fragment of blue sky and the slanting fronds of a shrub.

The woman in pink, too, is not simply an enigma. Not a cipher but a proposition: 'Look at me now and here I am.' Standing before the door of her home, she is poised between 'inside' and 'outside'. Her half-turned position might be a turning away or a turning towards. Her calm countenance, as it catches the sunlight, seems like an invitation.

Now and then

Beverly Hills Housewife, which Hockney painted between the summers of 1966 and 1967, is a picture with multiple subjects. Through the specific vision of Betty Freeman in her garden, the painting provides a window onto other realities: Hockney's discovery of Los Angeles, Freeman's past and future, the character of the city that they both loved (the light, the architecture, the sense of possibility). Then there are the numerous artistic forebears that glimmer within the prosaic 'present tense' of the Beverly Hills woman on her garden terrace. The painting is not simply a record of a moment in 1966, or of the era in which that moment occurred, but an element within a longer historical narrative.

The *now* of Hockney's encounter with Freeman conceals within itself the *then* of other such encounters, at diverse moments in the past.

The composition takes its viewer on a ranging, circular journey between its details – the antelope, the dark interior, the zebra-skinned recliner, the sunlight glancing off the glass – while keeping Freeman in central view. In an equivalent way, the account that follows moves between numerous subjects, while returning always to Betty: her life, her persona, her enduring bond with Hockney.

For Hockney, Freeman was an object of fascination, but ever to be contemplated from a slight remove. This book retraces the facts of their lives, but more than anything, it employs the painting as its primary source: a way of understanding who Betty Freeman was, who Hockney was at the time they met, who they were to each other. The book is a portrait, like Hockney's, that ranges in time and place, and one that stands back from Freeman to contemplate both the real woman and her timeless image.

'If Betty were alive, I don't know that she would love this project', says Stephanie Barron, senior curator at the Los Angeles County Museum of Art. 'Betty didn't like to be the centre of attention. She liked to be behind the camera instead of in front of the camera.'[2] Hockney's painting, which originated in a set of Polaroid photographs, was a rare instance of Freeman being cast in the spotlight. And yet her retiring instincts are palpable in the picture, not only in her fugitive gaze into space, but in the very title of the work. It was her idea to cloak it in anonymity. In doing so, she allowed it to become a representation of Hockney's vision of Los Angeles – less a portrait of her than a representation of a moment in his life. The mystery of the painting – its inscrutable subject, its pristine lucid space – turns it into a mirror. It is not simply a picture of a woman at home, but a covert portrait of the artist as a young man.

Premonitions

Long before he went to Los Angeles, Hockney harboured a powerful sense of the city. As a boy in wartime Bradford, he would go with his father and siblings to see Hollywood movies at picture houses such as the New Victoria – then the largest auditorium in Britain – as well as 'fleapit' cinemas that had nothing more than a suspended sheet for a screen. He was struck by the look of Los Angeles in those films, as much as by their stories, in particular by the hard-edged shadows and sharp sunlight in movies such as *Big Business* (1929), in which Laurel and Hardy play a pair of hapless Christmas tree salesmen. He would later say that he had grown up in both Bradford and Hollywood.

At the Royal College of Art in London – where he studied between 1959 and 1962 – a fellow student, Mark Berger, introduced him to Californian magazines such as *Physique Pictorial* and the *Young Physique*. Berger used to plaster pages from these titles over the walls of his painting booth. The photographs featured muscular youths in half-serious mythological guises (a near-nude Roman soldier, a master with a chain gang of slaves), or in more 'everyday' scenarios: climbing out of swimming pools, frolicking in showers.

Gay erotica veiled as body-building photography this genre was the invention of a Los Angeles-based photographer and publisher called Bob Mizer, who produced more than a million negatives in his lifetime. His magazine, *Physique Pictorial*, which had originally spun out of his agency for young male actors and models (the grandiosely named Athletic Model Guild, founded in 1945) became the epitome of the 'beefcake' aesthetic of the mid-twentieth century. Mizer's models were a diverse group: gay, straight, body-builders from Muscle Beach, hustling down-and-outs, college students, and returnees from the war. In the 1980s, he made a reported fortune from his 'session videos'. He once wrote to his mother: 'My ambition is everything.'[3]

Through the imagery of Mizer and his acolytes, California held out a promise of burnished men, cloudless skies, and glossy American brands. Hockney later told a French journalist: 'I was drawn towards California, which I didn't know [...] because I sensed the place would excite me. No doubt it had a lot to do with sex.'[4] 'No wonder Hockney was attracted to California and its beefcake magazines', the novelist Edmund White has remarked. 'In fact, the beach communities in California after the Second World War were the nurseries of modern homosexuality, and it is no accident that Hockney imagined the place before moving there.'[5]

Young Physique, October 1961, front cover

That imagining began to reveal itself in his work at the Royal College of Art, notably *Life Painting for a Diploma* (1962), which borrows the spectre of a hyper-muscular man from the cover of the October 1961 issue of the *Young Physique*. (In the original photograph, the model stands in a rock pool, hand on hip, grinning into the sunlight and wearing only a G-string 'posing strap' – the mid-twentieth-century answer to a fig leaf.) The painting was made to satisfy, but equally to ridicule, the RCA's demand that students complete a certain number of paintings from life. Technically, it is a painting of a photograph, although in a nod to the long tradition of drawing from anatomical specimens, Hockney also pasted a graphite study of a skeleton to the surface of the canvas: a memento mori to counterbalance the flexing everyman.

The following year, Hockney's premonition of Los Angeles was compounded by John Rechy's novel *City of Night* (1963), a lurid, lyrical evocation of the city's gay underworld, related from the perspective of a lonely hustler who travels from LA to New York to New Orleans:

> Each morning the pale sun rose in the imitation-blue sky of Los Angeles, and the endless resurrection of each new day began. Like the palm trees that lined the streets of the city, the world seemed to be shrugging indifferently.
>
> For me then there followed a period of untrammeled anarchy as I felt my life stretching toward some kind of symbolic night, as the number of people I went with multiplied daily. With those many people – only in those moments when I was desired – the moments before we became strangers again after the intimacy – I felt an electric happiness, as if the relentless flow of life had stopped, poised on the very pinpoint of youth; and for those moments, youth was suspended unmoving.[6]

Rechy's story provided a counterpoint to the utopian conception that Hockney had so far nurtured, shadowing forth the city's abject side and chiming with the journalist Carey McWilliams's cynical maxim that 'Los Angeles is the kind of place where perversion is perverted and prostitution is prostituted.'[7] That 'other side' was itself a source of dark seduction.

Domestic Scene, Los Angeles (1963), painted before Hockney left London, suggests a concentrated vision of his mind's eye. Two young men are shown in a make-believe interior. One is nude except for white socks and an apron-style loincloth; the other is naked and engulfed (haloed, almost) by the jet from a shower head. The first man touches the other's back – a sedate, strangely loaded gesture. The room around them is a mere expanse of unprimed canvas, populated only by a patterned armchair – an actual chair from Hockney's London flat – and a pot plant on a shelf. Hockney later remarked, in reference to the source photograph: 'There's a boy with a little apron tied around his waist scrubbing the back of another boy in a rather dingy American room; I thought that's what a domestic scene must be like there.'[8]

Naked city

America was not entirely unknown to him. He had first gone to New York in July 1961, while still a student, and had fallen in love with the twenty-four-hour city. He chronicled his time there in a sequence of etchings,

A Rake's Progress (1961–3), casting his experiences in the lineaments of William Hogarth's satirical paintings and etchings of the 1730s – the story of a young man's decline and fall. A photograph from that first trip shows Hockney, his hair dark and cropped, smoking a cigar on the beach as though trying on a mask of American machismo. He was staying with Mark Berger's family in Long Beach. As he and Berger watched television one evening, an advert for Lady Clairol hair dye flashed up: 'Blondes have more fun. Doors open for a blonde.' The random broadcast proved to be life-defining. Hockney announced that he would dye his hair – and did, to the consternation of Berger's father. A plate from *A Rake's Progress* commemorates the transformation: Hockney's naked upper body appears beneath a bottle of Lady Clairol. He gazes quizzically through an open door, into the future. Below the picture, a legend reads: 'The start of the spending spree and the door opening for a blonde.'

Hockney returned to New York in the spring of 1963. While exploring the city with his 'sexy friend' Jeff Goodman, a young American whom he had got to know the previous year, he met Andy Warhol, then at the height of his early celebrity.[9] They came across Warhol by a stroke of luck at the Stable Gallery, a cutting-edge space on East 74th Street where the artist had staged his first exhibition of Pop works the previous year. As Warhol related in his memoir *POPism*, the gallery's original space, on West 58th Street, 'had been an actual stable where rich people kept their horses, and in the spring when the wetness was in the air, you could still smell the horse piss, because that's a smell that never goes away'.[10]

Warhol invited Hockney and Goodman to his town house on Lexington Avenue that evening. Here, they met the actor Dennis Hopper, along with an ambitious young curator at the Metropolitan Museum of Art, Henry Geldzahler, one of Warhol's foremost proponents at that time. Soon afterwards, Hockney, Goodman, Warhol, and Geldzahler made a trip to the Harlem set of the TV police procedural *Naked City*, in which Hopper was starring.[11] A black-and-white photograph taken by Hopper shows the other four huddled together in moody fraternity: it might almost be a still from the hard-boiled drama. Warhol lurks behind, masked by sunglasses. The others stand in a line, their dark ties chiming. Hockney frowns behind his heavy-framed glasses; Goodman

Dennis Hopper, *Andy Warhol, Henry Geldzahler, David Hockney, and Jeff Goodman*, 1963

exhales a cloud of smoke; Geldzahler chomps on a cigar, the collar of his anorak tugged upwards.

Two years Hockney's senior, Geldzahler was free-thinking, sardonic, vastly knowledgeable, and exuberantly gay. In Hopper's snapshot, he eyes the camera with an air of impudent challenge. The following year, Warhol would shoot a film of Geldzahler smoking a cigar on a couch in Warhol's Factory studio over a ninety-seven-minute duration. In the course of the film, the urbane curator becomes increasingly fidgety – visibly self-conscious. 'Andy must periodically have felt the lash of Henry's educated and sarcastic tongue,' in the view of Warhol's biographer, Wayne Koestenbaum, 'and so this portrait becomes a roundabout way of paying Henry back, and of retaliating against every merciless eminence who thought Warhol a poseur, a hick, or a freak – even as this film is also his generous attempt to make Henry a beauty.'[12]

Hockney's chance meeting with Warhol at the Stable Gallery marked the beginning of a long acquaintance. Years afterwards, Warhol would remark: 'David's cute, he really is magic.'[13] But it was Geldzahler, of all the people he met in New York, who was to play the largest role in Hockney's life. He would become one of the artist's closest friends, as well as an intellectual mentor – one of the few people whose criticisms mattered.

Flight to Los Angeles

Hockney was back in New York in December 1963, seeing in the New Year in Times Square. This time, however, Manhattan was only a stop off. In early January, he flew to Los Angeles, acting on a deep-seated impulse. As he stated in an interview four years later:

> Somehow I instinctively knew I was going to like it. And as I flew over San Bernardino and looked down – and saw the swimming pools and the houses and everything and the sun, I was more thrilled than I've ever been arriving at any other city, including New York.[14]

Before leaving New York, Hockney had been put in touch by the art dealer Charles Alan with a sculptor living in Los Angeles, Oliver Andrews. It was Andrews who drove him, when he first arrived, to Santa Monica in a pickup truck: 'David immediately sent a telegram back to Charles Alan, telling [him] that he had safely arrived, and said, "Venice California more beautiful than Venice Italy".'[15] Hockney did not know anyone in Los Angeles, and had little comprehension of the scale or layout of the city. On his first night, he stayed in a motel at the base of Santa Monica Canyon. Walking onto the nearby beach, he saw some distant lights, which he assumed to be the city centre. He walked for two miles, only to arrive at a gas station.

The next day he bought a bicycle. He had noticed on a map that Wilshire Boulevard ran all the way from Santa Monica to Pershing Square in downtown Los Angeles, the place that Rechy had depicted as a crucible of gay nightlife. And so he set about cycling there, without realizing that it was a sixteen-mile distance. Arriving in the middle of the evening, he found only a deserted city park. He drank a glass of beer and cycled back.[16]

Andrews was alarmed to hear about the quixotic bike ride, and suggested that he teach Hockney how to drive. Within a week, Hockney had purchased a Ford Falcon and passed his driving test, although venturing solo onto American roads proved terrifying. 'He immediately got on the freeway and drove all the way to Las Vegas before he figured out how to turn around', according to Andrews. 'So he spent a couple of days in Las Vegas until he had gambled away all his money, and then he got in his car and drove back to Los Angeles.' Work was never far from his mind, however: Hockney's sybaritic facade belied an iron discipline, as Andrews observed: 'He appears to take life very easily and to lead a very full social life, but actually David works very hard.'[17] Soon after the trip to Vegas, Hockney rented an apartment in Santa Monica and found a studio in Venice, where the coastline boardwalk with its roller-skaters suggested a bare-fleshed version of Portobello Road.

Already, the attractions of the real city were supplanting his long-harboured fantasies. The otherness of America, the sense of being a world away from home, had an irresistible allure. But Los Angeles was more than simply an escape from the smallness and conservatism of England. It felt innately new – a city where everybody came from somewhere else – and almost boundless: the vast grid of streets and highways giving way only to mountains or ocean, or fading into the ether of the desert. Some of the city's freeways were still being built. In his first week, Hockney spotted an unfinished ramp soaring into the air, which inspired the thought (later a personal maxim) that the city needed a Piranesi.

He loved the light, the architecture, the sense of space, and the sense of possibility – not least the possibility of greater sexual freedom. West Hollywood boasted a large gay bar, the Red Raven on Melrose Avenue, that was unlike anything in New York. Here at last was a real-life counterpart to the underworld he had glimpsed in Rechy's novel:

> In Los Angeles, night comes like a blessing, even after the warmest afternoons. Soon, long shadows will protect the exiles, shelter them soothingly before the concealing night. And as it becomes later and the loneliness and the determination become hungrier, the frenziedness will increase.[18]

'The only time I was promiscuous was when I first went to live in Los Angeles', Hockney wrote in 1976. 'I've never been promiscuous since. But I used to go to bars in Los Angeles and pick up somebody.'[19] There was also – always – the lure of the beach, with its pageant of sculpted bodies.

AMG, not MGM

Two months after Hockney had arrived in Los Angeles, John Kasmin showed up. Hockney's London dealer was himself on the rise, having opened his eponymous gallery on Bond Street the previous spring. A high, skylit room with a Pirelli rubber floor, accessed via a dark tunnel entrance, Kasmin's was like no other gallery at the time. Its inaugural exhibition of 'target' paintings by Kenneth Noland in April 1963 had set the tone of the programme, which leant heavily towards American abstraction. As Kasmin puts it: 'Hockney stood out as the only figurative guy in an abstract gallery.'[20]

A would-be poet turned art dealer, the young Kasmin – known universally as Kas – was as striking a figure, in his way, as Hockney: preppily turned out and suavely bohemian. He also shared the artist's gift of self-belief. He had first come across Hockney's work in 1961 at the Royal Society of British Artists' annual *Young Contemporaries* show, while still working at Marlborough Fine Art in Mayfair, and had been struck as much by the artist as by his work:

> He had a sort of gifted impudence that really attracted me. And he obviously had an extraordinary natural talent. I knew he would be a success, and his open homosexuality and use of gay imagery would be to his advantage since gay people always have money to spend on art.[21]

At *Young Contemporaries*, Kasmin had bought the painting *Doll Boy* (1960–1) for £40. It was characteristic of the artist's work at this time, combining an expressionist-cartoonish figure (a boxlike body and bared teeth) with hard-edged panels of red. Its title was a coded allusion to Hockney's crush on Cliff Richard. Kasmin recalls: 'I found it enigmatic, to start with, because – what was it all about? Somebody was clinging to a cliff in the painting, and I had no idea that it was to do with Cliff Richard, because I wasn't the faintest bit interested in pop music.'[22] Soon after, Kasmin left Marlborough and became Hockney's main dealer, initially showing paintings in his Earl's Court flat and selling them for between £50 and £100.

By the time Kasmin arrived in Los Angeles in the spring of 1964, Hockney had moved into his rented apartment in Santa Monica. But he happily shared Kasmin's bedroom at Gene Autry's Hotel Continental on Sunset Boulevard, and spent much of his time beside the rooftop pool, against the backdrop of the Hollywood Hills. On one occasion, he persuaded Kasmin to join him on an expedition to Bob Mizer's studio, the headquarters of *Physique Pictorial*, in downtown Los Angeles. They arrived to find a compound sequestered behind a thick hedge. Mizer's operation had grown over the years to incorporate most of the block around his mother's house.

'David wanted to see the location', Kasmin recounts. 'So we went to a house in a backstreet, in an area we didn't usually go to. It was their offices. And there they had a sort of set: two walls with a shower.' There was also a papier-mâché mountain range and a pool that sometimes masqueraded as a tropical lagoon. It came as a thudding anticlimax to find that the source of *Physique Pictorial*, with its dazzlingly erotic visions of American men, was a workaday stage set. The pool was surrounded by plaster statues in a 'tacky' Grecian style; its concrete sides were stained. 'There was absolutely no glamour – David had thought there might be a bit of glamour. The most glamorous thing was penetrating through the hedge to get to the front door.'[23]

But even the seedy ordinariness of the place held a fascination. The poolside studio had revealed more luridly than ever the desolate flip side of Los Angeles – the striving and exploitation and mercenary compromise that lay just beneath the shimmering surfaces. Mizer was known for employing boys off the street or straight out of jail to pose for photographs, and liked to quip: 'Remember, we are AMG [Athletic Model Guild], not MGM.'[24]

In its odd diversity of wishful glamour and tawdry reality, Mizer's set-up reminded Hockney of the poems of Constantine Cavafy; he would go on to draw a set of illustrations for Cavafy's poems in 1966.

Indeed, Los Angeles seemed like a proxy for Cavafy's favourite city, Alexandria. As Hockney wrote a decade later: 'The hot climate's near enough to Alexandria, sensual; and this downtown area was sleazy, a bit dusty, very masculine – men always women are just not part of that kind of life. I love downtown Los Angeles – marvellous gay bars full of mad Mexican queens, all tacky and everything.'[25]

In his first Californian paintings, Hockney evinced that sensuality in less extravagant terms. His sparse compositions in acrylic translated the models of the Athletic Model Guild into a mundane register – one closer to the dreary reality of the AMG compound than the sexy spectres of Mizer's photographs. *Boy About to Take a Shower* (1964), based on one such photograph, shows a naked youth standing before the black-and-grey-tiled wall of a shower, his body abruptly pale around the buttocks. In *Man Taking Shower in Beverly Hills* (1964), a similarly built figure bends over beneath a torrent of spurting, dancing water. This latter piece was completed in Iowa during a six-week stint of teaching at the university – its coolly intimate scene a memory of what Hockney was already missing.

An English exile

'Isherwood is at Santa Monica
with miles of beach and a crazy pier.
He looks battered, but is smiling,
like an old seaman, nearly at rest.
A gold-haired boy twirls upside-down on rings.'

Alan Hollinghurst, 'Isherwood is at Santa Monica', 1975[26]

For Hockney, Los Angeles was its own cultural centre, 'the Florence of the twentieth century'.[27] It was the home of moviemaking, and had moreover garnered formidable intellectual currency through the arrival of wartime European exiles (forced or willing) such as the writers Thomas Mann, Bertolt Brecht, Lion Feuchtwanger, and Aldous Huxley, the composers Igor Stravinsky and Arnold Schoenberg, and the director Berthold Viertel and his screenwriter wife, Salka.

A key figure in this international circle was the English novelist Christopher Isherwood, who had emigrated to America in 1939 with W. H. Auden. Their departure on the cusp of the Second World War had prompted accusations of desertion, although the move had long been meditated. In part, Isherwood was following the example of the English philosopher and ascetic Gerald Heard, who had moved to California with Huxley in 1937. It was Heard who introduced him to Vedanta spiritualism – a guiding influence in his middle and later years, if also a burden with which he wrestled, long after abandoning the idea of becoming a monk.

Soon after arriving in Los Angeles, Hockney made a telephone call to Isherwood – to whom Stephen Spender (a mutual friend in London) had already sent a letter of introduction. He received an invitation to dinner. This meeting marked the beginning of a friendship that would endure until Isherwood's death in 1986. Some years earlier, Hockney had read *Mr Norris Changes Trains* (1935) and *Goodbye to Berlin* (1939), novels that evoke the sleazy glamour of Berlin in the pre-war years. (On a trip to Berlin in 1962, he had sought out the haunts described by Isherwood, finding only the Kleist Casino – a louche dive with pink lampshades.) Isherwood, then fifty-nine, seems to have perceived something of himself in the younger man. As he would later declare: 'Oh, David, we've so much in common: we love California, we love American boys, and we're from the north of England.' (Hockney wryly noted: 'Of course Christopher's from the opposite side of the north of England: his family was quite rich, mine is working class.'[28]) Isherwood had met his own American boy, Don Bachardy, on Santa Monica Beach in 1952, when he was forty-eight and Bachardy was eighteen.

That stretch of beach was visible from the house where Isherwood and Bachardy had lived since 1959, on a ridge above Santa Monica Canyon. The day that Hockney came to visit, Bachardy was away, and so Isherwood invited two friends to join them: Jack Larson, a screenwriter and poet who had achieved early fame for playing Jimmy Olsen in the television series *Adventures of Superman*, and his partner Jim Bridges, a polite 'Southern gentleman' (versus the more impishly gregarious Larson) who worked variously as an actor, writer, and director.[29] Both would become close friends of Hockney's. The artist arrived on his bicycle, and Isherwood was amused to hear of his early adventures in Los Angeles – including the ride to Pershing Square in search of the pleasures of *City of Night*. (John Rechy was also known to Isherwood, who considered him 'a very romantic writer'; Bachardy took the view that he was somewhat strange.[30]) A week later, Bachardy also met Hockney and took a similar liking to him.

Just as Isherwood discerned a kindred spirit in Hockney, the artist identified with (maybe already aspired to the status of) the renowned English exile. Isherwood had found in Los Angeles an escape from the class prejudices and narrow conservatism of much of British society – a more accepting climate for his pacificism and homosexuality. In his 1964 novel *A Single Man*, he reflected on his situation as an émigré writer, imagining a day in the life of a gay English professor teaching in California. Within the story, he depicts the close, confidential friendship between the fifty-something protagonist, George, and a fellow expatriate called Charlotte.

> And now, as George pours the vodka (giving her a light one, to slow her down) and the scotch (giving himself a heavier one, to catch up on), he begins to feel this utterly mysterious unsensational thing – not bliss, not ecstasy, not joy – just plain happiness. Das Glueck, le bonheur, la felicidad – they have given it all three genders, but one has to admit, however grudgingly, that the Spanish are right; it is usually feminine, that's to say, woman-created. Charley creates it astonishingly often; this doubtless is something else she isn't aware of, since she can do it even when she herself is miserable.[31]

Isherwood's fictionalized self-portrait held up a paradigm to the young Hockney, a mirror in which the artist might have caught glimpses of both the present and the future. This brief evocation of the dynamic between a gay man and a heterosexual women is prescient, in particular, of the intimacy that Hockney would later share with Betty Freeman.

California art collector

Around the same time that he met Isherwood and Bachardy, Hockney began visiting the homes of wealthy art collectors, often accompanied by Kasmin. 'It was impossible to make a living as a dealer in England if you didn't travel', Kasmin explains. 'You had to have collectors from somewhere else, because in England very few people collected contemporary art. So when I went to LA, it was to join in an art scene out there – to meet collectors. We'd go and visit collectors together.'[32]

Neighbourhoods such as Beverly Hills, Bel Air, Brentwood, and the Hollywood Hills were a revelation. Hockney took delight in their assorted architectural styles – that very quality of giddy variety that has appalled other observers. Woody Allen, in character as Alvy Singer in *Annie Hall* (1977), passes sardonic judgment on the mansions of Beverly Hills as he is driven through the neighbourhood: 'The architecture is really consistent, isn't it? French next to Spanish, next to Tudor, next to Japanese.' Nathanael West responded in similarly jaded terms in his 1939 novel *The Day of the Locust*, via the consciousness of the young artist Tod Hackett: 'Only dynamite would be of any use against the Mexican ranch houses, Samoan huts, Mediterranean villas, Egyptian and Japanese temples, Swiss chalets, Tudor cottages, and every possible combination of styles that lined the slopes of the canyon.' Hockney was fascinated not only by the metamorphosing styles of the houses, but also by their interior and exterior ambiences – their veneers, furnishings, and swimming pools. He was fascinated, too, by the archetype of the wealthy female collector: he and Kasmin were invariably hosted by the woman of the household, the husband being out at work.

California Art Collector (1964) was a picture of one such woman. It was painted in acrylics – a new discovery that allowed him to create a smooth, uniform, and thin surface that befits the picture's ambience of suburban calm. A middle-aged lady in a green dress sits on an armchair, underneath a minimalist canopy. She is captured in profile alongside two sculptures: a stack of rocks by William Turnbull and a stone head, redolent of the Moai of Easter Island, that mirrors her own inexpressive countenance. In the background, separated from the domestic scene by a rainbow-like band of colour and a grey patio, is a pool: a trapezium of blue, turning from light to dark, backed by a few items of scenery (sprigs of foliage, a pair of palm trees, the blanched silhouette of a mountain range).

The tableau of *California Art Collector* was invented. It was a homage to the planar architecture of Fra Angelico, Giotto, and Piero della Francesca as much as a portrayal of contemporary Los Angeles.[33] The only 'real' element was the pool, and even that was an image of an image: Hockney had copied it from an advertisement in the Sunday edition of the *Los Angeles Times*. The woman, rigidly erect on the edge of her seat, is an object rather than a subject, a type rather than an individual little different from the sculptures that surround her. Her face is no more than a notation: pinprick eye, pursed lips, brown hair tied in a double bun. A picture of respectability, of inexpressive restraint, she fails to inspire – refuses to communicate. More striking by far than the woman herself is the stage set of her house and yard, with the pool hovering like a mirage in the middle distance. The painting marked the first appearance of a pool in Hockney's art, followed by *Picture of a Hollywood Swimming Pool* (1964). These everyday marvels – first spotted from the plane over Los Angeles – would soon become his leitmotif. In 1965 he began to add figures – sunbathers or idling swimmers – and yet these remained mere details, not a subject in their own right.

California Art Collector was the second painting that Hockney completed in Los Angeles. The first, *Plastic Tree Plus City Hall* (1964), was similarly a 'collage' of resonant things: a plastic palm tree, the cut-out vista of City Hall's mausoleum-like tower, scattered orbs of reflected light, an impressionistic sky. Both works demonstrate how his early conception of the city was itself a sort of montage, and both seem to articulate a preconceived *idea* of the place, albeit that tangible details stand out with bracing clarity – the tufts of the shagpile carpet, for instance, under the chair of the unnamed collector. That carpet is the only thing in the picture with depth and weight.

Back and forth

Hockney's first stay in Los Angeles lasted only six months. He travelled through Iowa in the summer of 1964, and afterwards took a road trip with Ossie Clark, the British fashion designer, from Chicago back to Los Angeles to see the Beatles perform at the Hollywood Bowl. They were then joined by Derek Boshier, an artist friend from the Royal College of Art, on a drive to the Grand Canyon and on to the Deep South, where Hockney's and Clark's flamboyant appearances attracted a degree of redneck ridicule: 'Whenever we sat down in a restaurant, everyone thought Ossie was a woman, because he had long hair, and nobody had long hair in those days. None of us really cared when they made fun of us.'[34]

Boshier left them in New Orleans ('I said to the others I was going to go off and find some girls'), and Hockney and Clark drove to New York for Hockney's first American exhibition at the Alan Gallery, which included the paintings *California Art Collector* and *Boy About to Take a Shower*.[35] 'Hockney is primarily not a painter but an image maker,' declared Stuart Preston in the *New York Times*, 'both fun and funny in his deadpan, zanily satirical interpretations of modern life.'[36] Hockney returned to London in the autumn, where he worked on *Picture of a Hollywood Swimming Pool*, a study of serpentine blue lines running back and forth across the surface of water – the first painting in which a pool forms the dominant subject. In subsequent works, male figures began to appear, levering themselves from the water or buoyed on lilos, but they were minor characters within the scenes – details rather than protagonists.

He was back in Los Angeles for the summer of 1965. One night at the Red Raven, he met 'a ravishingly beautiful boy' called Bobby Earles, a classic Californian youth with blond hair. 'It was lust on my part, sheer lust, and I thought, fantastic!'[37] Hockney sketched Earles as the young American lay asleep on a transatlantic liner back to Europe at the end of the year. The pencil drawing, titled *Bob Aboard the 'France'*, shows Earles face down on a cabin bed, his full buttocks shaded in scorching pink. It is the only relic of a transient affair. Earles quickly grew tired of London, seeming only to want to find gay bars (of which there were none comparable to those of Hollywood). He later wound up as a go-go dancer in Laguna Beach, and died tragically young from a drug overdose.

Hockney returned to LA in 1966. He had been invited to teach painting for a summer school at the University of California, Los Angeles, having been introduced to the university's circle of artists by Oliver Andrews. The painter Bill Brice was also a nodal point of connections at UCLA and beyond (the son of the Hollywood star Fanny Brice, he was well acquainted with actors and writers). Ahead of his stint at the university, Hockney drove across the States from the East Coast in an open-top

Triumph that he had had shipped from England. '[A]s I got further and further West,' he wrote a decade later, 'I became more and more excited, imagining my class would be full of young blond surfers.'[38]

Peter

To his disappointment, the class at UCLA consisted mostly of housewives and middle-aged would-be teachers. Many were attending to gain credits that would entitle them to higher pay. But one member of the class was different: a diffident, good-looking boy of eighteen with flowing hair. Peter Schlesinger came from a middle-class Jewish family living in Encino, a suburb of Los Angeles in the San Fernando Valley. His two brothers had gone to Israel to further the cause of Zionism, but Schlesinger had a different ambition – he had wanted to be an artist since he was a young child – and a distinct sense of self. He had enrolled at the University of California, Santa Cruz the previous autumn to study world history; but as there were no art classes, he had come back to LA for the summer.

Over the six weeks of the course, a tentative intimacy developed between Schlesinger and his teacher. They went out to see a film and to dinner, then began to meet more regularly. Schlesinger started to rely on Hockney, telephoning for help, one evening, after a man he had hitchhiked with tried to solicit him. It took Schlesinger time, even so, to reciprocate Hockney's desire, and time to acknowledge his own sexuality. Hockney agonized over his depth of feeling. For hours, he would pour out his longing to Nicholas Wilder, a new friend at whose West Hollywood apartment he often stayed.[39]

But late in the summer, something shifted. A love affair began. It was the first that either Hockney or Schlesinger had experienced. Hockney was amazed to have come upon the 'sexy boy' of his fantasy, and moreover a boy – contrary to Earles – who was curious and intelligent.[40] Schlesinger spent most of his time outside class in the company of Hockney and his growing circle. 'We all went out to dinner,' recalled Jack Larson, 'and there was no doubt in my mind that David was absolutely smitten with this wonderful-looking, quite charming young boy who wasn't a Beverly Hills swinger, but a boy from the Valley.'[41] A picture taken that August shows the pair of them at Liberace's antique store on La Cienega Boulevard, next to a giant candelabra. In another, they stand in Forest Lawn cemetery beside a replica of Michelangelo's *David*.

Hockney found an apartment cum studio in a converted garage close to the junction of Pico and Crenshaw Boulevards, 'a really run-down little house in a very seedy part of Los Angeles'.[42] It consisted of a bedroom, a poky kitchen and bathroom, and a room at the front in which he could paint. It was here, when they were not staying at Wilder's place, that Hockney and Schlesinger first lived. Whenever Hockney put the gas on, cockroaches scuttled out of the oven, but it hardly seemed to matter.

What Betty knew

In *Beverly Hills Housewife*, Betty Freeman's level of awareness is impossible to measure. It is easy to intuit a silent, telepathic sort of interaction between her and the objects in the picture – as though she were conscious of how those objects frame her, and consciously framing herself for the artist's eye. It is equally possible to perceive a total disconnect. In the moment of being captured, she might have been aware only of herself.

How much she knew of David Hockney before they met is unclear, but that she did know of him is certain. On a visit to London with her friend Virginia Dwan in 1966, she went to the production of *Ubu Roi* at the Royal Court theatre that Hockney had designed – 'a very tough, mean show', as she remembered, 'a little bit like Beckett' – as well as an exhibition that same summer at Kasmin's gallery, featuring drawings for the sets and costumes.[43] She had bought one of the drawings, a crayon piece titled *Ubu Collecting his Taxes*: 'I would have bought more, but they'd already been sold.'

It was Jack Larson who first introduced Hockney to Freeman. She had become friends with him after he began writing librettos for contemporary operas in the early 1960s. Larson was serially connected in Los Angeles, and as Schlesinger recalls, 'knew all the Hollywood gossip – all the secrets'.[44] In Kasmin's account: '[Jack] told me he'd got this new friend who was *avant-garde*. "The stuff on her lawn looks like grass," he said, "but it's more like watercress."'[45] Hockney, who was at the peak of his fascination with swimming pools, asked Freeman if he could come over to her house to take photographs. 'What I didn't know', Freeman related in 2005, 'is that David is a little more complicated than that.'[46]

Polaroids

Number 703 Hillcrest Road is a classic Los Angeles residence. Set back from the sidewalk by a swathe of lawn, it embodies the colonial style of the 1930s. The cream-painted brickwork is inset with modestly sized windows, each flanked by wooden shutters. A long balcony runs along the upper storey. Less palatial than many houses in the neighbourhood, its appeal lies in its understatement, not that any Beverly Hills residence is a shrinking violet.

The typicality of the home, the elegance of the setting, were what Hockney wanted to capture with his Polaroid camera. He was looking for a reality to mirror, perhaps to confirm, some of the fantasies expressed by his earlier paintings, although he was unlikely to find showering youths in the home of Betty Freeman. And so, on that day in the late summer of 1966, he took along Peter Schlesinger.

It is intriguing to imagine how the artist and the collector might have struck one another. Each was an exotic curiosity, probably, in the eyes of the other: the bleach-blond Yorkshireman, cheerily beguiled by the house and its glamorous inhabitant; and the forty-five-year-old American woman – also newly blonde – who had picked out a long fuchsia dress of Indonesian silk to wear that day. She may have been bemused more than beguiled, but was happy to play along.

At the same time, it is likely that each discerned a kindred seriousness in the other. Freeman was a deeply serious person, little given to small talk. Hockney had already applied himself with an almost monastic discipline to his art. Beneath their overt differences – and in tandem with their shared blondness – was a common sincerity of purpose. As Stephanie Barron puts it: 'There was nothing flighty about her. I think there was a rigour to her that David also has. Even with all his enjoyment of the atmosphere of Los Angeles, that seriousness came through. My guess is that they responded to that in each other.'[47]

Freeman guided her visitors through the art-filled interiors of the house: the double-height hallway, the breakfast room with its tiled

David Hockney, preparatory photographs for *Beverly Hills Housewife*, 1966

David Hockney, preparatory photograph for *Beverly Hills Housewife*, 1966

floor and table in the design of a white poppy, the smaller bar room. Schlesinger remembers above all the impact of the living room, a large salon adjoining the bar room where the walls were coated in black velvet: 'You couldn't see the walls. There were these two huge paintings – a Clyfford Still and a Sam Francis.'[48] In the garden, the house presented a different aspect. A curving portico, supported by slender columns, ran along the ground floor, introducing a note of California Modernism. A short distance away, across a grid of paving stones and a stretch of lawn, was the pool – always heated and ready.

Hockney tried out a number of possibilities for a picture. Freeman recalled how he asked her to stand at the top of the garden while Schlesinger emerged from the pool. 'He came out naked. I was a little bit startled by that. I tried not to show it, but I was startled.'[49] She succeeded in hiding her surprise, if the photographs that Hockney took of her are anything to go by. In two Polaroids, the composition is practically the same (p. 29). Freeman stands on the terrace, underneath the portico. She looks into the middle distance. The setting is outside, and yet the canopy and the wall to her right create the sense of a dismantled interior, as though a fourth wall had been removed to allow the daylight in. In one shot, she is slightly out of focus; but in the second, the sun has come out to transfigure her. Slim and solemn, she holds herself with a touch of self-consciousness, somewhere between casual and statuesque. On the wall of the alcove in which she stands, three stuffed antelope heads rear outward: a static herd. Her husband, as she no doubt mentioned to Hockney that day, was a 'great white hunter'.

In a third Polaroid, Freeman appears in the same pose but from a closer remove. It is clear, from this image, that she is still youthful. Behind her, the details of the house have blackened out. Palm fronds fan out to her right, catching the light. She looks away, as before, but her expression is clearer now. It suggests a dreamy sort of concentration, perhaps the very beginning of a smile. Her hair, coiffed into a bob, matches the pale tone of her dress. She is every inch the affluent collector, but she is also simply *herself*: Betty Freeman at the entrance to her garden. For a while, as Hockney took snapshots of the garden and the outside of the house, Freeman busied herself by dusting the stuffed animal heads.

'There's only one thing you can call this painting,' she told him, 'since I'm dusting: *Beverly Hills Housewife*.'[50]

There are two additional photographs from the trip to Hillcrest Road. They show Hockney and Peter Schlesinger taking turns to lie on Freeman's zebra-skin lounger. Schlesinger is clothed and lies back with his eyes closed. He is rendered almost featureless by the sunlight: a white shape. Hockney, by contrast, sits up a little – his arms folded. He returns the camera's gaze.

His portrayals of Los Angeles would soon acquire a new focus. In place of generic beautiful boys and impersonal settings, he would paint the people close to him, Schlesinger foremost among them. In the summer of 1966, he had begun sketching his new friend whenever they were together, such as the occasion when Schlesinger first invited Hockney back to his parents' house in Encino. And yet these sketches have a provisional aspect: they are diary entries, exercises. One senses that the romance was somewhere beyond the reach of Hockney's artistry. The pencil drawing *Peter, Swimming Pool, Encino, California* (1966) is more concerned with the play of light on water than the boy who sits reading in a pair of red trunks. For his first major painting that he began that summer, Hockney settled upon another subject. The Polaroids that he had taken at Hillcrest Road had fixed for him a vision of his first Californian muse.

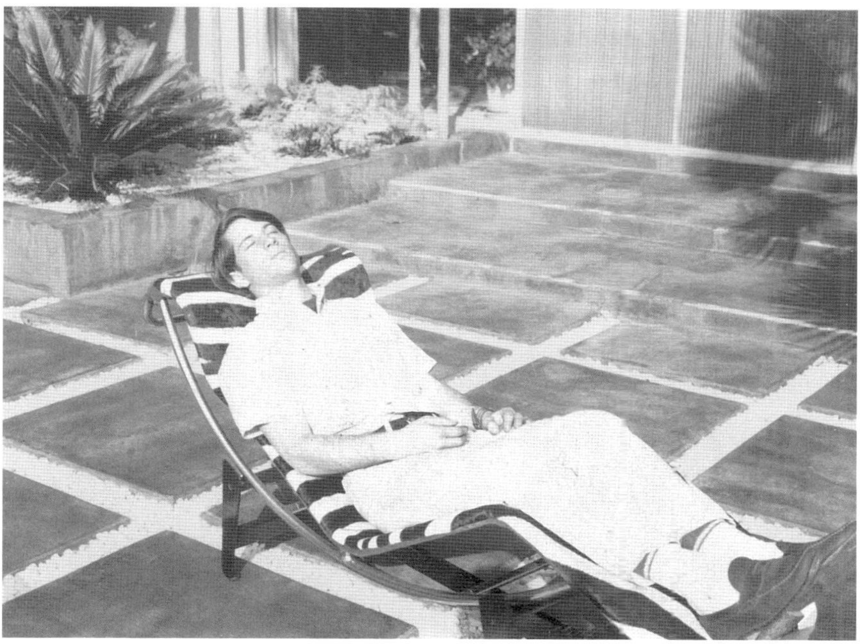

David Hockney and Peter Schlesinger at Betty Freeman's house, 1966

2

Mrs Stanley Freeman

'All right, Mr DeMille, I'm ready for my close-up.'

NORMA DESMOND, PLAYED BY GLORIA SWANSON, IN *SUNSET BOULEVARD*, 1950[1]

Freeman and Hockney each came with a well-formed identity: the artist on the rise and the Beverly Hills collector. But it is possible that they each betrayed (each perceived in the other) a more labile interior.

Who was Betty Freeman? The paradox of the painting is that she is plain to see, *there* at the centre of her domestic world, even as the title consigns her to a stock role. Hockney has amplified that paradox, hinting at the life she leads while holding her at an inexorable remove. She stands alone, somehow apart from the house and its assorted objects, as though it were a film set that had little to do with her; as though her life had dropped into place around her.

Betty Wishnick

She was born Betty June Wishnick on 2 June 1921 in Chicago, the daughter of Robert and Freda Wishnick.[2] In 1924, when she was three, the family moved to Brooklyn, and it was here, in the shadow of the Brooklyn Museum, that she lived until the age of eleven. Her father was a self-made businessman. A Russian-Jewish émigré, he had come to Chicago in 1898 at the age of four, the youngest of eight children. His own father, Eli Wishnick – a horse trader, merchant, and part-time rabbi – had emigrated from the Kaluga Governorate of the Russian Empire to the United States two years earlier, making his way to Illinois and setting up a small country store with his eldest son. Robert and his mother were sent for last of all, travelling over on the cattle boat.

Chicago was undergoing a boom at this time; its population had doubled in the space of a decade to more than a million people by 1890, of whom eighty per cent were immigrants or the children of immigrants. Already an important centre of trade in grain, cattle, and lumber, the city was also becoming a force in diversified manufacturing.[3] Jewish immigrants settled mainly in the city's west side, around Maxwell Street, where many of them worked as pedlars. While he was still a boy, Robert would spend the summer months travelling through the countryside by horse and wagon, selling old pots and pans. Occasionally he would sleep beneath the wagon in an open field.

Robert's life and character were shaped by an early catastrophe. Aged seven, while playing 'follower the leader' with older boys, he fell off a high fence and broke his right arm. There were no specialists in the neighbourhood to treat the complex fracture, and so the lower part of the arm had to be amputated. It was a devastating loss, but in time he learned to compensate for the disability. He would always credit his mother for her refusal to help him. As a teenager, he dedicated himself to academic success, winning a scholarship to the prestigious Armour Institute of Technology, where he studied chemical engineering, before taking a law degree from the Kent College of Law in 1917.[4]

In 1919, while working as a salesman for a local chemicals firm, Robert met Freda Monica Frankel, a striking, dark-haired woman of twenty-six. The occasion was his boss's wedding: Freda had been the college roommate of the bride. She was a graduate of the University of Wisconsin, where she had studied mathematics. She had since returned to her rural hometown of Eagle River in north Wisconsin to teach maths in the local school – and numerous other subjects besides: according to her son, 'When somebody would ask a question in, say history, she'd always say: "That's ahead of the class. You've just got to wait and we'll cover that next week." That night she'd go home and study that.'[5] A 1916 copy of Freda's university yearbook features a photograph of her among the seniors of that year: she regards the camera with the hint of mirth that would become her daughter's hallmark.[6]

Robert and Freda fell immediately in love: they were married forty-one days after their first meeting. In 1920, Robert cofounded the Wishnick-Tumpeer Chemical Company on East Illinois Street. Over the next three decades, it would grow to become a worldwide producer of specialty chemicals and petroleum products. Known after 1944 as Witco, it was one of the giants of twentieth-century manufacturing, an emblem of the American Century. A clever, determined, and stoical character, Robert was an inspiration to Betty from her early childhood. It was easy to forget that he had only one arm: he enjoyed golf, tennis, and horse-riding; and when driving, he held the steering wheel with his elbow in order to shift gears. She admired the way in which he had overcome adversity, put himself through college, and built a successful business. More than anything, she was proud of his philanthropic work for organizations such as the Jewish Child Care Association of New York and the Hebrew University in Jerusalem (where a dormitory was named after Freda), as well as New York cultural institutions including the Lincoln Center and the Metropolitan Opera.

Robert took over the New York branch of the company in 1924, and that summer the family moved to Brooklyn. Betty recalled it as a wonderful place to grow up – quiet, suburban, and still pleasantly provincial at that time, with broad avenues that gave it a European atmosphere. She was artistically inclined, unlike her parents or two younger brothers, and used to pore over the Egyptian collections at the Brooklyn Museum.[7] Even as a child, however, she gravitated towards the new. 'From an early age (I think I was about 15), I liked all the arts of my time', she said in 1998. 'It seemed to me to be perfectly natural – and still does.'[8] Around the age of twelve, she learned the piano for a while, but did not enjoy it.[9] As a teenager, she met the modernist architect Frank Lloyd Wright, whom her father knew – an encounter that she never forgot. As she later described it, Wright was seated in a thronelike chair, and it felt like meeting God. (Notorious for his egotism, Wright once admitted that this quality was 'more armor than character, more shell than substance ... the usual defence for exaggerated sensibility'.[10])

In 1937, the Wishnick family moved to the New York suburb of New Rochelle. After high school, Betty attended Wellesley College, a leading women's university in Massachusetts, where she majored in English literature and minored in musicology. Her lecturers included Vladimir Nabokov, then newly arrived from Europe. 'His aura was one of casual, relaxed manliness', according to a near contemporary of hers. 'He had the pleasing smell of a man who smokes, and an air of a certain diffidence, of a natural aristocratic dignity.'[11]

Another teacher was Victor Zuckerkandl, a Jewish-Austrian musicologist whose emphasis on the mystical properties of music left a deep impression on Betty. Some years later, Zuckerkandl would write, in his book *Sound and Symbol* (1956): 'Hearing a melody is hearing, having heard, and being about to hear, all at once. [...] Every melody declares to us that the past can be there without being remembered, the future without being foreknown.'[12] It is tempting to wonder whether, in her student years, Betty had already developed her singular capacity for living in the present, unburdened by memory.

It was at Wellesley College that Betty took up the piano once more. As she walked along the corridors of the music centre, Billings Hall, she heard other girls practising. 'I got this real urge to study piano. My parents had told me they wouldn't help me financially [...] so I took it out of my allowance and studied for three and half years, two hours a day.'[13] Music then became a near obsession. Since her mid-teens, she had gone up to the Boston Symphony Orchestra's summer festivals at Tanglewood. Every Friday, she would attend the Symphony Hall to hear the orchestra, then under the directorship of Serge Koussevitzky, whose humble origins in Russia paralleled those of her father. In this way, she became acquainted with contemporary composers such as Béla Bartók, Igor Stravinsky, Sergei Prokofiev, Sergei Rachmaninoff, and Dmitri Shostakovich.

While still at Wellesley, she had met Stanley Milton Freeman, a graduate in chemistry from Massachusetts Institute of Technology. Their first encounter was a blind date: Freeman's father had advised her that Stanley would be a good match. Robert's approval of the union seems almost to have predetermined it. Stanley and Betty married in 1942, soon after she had graduated. During the war, he served as a lieutenant commander in the US Navy. She worked at CBS, the leading radio

Betty and Stanley Freeman, 1944

network, which was nicknamed the 'Tiffany network' because of its reputation for excellence.[14]

Not yet a blonde

A photograph taken in 1944 shows the young couple side by side in a garden: Stanley in his uniform and beret, Betty in a matching jacket and skirt. He looks away from the camera with saturnine calm, into the sunlight, while she looks straight back at the lens, smilingly confident and yet gauche, as though she were repressing a giggle. They stand together and yet remain in some sense disconnected, their hands resolutely by their sides. The buttons of her jacket are tightly fastened; her feet are neatly together. Strikingly, her hair is dark in this picture. She had not yet become a blonde.

Betty continued her piano training wherever she found herself: at the Juilliard School, New York, during the summer; while living in Virginia during the war; and at the New England Conservatory of Music in Boston, where the couple subsequently settled. In the meantime, more conventional obligations took priority. She had four children within six years (Robert in 1945; Corey in 1948; Shelley in 1949; Claudia in 1951). Stanley began working for Witco in 1949 as a sales manager. The following year, the family moved to Los Angeles after he was assigned to the company's regional office. Soon after, he received a promotion to run Witco's operations on the West Coast.

Their life was in many respects typical of an affluent bourgeois family in Beverly Hills in the mid-twentieth century. They lived in a detached 1920s house on a leafy corner of North Maple Drive. The children, while they were still young, attended the local Hawthorne Elementary School, where fellow pupils included the offspring of Groucho Marx, Rita Hayworth, and Veronica Lake. At home, an African-American maid called Elvina worked for the family, and used to entertain the children with stories of Elvis Presley, for whom she had previously cooked. (Presley had been so devoted to her that he bought her a house.) Stanley was a committed supporter of the Republican Party. A photograph exists of him, Betty, and the children meeting Richard Nixon around the time of Nixon's unsuccessful campaign for the governorship of California in 1962; Shelley Freeman recalls how she and her younger sister were made to wear their best dresses for the occasion.[15]

All the while, alongside her roles as wife and mother, Betty's determination to improve at the piano persisted. She took lessons from Joanna Graudan, a Jewish émigré from Germany and 'a wonderful teacher', and practised for six or seven hours a day.[16] After moving to Hillcrest Road in 1959, she played in the large room at the front of the house, surrounded by walls of black velvet.

Art and the martini set

Betty Freeman's other great love in these years was contemporary art. Her first purchase, made in 1952, was a group of three sculptures by Henry Moore. In the late 1950s, she attended classes taught by Walter Hopps, the director of the Ferus Gallery on La Cienega Boulevard, and his wife Shirley. Entitled 'Looking at Modern Painting', Hopps's classes were organized through UCLA, but took place in the house of Freeman's friends, Fred and Marcia Weisman.

The Weismans were a wealthy middle-class couple who had cultivated a reputation as trendsetting collectors. Their home was a nucleus for the Beverly Hills 'martini set', and other members of the classes included the real-estate baron Edwin Janss and Monte and Betty Factor, owners of a men's clothing business. The Weismans' marriage, however, was a difficult one. As the president of Hunt Foods, Fred worked for Marcia's industrialist brother, Norton Simon, and the resentment arising from the arrangement had once come to a head when Marcia acquired a Francis Bacon 'screaming pope' and installed it at home without her husband's knowledge. Upon arriving home and seeing the painting, Fred was horrified by the resemblance to Norton Simon. 'I work with this man every day of my life', he is said to have shouted. 'Get it out!'[17] The painting was removed the next day.

Hopps's art classes were primarily an exercise in cultivating a network of collectors, but he was also an inspirational teacher: clever and iconoclastic. (Shirley recalled how on one occasion, 'He was showing a Barnett Newman slide and Fred Weisman said, "You've got me there, kid", and walked out of the room. This stuff was Latin to these people.'[18]) He had founded the Ferus Gallery with the artist Edward Kienholz in 1957, when he was only twenty-four, and the space had fast become one of the edgiest on 'Gallery Row', a three-block stretch of La Cienega Boulevard that formed the epicentre of the city's art scene. Hopps had drawn the name from the Latin for 'fierce' or 'wild'; the word 'seemed to describe somewhat ironically the kind of people we were and the kind of people we were dealing with', he said in 1975.[19] The name is also said to have derived, at least in part, from that of Jim Farris, an art student at Eagle Rock High School (where Hopps had studied) who had committed suicide.

Hopps hung out with artists, but cut a maverick figure among them, turned out like a matinee idol in suit and tie. He was an enigma, with a knack for disappearing at critical moments (such as on the night before Ferus opened, when he vanished for hours, only to turn up with a Clyfford Still painting for the launch exhibition). When looking closely at a work of art, he would wear a second pair of glasses. He was also famously voluble, his pronouncements soaring into rhapsodies.[20] In the words of Betty Asher, a collector who was part of the social set that orbited the Weismans – speaking in 1980 – 'Walter is very elusive.'[21] After dropping out of Stanford and then the biochemistry department of UCLA, Hopps had made a name for himself by staging an exhibition of contemporary masters (Still, Mark Rothko, Richard Diebenkorn, Jay DeFeo) on Santa Monica Pier in 1955. Entitled *ACTION 1*, this had involved stretching canvas around the merry-go-round on the pier to create a circular enclosure, within which twenty-four paintings and drawings were mounted; the entire display revolved in tandem with music by John Cage and Dave Brubeck. Although visitors consisted mainly of beachgoers, tourists, and patrons of a nearby transvestite bar, *ACTION 1* did attract the notice of Beat writers such as Allen Ginsberg, Neal Cassady, and Jack Kerouac. It quickly became part of Hopps's personal mythology. Irving Blum, who joined Ferus as Hopps's partner in 1958, attended some of the classes and later described their 'very personal, very tailored' format:

> Walter would bring slides [...] and talk at length about
> the work of a particular artist, or talk about the New York
> experience and what his understanding of that experience was

> at that time – with the focus maybe on de Kooning, or the focus maybe on Barnett Newman another evening, maybe on Pollock [...] and talk at great length and answer whatever questions these people had, and show transparencies of their past work, their current work.[22]

Through Hopps, Freeman became aware of several of the artists she went on to collect. Key among them was Still, to whom she wrote for the first time in late 1959:

> I consider you the finest painter in the world today and I would give almost anything to own a picture by you. I have never seen your work in the original and am always asking, either out here on the Coast, or when I visit New York. Your No. 2, 1949 I know in my heart and think it is the finest picture of this century.[23]

It soon became clear that one never 'bought' a painting from Still, but rather traded money in return for a work that he selected.[24] Freeman continued to write; finally, in the autumn of 1960 – after a cordial meeting with Betty and Stanley in New York – Still agreed to sell them the monumental painting *1955-H* (1955) for $15,000. The work had featured in his exhibition at the Albright Art Gallery, Buffalo, a year earlier, a display that Freeman called 'the most awesome experience of all the years of my loving art'.[25] Still had picked out her favourite painting from the show, she said, without her ever having told him.

Still's *1955-H* consists of nine 'columns' of paint, applied to a raw canvas using a pallet knife – 'lifelines', as the artist called them, in ragged-edged black, white, and red. Each traverses the work from top to bottom. After Still had proposed the painting for Freeman, she wrote back with a transcription of the opening stanzas of Percy Bysshe Shelley's 'Ode to the West Wind' (1819), which seemed to her to encapsulate the power of the composition:

> O, wild West Wind, thou breath of Autumn's being,
> Thou, from whose unseen presence the leaves dead
> Are driven, like ghosts from an enchanter fleeing,

Yellow, and black, and pale, and hectic red
Pestilence-stricken multitudes: ...[26]

Early the following year, she installed Still's painting in the dark, velvet-walled space of her living room (also the 'music room'), where the only furniture at that time was her Steinway piano. Hopps and Blum came to help with the stretching and hanging of the canvas, together with artist Bob Irwin. 'They were in shirt-sleeves and the perspiration running', Freeman recounted, 'but would stop for nothing.' When the picture was finally up against the wall, she opened a bottle of her best champagne and her four children came downstairs to admire it. Robert, then fifteen, and her elder daughter Shelley, eleven, were especially thrilled. 'Their reaction was beautiful. An immediate joyful acceptance. Then back to bed they had to go.'[27]

As she later claimed, there was not much of a following for contemporary art in Los Angeles at that time. Freeman's gallerist friend Virginia Dwan kept a list of clients in the city that numbered only twelve in 1965 (and Freeman was not among them).[28] The year after the Freemans' acquisition of *1955-H*, Still acknowledged: 'It is the only work of mine I know of in your part of the country, and its presence at your place will become swiftly and widely known.'[29] Things began to change noticeably only after the American Pavilion in Venice won the Golden Lion in 1964, with an exhibition by Robert Rauschenberg.[30]

Freeman would always claim not to have been a patron of the arts – only a collector; but her status in the city's art scene is evidenced by the fact that she and Stanley were one of six couples, along with the Weismans, who were invited in 1961 to advise the Los Angeles County Museum of Art on its fledgling contemporary programme. By 1965, however, as she admitted to the artist Sam Francis: 'I really am through with the L.A. County Museum Council, as I consider [Maurice] Tuchman a complete fool.'[31] Tuchman was the museum's first curator of contemporary art, a dandified and sociable figure who tended to polarize opinion. But his very position (inaugurated in 1964) was a testament to the city's growing cultural profile – a profile that Freeman and only a handful of others had helped to shape.

Artist friends

Freeman's preference, then and later, was for abstraction. She was one of the first Los Angeles collectors to acquire the work not only of Still, but also of other leading painters of the New York School, among them Philip Guston, Franz Kline, and Willem de Kooning. Many of her early purchases were subsequently sold, but she gradually built up a pre-eminent collection of contemporary art. In the 1960s and 1970s, her taste expanded beyond the New York abstractionists to include the work of Roy Lichtenstein, Bruce Nauman, Claes Oldenburg, and Dan Flavin. Hillcrest Road became a showcase for works by these and many others.[32] By the 1980s, her collection numbered more than one hundred objects.

Before Hockney, the two artists who mattered most to her were Still and Sam Francis. Both were American abstract painters, and yet their distinctness from one another – Still's epic monumentality, Francis's lyricism – attested to the breadth of her tastes. Still was a titan of American art, as reflected by his landmark exhibition at the Albright Art Gallery. He was also notorious for his implacable disdain for the art establishment. Francis, on the other hand, had come to artistic maturity while living Paris in the 1950s (he remained better known in Europe and Japan than in the United States), and his work combined abstract motifs with suggestions of Asian calligraphy and the shimmering colours of Monet, Matisse, and Bonnard. Freeman became friends with both.

For close to a decade, Still and his wife Patricia were regular correspondents of Freeman's. A genuine warmth, always somewhat formally expressed, evolved between them. In the spring of 1961, Freeman hosted the Stills on a visit to Los Angeles. Photographs show her, Clyfford, and Patricia at the Los Angeles County Museum of Art – then housed within the Museum of History, Science and Art – viewing a display of his work. Still dark-haired at this time, Freeman is dressed in a chic pullover and sharp-collared blouse, her hair styled in a bob. Standing in the doorway to the gallery, clutching a floral-patterned handbag, she looks into the room with a smile. Still grins alongside her.[33]

As they wandered around the museum that day, Freeman admitted to Still that the work of Georgia O'Keeffe left her cold. He replied that that

Betty Freeman, Clyfford Still, and Patricia Still at the
Los Angeles County Museum of Art, 1961

was because O'Keeffe was cold herself. He was not short of strong opinions: Bauhaus had ruined New York; Rothko was a 'feminine' painter; Mondrian had taken painting to a dead end. The Stills' visit seems to have passed cordially, except for a dinner at Romanoff's, a trendy night-spot in Beverly Hills. Among the party was LACMA's curator Jim Elliott, who after several drinks voiced a desire to curate a show illustrating the history of art since the Enlightenment era, from Fragonard and Boucher through to Still. Appalled to be bracketed with the frothiest of Rococo painters, Still became so 'excited' – despite having lost his voice – that Stanley Freeman insisted they all leave.

Freeman's notes from the Stills' trip to Los Angeles attest to her profound admiration for him. 'When other people are around S. becomes a giant', she observed. 'Even his height increases & he really appears awesome.'[34] But their relationship was not without tensions. Letters between them increasingly betray an undercurrent: Freeman appearing to tread on eggshells; Still veering between warmth and hauteur. On one occasion, she telegrammed him to ask about showing *1955-H* to a visiting curator from the Museum of Modern Art, New York. Still shot back: 'I would not permit agents of the Museum of Modern Art to enter my hallway. [...] I know them to be arrogant, malevolent and deceitful in all their machinations.'[35]

By this time, Still had angrily renounced the entire professional art world – artists, curators, writers, and dealers; the contemporary scene was, he felt, a 'farce' run by socialite critics.[36] 'I am completely disgusted with the pseudo-scholarship and historical lies or ignorance being set down by the international critics', he fulminated to Freeman. 'And not a little in contempt of the arrogant political gestures in which the art world has involved itself. Especially is it prevalent among the painters as they degenerate into alcoholic insolence, and their reviewers who perform like jockeys on their backs in their race for notoriety.'[37] Patricia Still seems to have rallied to her husband's cause. Referring to the art dealer Leo Castelli, she once told Freeman: 'Mr Still has refused always to have anything to do with him or his gallery. In fact, he has always considered him something approximating vermin on the art scene.'[38]

In the early years of the friendship, at any rate, Still's trust in Freeman was such that he invited her to write a monograph on his art.[39] 'You write very lucidly of your impressions of my work', he wrote in 1961. 'It is unfortunate the professional writers seem totally incapable of approaching the work with comparable clarity and understanding.'[40] Two years later, Freeman sent him a preliminary draft, which he greeted with numerous reservations. Despite his earlier praise, he felt that Freeman did not fully comprehend his art or the intentions behind it indeed, Still seems ultimately not to have trusted in the ability of anyone to communicate those intentions.

Freeman's response was undramatic. 'I have no interest in a career or publicity of any sort. [...] If you don't want the book there will be no book. I only considered myself an instrument to organize and present your ideas.' While admitting that she did not have the grounding of an art historian, she ventured that she was capable of dealing with the impact of the work with 'conviction and awareness'.[41]

They revisited the plan the following year, but Still's misgivings persisted and his increasingly fanciful demands proved impossible to accommodate. In a megalomaniac flourish, he proposed that the book should explain how he had altered the course of nineteenth- and twentieth-century thought, as exemplified by Einstein, Kierkegaard, Kant, Freud, and others.[42] Freeman did not feel able to produce the desired hagiography. Still told her plainly that he did not believe she understood his art ('its issues, it purpose, and meaning').[43]

There was a final attempt to revive the plan in 1968, when Freeman sent Still yet another reworked manuscript. His response confirms his reputation for vitriol, or perhaps outright paranoia

> I must tell you that from my knowledge of your life and attitudes a reading of your manuscript convinces me that you are sharply aware of what you are doing and that a large section of it implies that you are engaged in a calculated and ill-concealed assault on my motives, my person, my work, and my integrity. As I read your paper, the contradictions, your choice of many of your sources, your judgments, your

> omissions and your falsifications of known and provable facts about me and my work suggest a very considerable amount of irresponsibility and malicious intent to injure me.[44]

Freeman wrote back in characteristically restrained style, but with a clear hint of hurt. 'If you had said my manuscript was inept and unintelligent I would accept your judgment. But to accuse me of purposely falsifying for evil ulterior motives is a charge I unequivocally deny.'[45] Her letter is marked 'unopened' in the Clyfford Still archive. There were no further exchanges between them. More than two decades later, Freeman remembered Still, who died in 1980, as a legendary artist but a very difficult man.[46] In 1991, she donated *1955-H* to the Los Angeles County Museum of Art, requesting that it be identified as 'Anonymous Gift, Los Angeles'.

In the early 1960s, Freeman had begun work on a book on Sam Francis, who had moved to Santa Monica after spending a decade away from California, primarily in Paris but also in Tokyo, Mexico City, Bern, and New York. In her manuscript, she described Francis in terms that might equally have applied to herself: 'As much as ideas and dreams of the extraordinary attract him, so do all kinds of concrete objects. The whimsical appeals to him as does the strange, curious or exotic.'[47]

Unlike Still, with whom Freeman always maintained a certain formality, Francis became an intimate confidant in the same way that Hockney later would. Freeman was able to confess her feelings freely, even to admit a level of naïveté that she might not have done with many others. Writing after a trip to New York in October 1966, at which she had attended the sale of a prestigious modern art collection, she remarked:

> I went to the Ross auction with Martha J. [Jackson, the gallerist] and that was interesting in a curious way, as all the men had beards and looked European and all the women looked quite forbidding. I had never been to an auction before where pictures became objects for [the] highest bidder.[48]

1 David Hockney, *Beverly Hills Housewife*, 1966–7 (detail)

2 David Hockney, *Beverly Hills Housewife*, 1966–7.
Acrylic on two canvases, 182.9 × 365.8 cm (72 × 144 in.)

3 John Kasmin, *David Hockney, Venice, California*, 1964
4 John Kasmin, *David Hockney driving his new car, Venice, California*, 1964

5 David Hockney, *Study for 'Beverly Hills Housewife'*, 1966.
Pencil and coloured pencil on paper, 35.6 × 43.2 cm (14 × 17 in.)
6 David Hockney, *Study for 'Beverly Hills Housewife'*, 1966.
Pencil and coloured pencil on paper, 34.9 × 42.5 cm (13¾ × 16¾ in.)

7 Fra Angelico, *The Annunciation*, 1440–5.
Fresco, 230 × 297 cm (90½ × 116⅞ in.)

8 Piero della Francesca, *Flagellation of Christ*, 1459–60.
Oil and tempera on panel, 58.4 × 81.5 cm (23 × 32⅓ in.)

9 David Hockney, *Beverly Hills Housewife*, 1966–7 (detail)

Francis and Freeman shared not only a love of the odd and the offbeat, but an enthusiasm for innovation that bordered sometimes on the messianic. In 1966, Francis was one of a gathering of friends hosted by Freeman in her living room, together with Walter Hopps and a lawyer named Bill Elliott. This was the catalyst for a radical scheme, the New Arts Society – whose aim, as the inaugural statement declared, was: 'To act as a stimulator not only to art but also to science, industry, urban life, even mental health and civil rights.'[49] Behind the sweeping proclamation was a concrete intention to help artists in Los Angeles to forge closer communities. Artists, writers, and figures from other disciplines such as science and philosophy would run the organization, staging exhibitions, happenings, and demonstrations in an anti-institutional spirit. Within eighteen months, the utopian venture had fizzled out. As Francis admitted, there had been a 'slight megalomania' to his vision.[50]

Grand plans were simply part of his nature, and for the most part they centred on his own work, of which Freeman was a staunch supporter ('I should tell you that I am full of the most enormous enthusiasm for your work. Wow!'[51]) She purchased several of his paintings and paid for the restoration of his seminal *Basel Murals* (1956) – a triptych of three vast canvases, each one a clustering, ethereal nebula of colour – after two of the three canvases were damaged by water in the late 1960s. (*Basel Mural III* was broken into four smaller panels, under her direction, two of which are now in the Norton Simon Museum, Pasadena.)

Freeman's book on Francis was completed in 1969. Like that on Still, it remained unpublished and largely unseen. The obstacle was more mundane: the release of a rival publication meant that there was no interest in publishing hers. But on this occasion, the abortive project does not seem to have impacted the friendship. Freeman remained close with Francis until his death in 1994, visiting him regularly at his home and various studios in Santa Monica, and often indulging in their shared love of sushi.

A photograph taken by Francis in 1976 shows Freeman at his house, grinning as she tries on a towering basket hat (p. 51). The headwear is a bizarre foil for her dark jacket and turtleneck sweater, but it seems peculiarly apt – peculiarly *her*. The formality in her (the quality that had pervaded her letters to Still) was always looking, one senses, for an outlet.

Betty Freeman and Sam Francis in the artist's home studio,
Santa Monica, 1976

Betty Freeman, 1976, photo by Sam Francis

Closing the piano

Into the 1960s, Freeman's piano playing had remained a private affair. At home or at her teacher's house, she sometimes performed works by Schubert, Beethoven, Mozart, or Bartók. But despite her countless hours – years – of practice, she never played professionally, only for 'captive friends'.[52] In a letter to Still in the spring of 1961, she related:

> I made my recital tonight and am feeling very happy as it seemed to go well and gave pleasure to the listeners. It was small, about 30 people at my teacher's house, but I worked hard for it and it was important to me, being an unprofessional. Now I am free to start some new pieces & am going to concentrate on Schumann, whom I love & who is very difficult to perform.[53]

One day in 1964 came a realization. She had just finished playing, and felt that she had played more brilliantly than ever. It was the kind of virtuoso performance that she had dreamed of for twenty years. And yet she knew, in that moment of triumph, that she would never be better. 'I closed the piano and never touched it again. I was very conscious of what I was doing. I was practicing six to eight hours a day, and all of a sudden I realized what it takes to be a concert artist. And I knew I didn't have it. You *know* these things.'[54]

Concealed within that single emphasis is the weight of her unrealized ambition, her years of dedication, and her dawning awareness that she was not – would never be – as good as she wanted. Throughout the early years of her marriage, piano playing had been the one facet of her identity that was not annexed to her role as a wife and mother, the one respect in which she was more than just a Beverly Hills housewife. And yet whatever pain the decision involved seems rapidly to have dispelled. In the spring of 1964, she was invited to take part in the organization of a new series of avant-garde music performances at the Pasadena Art Museum. She soon decided that the collaborative work of producing music events was preferable to her lonely existence at the piano; and so the piano stayed closed.[55]

Over the years, Freeman would frequently retell the story of her decision to give up playing. Sometimes she mentioned that having four children was a factor.[56] But the account was never rueful, always matter-of-fact. 'It didn't seem to bother her one bit', says her elder daughter, Shelley. 'She was like that with everything. When it was done, it was done. It was as nothing.'[57]

She was equally unsentimental about the unpublished books on Still and Francis. Even so, the knowledge of her private failures adds a new layer of import to *Beverly Hills Housewife* – painted two years after she gave up the piano, and while she was still striving to complete the two books. There is an edge of melancholy to the picture's luminous calm. In the still, poised figure of Freeman, it is possible to imagine a deepening stoicism, almost as though she were expressing a silent response to the injunction expressed by W. B. Yeats in his 1913 poem 'To a Friend Whose Work Has Come to Nothing':

> Bred to a harder thing
> Than Triumph, turn away
> And like a laughing string
> Whereon mad fingers play
> Amid a place of stone,
> Be secret and exult[58]

Freeman's writings of the 1960s, both her letters and her books, mark out a hinterland that *Beverly Hills Housewife* fails to disclose. Her talents went largely unrecognized at this time, although Hockney's painting constitutes its own unspoken recognition. Perhaps, after all, he discerned the 'laughing string' in her – the secret exultation.

Mrs Stanley Freeman

In a letter to Still in 1961, Freeman's printed letterhead describes her as *Mrs Stanley Freeman*. 'We have just come back from four weeks at Bass Lake, where we have been going each summer. We bought a little house there the end of last year. It means so much to the children to have something of the simple life.'[59] A rural town in the Sierra National Forest, two hundred miles from San Francisco, Bass Lake offered an idyllic retreat with its ponderosa palms, clear lake, and uneventful days.

Freeman did not talk much about her feelings, and it is impossible to know what emotional undercurrents the sunny tone of the letter might have concealed. But from the start of her marriage, she had known that she and Stanley were ill suited. He loved shooting and hunting – their son Robbie recollects being taught by his father how to clean and shoot a rifle. Betty did not share this love. Only once did she accompany her husband on an expedition to Africa – and she hated it. Stanley was similarly disenchanted when she took him on a tour of museums in Europe.[60]

Towards the end of Freeman's letter to Still, the focus shifts – as if to the true point of interest: 'I was sorry not to be in town when you and Mrs Still were out here. But I do go often to the museum to see your paintings. They are a constant source of beauty and magic, always changing yet always there.' In this simple statement, one gains a sense of the deeper sensibility of Mrs Stanley Freeman – her belief in the potential for a work of art to become 'new' again in every moment of viewing.

3
Turning

'The philosopher Attalus used to say that it was more of a pleasure to make a friend than to have one, "in the same way as an artist derives more pleasure from painting than from having completed a picture".'

SENECA, 'LETTER IX', 1ST CENTURY AD[1]

When *was* this? All we see is a woman in a domestic setting, arrested midway between somewhere and somewhere else. We have not yet learned any of the particulars of her life. At one level, the picture has a precise coordinate in space and time – in the garden, on the day that Hockney visited. But time also dilates, abstracts, in the picture; it seems alternately to condense and expand. The art historian Andrew Causey once remarked that Hockney's early Californian art is 'concerned with the removal of art from temporality – the way representational images, as well as abstract ones, can be made to seem, at least, both "prior and final"'.[2] The fragment of reality presented by *Beverly Hills Housewife* might equally be a beginning or an ending, or neither.

Looking at the painting, all the same, one senses Hockney's presence. There is a powerful, persistent impression that we are seeing what he sees, observing Betty Freeman from his vantage point across the garden. It was an effect, coincidentally, that Freeman had herself noticed in a work by Clyfford Still: 'It is almost as if you were at the center, at the vortex of life itself, and are able to reveal this to us who are further away.'[3] The effect is common to a number of Hockney's paintings from this period, including those where a blank-canvas 'frame' surrounds the picture, such as in *A Bigger Splash* (1967), reminding us always that this is a mediated view, Hockney's view, for all its pugnacious directness. In subsequent pictures, he amplified the effect by having the subjects of his portraits look back – at the artist, at the viewer – as, for example, Don Bachardy does in *Christopher Isherwood and Don Bachardy* (1968), or Celia Birtwell does in *Mr and Mrs Clark and Percy* (1970–1).

'From the very beginning, literature, homosexual desire and tributes to friends came together as sources in David Hockney's work', according to Edmund White.[4] In an unstrident way, *Beverly Hills Housewife* bears out the statement. The sense of a passing moment, a snapshot from a larger narrative, endows the painting with the quality of a short story – or the opening sentence of one. The fascination with a glamorous woman was itself a kind of corollary of homosexual desire – compare Andy Warhol's apotheoses in silkscreen of Marilyn Monroe and Elizabeth Taylor, or Francis Bacon's more earthly fascination with Henrietta Moraes, the fabled 'Queen of Soho' in the 1960s. And, perhaps unconsciously, the painting was the expression of a dawning friendship.

Case study

Hockney's original idea of depicting Freeman's pool had quickly given way to a desire to paint the collector herself, framed by her domestic habitat. The Polaroids that he took in her garden reflect an ongoing fascination with scenic details: the palm fronds, the sliding door, the outdoor sculpture. But those details were secondary, now, to the human subject. In Freeman, perhaps for the first time, he had found a protagonist. A subject rather than a generic object.

The drawings he made in advance of the painting emphasize the dynamic between the human figure and the architectural setting. In two sketches in coloured pencil (figs 5 and 6), Freeman is set within a simplified geometric rendering of the house and garden. Hockney homes in on the porticoed structure at the back of the house – isolates it and pares it down to suggest a far more modernist property than Hillcrest Road in fact is. In this regard, he was adjusting reality to his own ideal. Before coming to Los Angeles, he had been intrigued by the 'Case Study' photographs of Julius Shulman – a series of shots that documented new homes by leading architects in LA and its environs; the constructions

Julius Shulman, *Richard Neutra, Kaufmann Desert House, Palm Springs, California*, 1947

were conceived as high-end prototypes for affordable postwar housing. In *Beverly Hills Housewife*, the back terrace of Hillcrest Road became a slice of 'mid-century modern', akin to the starkly rectilinear structure of the Kaufmann Desert House designed by Richard Neutra, a Palm Springs retreat photographed by Shulman in 1947.

The house plays a certain version of itself; its more classical elements are banished from sight. The same holds true for the picture's human subject. Hockney wanted to portray Freeman, but he also wanted to make her a character. Perhaps it was this impulse that Stuart Preston had perceived when he wrote that the artist's pictures were 'intensely literary, meant to be "read" like little stories'.[5]

In the studio

Hockney painted *Beverly Hills Housewife* in his studio on Pico Boulevard, near the junction of Crenshaw Boulevard. The space was so small that he could never stand back more than five feet from the work. Measuring six feet by twelve, it was the largest painting that he had made since *A Grand Procession of Dignitaries in the Semi-Egyptian Style* (1961; a frieze-like vista of box-shaped marionettes), so large that he was forced to divide it into two panels, painting each half separately. He worked in acrylic, having lately decided that American acrylic pigments were superior in quality and brighter in tone than the ones he had tried in England. The flattened, simplified forms of the pink dress, the grass, the sky are all indicative of the lucid style that he developed in this medium, which would dominate his output for the next decade.

It was the first painting that he had based on black-and-white photographs. Freeman recalled having asked him, that day in her garden, how he could work from colourless pictures. 'I can only work from black-and-white photographs', he told her, 'because the colour of photography is never the same as real life.'[6]

It took many months to paint, extending into the spring of 1967. During this time, Freeman was contacted by Hockney's dealer Felix Landau, who planned to exhibit the picture in his New York outpost once it was finished, asking if she wanted to view the work in progress. She paid a visit to the studio, later remembering the run-down surroundings and the cramped confines of the space, where a beaded curtain separated the studio proper from a back area with a stove. Upon seeing the painting, she was not impressed. She had not commissioned the work and felt no obligation to like it. 'It was all monochrome, there was no colour.'

A month or two later, Landau called her again and suggested a return visit to Hockney's studio. 'I didn't like it again. He'd added some details but it was still flat, with no colour – no special interest.' Another month passed, and the painting was finally finished. She agreed to go once more. 'And then, I flipped.'[7] Everything had changed. Hockney had introduced the luminous green grass, the blue sky, and numerous other details, including her Le Corbusier recliner. She bought the painting for $4,500, in advance of its exhibition at the Landau-Alan Gallery on

Madison Avenue. There it appeared alongside Hockney's other recent scenes of California, including *Portrait of Nick Wilder*, *Hollywood Garden*, and *Sunbather* (all 1966).[8]

Freeman's reactions to the painting as she visited Hockney's studio offer an insight into her nature – decisive and insistent, but not incapable of change. Those reactions would be mirrored, some forty years later, when she commissioned an opera from the Finnish composer Kaija Saariaho, *L'Amour de loin*, which was performed in Salzburg in 2000:

> Peter Sellars directed it on water: somebody loved somebody from afar and had to paddle across this water to get there and when he got to her he died. [...] Since I was the commissioner, I saw it the first time: I didn't like it. The second time, I didn't like it. The third time, I didn't like it. The fourth time, I didn't like it. The fifth time, I began to like it.[9]

One senses, in her eventual warming to the painting, a deepening regard for the artist. Her suggestion for the title had been made in a spirit of gentle irony – and she came to admire the same faculty in Hockney, stating towards the end of her life that 'what I respect is his outlook on life: he's a great satiric person. [...] He sees the foible in human life.'[10]

Freeman's final admiration for the picture was in pronounced contrast to the response of Fred and Marcia Weisman to a portrait that Hockney made of them the following year (fig. 17). Like Freeman in *Beverly Hills Housewife*, the figures in *American Collectors (Fred and Marcia Weisman)* (1968) are imagined somewhat as objects – surrounded by actual art objects in their backyard (a Henry Moore sculpture, a totem pole, a William Turnbull stack of boulders). Fred stands in rigid corpselike profile, his hand surreally bleeding flesh-coloured paint; Marcia gazes out with a grimacing smile. The difference, here, is that there is no longer an inkling of the subjects' interiority. Marcia's expression is a double of that of the painted face of the totem pole. Life has become a faintly gruesome mimicry of art.

American Collectors shares with *Beverly Hills Housewife* a fascination with surface and externality. We find the same even light, the same sleek panes of glass and perpendicular structures. But surface is all there is. It is little wonder that the Weismans did not like the picture.

According to Peter Goulds, Hockney's long-time dealer in Los Angeles: 'Marcia didn't like the fact that he depicted her face to look like the totem pole. And you notice that she's dressed not in the same elegant way as Betty in her painting. Marcia is the quintessential "bathrobe woman", and next to her is the businessman who goes out to earn the money to pay for her pleasures.'[11]

Turning

'Dark and dull is your distraction:
Walk then, come,
No longer numb
Into your satisfaction.'

W. H. Auden, 'Underneath an Abject Willlow', 1936[12]

Betty Freeman is half-turned, seen at a three-quarter view. She seems to be pivoting into her own private dimension, indifferent to her immediate surroundings. She stands apart from the contained world of the picture; apart, too, from the world beyond it. Aged forty-five, Freeman was at a pivotal phase in her life. The objects around her are the attributes of her married life, specifically the antelope's head and the zebra skin that has been stretched over a recliner designed by Le Corbusier – trophies of her husband's hunting expeditions. The house itself, which she had lived in since 1959, is a statement of their life together. And yet Freeman is looking away from these objects, from the entire setting, into an unspecified middle distance – or perhaps into some remoter place of the mind.

Freeman and her husband would not divorce until 1971, but they were living largely separate lives by 1966. At the centre of the picture, in the midpoint of life, she is simultaneously rooted in time and place – defined by everything around her (house, possessions, the very brilliance of the day) – and aloof, cut adrift: a ghost in her own home, floating over the manicured surfaces.

Freeman would remain at Hillcrest Road for the rest of her life. She was always, in a literal sense, a Beverly Hills housewife; and Hockney's painting saw to the fact that she would always retain this identity. But one way or another, she reinvented that identity. At times, she reduced it to an ironic moniker; at others, she imbued it with new possibility.

In the years following the painting of the picture, she cast off the role that had defined her for the past two decades. She became, instead, one of the most energetic and original champions of contemporary classical music of her era. Newness and reinvention became precepts for her, both in the music that she supported and in her larger outlook. 'She did the standard thing of marrying a nice man who had money, and having four kids', says Julia Bloomfield, former head of publications at the Getty Center, who befriended Freeman in the late 1980s.

> She hadn't developed quite into being who she became. After that, she was free to become herself. She became Betty Freeman. A most unusual and powerful and wonderful and wacky woman, with this extraordinary love for the contemporary. She was totally bored by old music or old paintings. She appreciated them but was bored by them. She wanted something new.[13]

At the time of *Beverly Hills Housewife*, this new life lay mostly in the invisible distance – somewhere beyond the gaze of the woman in pink – although its course had been set.

In the painting, Freeman is viewed in a single moment. But the moment itself appears to have slowed or expanded. Time has taken on an elastic quality in the transition from the original Polaroids to the final portrait. Rendered at the scale of a history painting, the scene has become obliquely historical, almost in spite of itself.

Within that single moment, lurking like ghosts, are other moments – longer histories. Glancingly, she resembles a caryatid, one of those stone maidens in fluted drapery, holding up the roof of an ancient temple. Her lowered arms, squarely planted feet, and erect bearing equate her concurrently with the archaic *korai*, or carved female figures, of ancient Greece in the sixth century BC, which originally functioned as tombstones. Such statues were generic in appearance, their bodies rigid and their faces composed of featureless planes. They served as representations of known people, but were equally monuments to womanhood in a universal sense. By contrast, the classical sculptures of the following century took on a more individual – more 'lifelike' – aspect. They are possible to interpret as particular characters: Apollo, say, or Aphrodite. Not just anybody, but a goddess.

Hockney's portrait, which conflates the type of the Beverly Hills housewife with the particular subject of Betty Freeman, expresses an ancient gravitation. Like a statue at the cusp of the archaic and classical epochs, it equivocates between rigid anonymity and trueness to life. In the same way as the bronze Charioteer of Delphi, say, a lone figure (cast around 470 BC) that was once part of a larger group, Freeman seems to have been extracted from a narrative context – becoming an image of nothing more than herself, while also demanding to be placed in context, reintegrated with the world. She is caught between the status of an epitome and the ordinariness of a woman who has just stepped out of her house.

Then there is her averted gaze. That semblance of obliviousness or dreamy inattention. Her *looking away* is another way in which she invokes an ancient, specifically classical, paradigm. It is the same gesture that characterizes many Greco-Roman statues: the figure (usually a deity) turns away, as though to evade or resist the gaze of the viewer. The classicist and art historian Jaś Elsner has theorized the development of naturalism in classical art in relation to the averted 'glance' of a statue such as the Kritios Youth (*c.* 480 BC). Unlike the frontal 'gaze' of earlier statuary, the glance is retiring. It alights 'on a specific point which happens not to be in the world of the viewer's experience but in that of the statue's own imagined experience'.[14]

Freeman looks away to a point in her own imagined experience – a point that the painting's viewers cannot access. Her glance prefigures that of Christopher Isherwood in the portrait that Hockney made of the novelist and his artist partner Don Bachardy in 1968. Here, however, we find a triangulation of gazes: Isherwood turns to look at his lover, who looks back at Hockney/the viewer. In *Beverly Hills Housewife*, there is no such closure. The scene is a record of Hockney's perspective, but he is not part of it.

And yet, the longer one looks at the picture, the more it seems that there *is* an affinity between the two of them, a kind of telepathic wavelength. Freeman's gesture of looking away does not simply connote inattention. She is aware – she has to be – of Hockney. Her averted gaze comes to seem like an expression of that awareness, a covert form of response. It is impossible, after seeing the source Polaroids, to regard the painting in any other way. She was nothing if not aware of the camera's presence, and that veiled awareness persists in the final picture.

Clearly, though, being seen by Hockney was a matter of remaining aloof. Becoming a muse entailed a standing apart and a retreat into herself. Perhaps it was a way of shoring up her sense of self, asserting her separateness from everyone and everything.

Hockney himself was at a turning point in 1966. His affinity for America was growing. So, too, was his preference for a sharp realist style that marked a new departure after the faux-naïve expressionism of the paintings he had made in London, whether *Doll Boy* (1960–1) or *We Two Boys Together Clinging* (1961). In these pictures, human characters were no more than cardboard cut-outs: ciphers through which he might convey a mixture of desire, daydreaming, and small-scale reality.

Hockney's portrait of Freeman at a decisive moment in her life was inevitably self-reflexive, an expression of his own 'becoming'. He was crystallizing his early Californian style and, with it, his sense of self. This reorientation was also a matter of looking back, particularly to the artists of the Italian Renaissance, artists who in turn had looked to ancient models in their cultivation of a new mode of realism. *Beverly Hills Housewife* might indeed be regarded as an annunciation in the mould of Fra Angelico's fresco of 1440–5 (fig. 7), with Los Angeles taking the place of the angel: a disembodied, everyday epiphany.

For Hockney, then, 1966 was a moment of personal and artistic rebirth, a turn towards the 'real' that carried echoes of older artistic rebirths, whether those of the ancient world or the Renaissance. And he found a mirror and a catalyst for his own renewal in the woman who stood across the garden from him, that day at Hillcrest Road. In his painting, she looks away, but not necessarily into empty space. If anything, she might be gazing into the depths of her imagination – and at the boundlessness of her future.

Quasi religious

Beverly Hills Housewife has an air of incipience, of things being about to occur. Freeman is stepping away from an old life, an abandoned ambition. Seen in hindsight, it is also a painting of the forming of a bond. It records Freeman at a time when she was essentially unknown to Hockney, and he to her. Yet the blank planes and pools of colour somehow embody their reciprocal intrigue. Freeman's stance, her dress, her immaculate house all present themselves to the artist's eye. That eye is felt, in turn, in the colour and clarity of the scene.

John Kasmin has a vivid memory of first meeting Freeman: 'She had a sort of atmosphere about her, as though she were in Cloud Cuckoo Land. She wasn't dumb, but she appeared to be – she had a slight appearance, on first meeting, of a dumb broad with money. But she knew what she wanted to do. She seemed genuinely to like the things that she liked.'[15] What Kasmin first perceived as ditziness was really a manifestation of her self-assurance. Her deep-seated knowledge of what she wanted to do, of the things she liked, endowed her with an air of separateness from the world, a quality of belonging to some other time. He adds: 'She was unpushy: entirely different, say, from Marcia Weisman. She didn't seem clued-in, on the ball, pushy, driven by ambition. She was really unlike most art-buying women. She was quiet; she reserved her opinions.'

It was that appearance of holding herself in reserve that Hockney translated into his painting. There is an impression in *Beverly Hills Housewife* that the woman might just have drifted in from a different realm. She has the bearing of a figure in an early Renaissance altarpiece: a saint or an angel. The sparing execution of the scene recalls the works in tempera of Duccio and Giotto. Freeman's face is composed from a thin, near-uniform layer of flesh-toned acrylic, interrupted by the reflective pink of her lip and the flecks of darker tone that signify her eyes, eyebrows, and cheekbone. And at one level, the picture *was* a piece of hagiography: Hockney's idealized version of the woman he had photographed at home. A saintly version of the actual Betty.

Flagellation

Between the summers of 1966 and 1967, in his studio on Pico Boulevard, Hockney painted a series of pictures that his friend Henry Geldzahler would later hail as the high point of his 'California manner' and indeed of his entire career to date. 'All are detached and crystalline,' Geldzahler wrote in 1976, 'all breathe a clarity of light, perception and realized intention that mark Hockney's new and greater intention to paint the world of today head on.'[16]

That desire to confront the 'world of today' was charged nonetheless by a historical awareness. When he had first visited the National Gallery in London, Hockney had been struck by the near-shadowless vision of Piero della Francesca's *Baptism of Christ* (1437–45), a painting he already knew from a reproduction in a book.[17] His works of the 1960s and 1970s allude in numerous ways to the tempera masterpieces of the early Renaissance painter. The showering man in *Domestic Scene, Los Angeles* (1963), for example, bears a faint resemblance to a background figure in Piero's *Baptism*, who bends over – pulling his shirt off – in readiness to be baptized.[18] Other references are more explicit. *Looking at Pictures on a Screen* (1977) depicts Geldzahler inspecting a series of taped-up reproductions of Western masterpieces, Piero's painting among them. The *Baptism* also appears fragmentarily, in the style of a footnote, in *My Parents* (1977): a reproduction on Hockney's studio wall is partially captured in a mirror between his elderly mother and father.

In addition to direct allusions of this kind, Piero's meticulous style and careful staging of elements are a pervasive presence in Hockney's paintings of the mid-1960s. He was pursuing a more 'realist' register, but also aiming for a sense of the theatrical – static, staged, and vaguely 'primitive', in the manner of the early Renaissance. His tendency, in the double portraits, for presenting one figure frontally and the other side-on – as, for instance, in *Henry Geldzahler and Christopher Scott* (1969) – can be related to the positioning of Christ and John the Baptist in Piero's *Baptism*.

Beverly Hills Housewife refrains from direct allusion to any particular model, but the picture's stilled ambience and compositional structure carry echoes of a painting of the mid-fifteenth century by Piero, the

Flagellation of Christ (1459–60), now in the Ducal Palace of Urbino in the Marche region of Italy (fig. 8).[19] In this small work in tempera on panel, the main subject has receded into the background. On the left of the picture, in the middle distance, Christ is seen being scourged by three men within the recess of a classical portico. Pontius Pilate looks on from a throne: the Passion is playing out in the governor's palace, framed by Corinthian columns and a coffered ceiling, although the setting is redolent of Piero's fifteenth-century Italy rather than Jerusalem in antiquity. To the right, in the foreground, stand three men in shin-length robes. The first of these men is bearded and middle-aged. The figure in the middle is angelically youthful. The third is stocky and balding, and wears a more ostentatious robe, an elaborate gold brocade on deep blue. Their identities have never been securely established. They seem dislocated from one another, juxtaposed rather than interacting. The declarative import of the background scene is countered by the ambiguity of their personae and placement. Some interpretations have proposed that they are allegories (the three ages of man, perhaps); others, that they have specific identities, for example as rulers of Urbino.[20] The art historian Kenneth Clark conjectured that the bearded figure represents 'a type of learned Greek'.[21]

Beverly Hills Housewife echoes the spatial arrangement of Piero's painting: the interior scene opens out into the picture plane while also receding by staged intervals, through a sequence of thresholds, with a vaguely Hitchcockian sense of zooming in and zooming out simultaneously. There is the impression, as in Piero's *Flagellation*, of the viewer's space extending into the picture, as though the same ground were shared between the two; and the same air of pristine stillness, condensed above all in the central figure. The various minor details of Piero's scene – the tree above the young man, or the gilded statue of a nude god on the column above Christ – themselves anticipate the various disconnected things of Hockney's picture, arrayed in static interplay.

'No other painting which has come down to us, except perhaps Uccello's *Deluge*, is so complete an expression of the Renaissance *mystique* of measurement, known as "perspective"', writes Clark of Piero's *Flagellation*.[22] *Beverly Hills Housewife* expresses its own mystique of measurement. The exactitude of the scene works to amplify the pervasive air of enigma. Hockney's picture channels not only the structure of Piero's, but its guiding tension – that between serene clarity and pregnant ambiguity. The very lack of dramatic action in Piero's painting, aside from the vignette of the flagellation itself, has been felt to invest it with a peculiar charge. In the absence of a 'main event', the effects of light – the very interplay of space and volume – achieve their own dramatic agency. For Bernard Berenson, *the* connoisseur-art historian of the Italian Renaissance, the work was 'of all Piero's paintings the one I love best, the one that satisfies me more and more every time I see it', precisely for its lack of incident: 'No feeling appropriate to action [is] expressed in face or figure, whether of the participants or of the mysterious trio in the foreground.'[23]

Hockney restates the numinous inactivity of Piero's scene. The stasis of his central figure, the dissected facets of the house, the precision of the palm fronds – all express an abstraction away from narrative, an eschewal of any single emotional register. Like the biblical sideshow in Piero's painting, the particulars of Freeman's existence (the objects around her, the interiors of her house) have become secondary. *She* is the dominant, unanswerable fact of the picture. Her isolation and self-possession, even the three-quarter angle at which she stands, evoke the young man at the centre of Piero's trio, who is part of the scene and yet in some deeper sense apart from it, gazing into space.

Art historians have sought repeatedly to impart identities to the three men in the *Flagellation of Christ*. The desire to assign meaning operates in a similar way for the viewer of *Beverly Hills Housewife*. Who is she? Where has she come from? In giving the painting its title, Freeman was at one level asking these questions of herself: stepping back from her own image, anonymizing and ironizing it.

The youth at the centre of Piero's group has the same aura of separateness, the same quality of beauty crossed with unknowability. In *Beverly Hills Housewife*, Hockney cast Freeman in this character's image, albeit obliquely – possibly unconsciously. In doing so, he was also projecting himself into the scene. The blond quattrocento youth – captured on the brink of maturity, striking a steadfast pose – might indeed be regarded as a double of the artist, a kind of second self. To discern the Renaissance revenant within the figure of Freeman is also to apprehend how Hockney and Freeman are mutually present in the picture, indivisibly connected, as though one were a reflection of the other.

One theory posits that the three men in Piero's *Flagellation* represent Saint Augustine's 'three modes of the consciousness of time'. According to this idea, past, present, and future exist only in our present-tense apprehension of them: the past in the form of memory, the present in the form of live concentration, and the future in the form of anticipation. Art historian Franck Mercier argues that the bearded man represents consciousness of the past; the right-hand figure in the brocaded robe represents consciousness of the present moment, his profile view 'reducing any depth, spatial or temporal, to a minimum'; and the young man at the centre 'looks toward a "beyond" of the picture that merges with the future'.[24] It is also possible to see the scene of the Passion in Piero's painting – as many commentators have – as a scene from *back in time*. That was then, in the irretrievable biblical past; this, the gathering in the foreground, is now. In this way, the spatial recession of the picture is also a temporal recession.

It is possible to infer a similar telescoping of time in *Beverly Hills Housewife*, even if the movement from past to present is subtler – more incidental. The picture seems, at first glance, to dispense with any sense of past or future: Freeman is seen in the present moment, and only there. And yet the house itself, as the scene of her married life, is not merely a backdrop but a marker of time. The receding interiors are indicative, perhaps, of the relative unimportance of the past, so far as Freeman was concerned – but they are reminders too of the past's ineluctable *presence*, its indelibility. Seen another way, those same narrowly glimpsed interiors might also be felt to connote the future. Half of her life lay ahead of her, in 1966, and Hillcrest Road would remain her home for all that time.

David Hockney and Betty Freeman at 703 Hillcrest Road, 1967, photographed by John Kasmin

Youth and age

For all its specialness, Hockney and Freeman's bond evoked a pattern – that of the mutual fascination of the gay man and the older, sophisticated woman. Andy Warhol thrived on such fascinations; he once said of Eleanor Ward, the art dealer in whose gallery he first met Hockney:

> Eleanor was an absolutely beautiful, aristocratic woman. She could easily have been a model or a movie star – she resembled Joan Crawford – but she loved art so much, and she just lived for it. She felt every artist in her gallery was one of her babies, and she called me her Andy Candy.[25]

There are countless precedents for this kind of attraction, whereby the woman occupies a curious place between the individual and the ideal, by turns herself and an image of something other than herself. A notable – notorious – example was Truman Capote's enthralment with the New York socialites whom he called 'the swans', although in Capote's case, adoration sometimes came with a hidden sting of irony. When the writer threw his legendary Black and White Ball at the Plaza Hotel in 1966, the guests included Gloria Vanderbilt, Lee Radziwill, and Barbara 'Babe' Paley.[26] Capote would go on to alienate these friends through his too-thinly fictionalized portrayal of them in 'La Côte Basque, 1965', an article in *Esquire* magazine in 1975. But in the gilded era of the mid-1960s, the affection in both directions was powerful, and the idealized vision still held strong. Of Paley, he wrote in his diary: 'Mrs. P. had only one fault: she was perfect; otherwise, she was perfect.'[27] In a more serious vein, he told his biographer, Gerald Clark: 'She was the only person in my whole life that I liked everything about. I was her one real friend, the one relationship she ever had. We were like lovers; she loved me and I loved her.'[28] Something of the same love and understanding came to define the relationship between Hockney and Freeman. At first, each represented the allure of the exotic – the other. Over time, fascination deepened into familiarity.

In her unpublished book on Sam Francis, Freeman had quoted some verses written by the artist upon the death of his older patron, Franz Meyer, in 1962. In these, it is possible to discern something of the dynamic that would evolve between her and Hockney:

> We had the meeting of youth and age.
> He let me share my secrets with him and refrained
> From old man's laughter which is most often tragic.
> I found him someone to give my blooms disinterestedly.
> My youth pretended to be old and he used my
> dreams to hide his deepest lines.
> Exchanging masks we looked over that middle ground.[29]

A specific portrait

The woman in the pink dress, like each of the figures in the foreground of Piero's *Flagellation*, or equally like one of Warhol's objects of worship (the models and movie stars and those who resembled them), wavers between the status of a personalized portrait and a type. Like those figures, she stands subtly apart from her setting, framed by it but not entirely part of it – as though nothing could wholly contain her.

Los Angeles itself had begun as an idea, its own sort of dream image, hardening only later into a concrete entity. In an interview in 1975, Hockney claimed that his earliest stay in the city marked 'the first time I had ever painted a place [...] In London, I think I was put off by the ghost of Sickert, and I couldn't see it properly'; but for a period, that portrayal took a generic form.[30] *Building, Pershing Square, Los Angeles*, which he painted in February and March of 1964, expresses the blank civic reality that he found in downtown LA in place of the seething hedonism of John Rechy's *City of Night*. The Pacific Mutual Life insurance building appears as a stack of dappled grey boxes, one atop the other, fronted by papery palm trees. A single pedestrian on the street features as a dark blot.

A similarly deadpan vision is conveyed by *White Building and Clouds* (1965), in which a white tower block seems to hover against a blank sky, with three clouds streaking far above like Matissean cut-outs. The painting, which was based on the 1965 Sierra Towers building that looms over Beverly Hills, was later acquired by Freeman. Here again, a simplified and flattened architectural structure invests the work with a modernist air – as though geometric abstraction had sprung into the guise of a representational picture. 'In 1965 I painted probably some of the most abstract pictures I'd done,' Hockney remarked a decade later, 'influenced, I think, by American cool abstraction. [...] But of course what made it very different was that I was using abstraction as my subject, commenting on it.'[31]

Human characters, where they occur in the early Los Angeles pictures, tend to be either clichés or ideals: the puppetlike woman in her stage-set house, the lithe nude (his head a pearly blank) of *Man in Shower in Beverly Hills* (1964). In the autumn of 1965, Hockney produced a set of six lithographs, 'A Hollywood Collection', that distilled his impressions of the city's art-collecting tastes into wry emblems.[32] Most were satires of heftily framed art he had seen on the walls of LA collectors, for instance, *Picture of Pointless Abstraction Framed Under Glass*, a tongue-in-cheek generalization.

Hockney therefore overlaid the reality he had encountered with a veil of ironic impersonality, or with his own imaginative preconceptions (compare Warhol's perception of his road trip to Los Angeles in 1963: 'The farther West we drove, the more Pop everything looked on the highways'). A similar overlaying can be discerned in a series of prints that he went on to make in 1966, responding to the poems of Constantine Cavafy; these extend Cavafy's own treatment of Alexandria as a mental (as much as a physical) entity. Having already visited Alexandria in 1963, Hockney went to Beirut 'to get some atmosphere [...] I found more in Beirut than in Alexandria because when I went to Alexandria all the Europeans had left.'[33] In Los Angeles, too, idea and actuality became meshed – impossible to separate. The myth and the real-life experience were concomitant.

With *Beverly Hills Housewife*, impersonality began to give way – if only a little – to a sense of the particular. Hockney simplified the scene as it appears in the photographs (in reality, there were three antelopes' heads); he abstracted away from the minute detail of 'real life'. The bars of light on the glass, rather than being based on direct observation or the evidence of photographs, were copied from illustrations in catalogues – a notation of the type used by Roy Lichtenstein in his 'Mirror' paintings of 1969–72.[34] And yet, as he put it in 1976: 'I did a specific portrait and a specific house, a real place that looks like that.'[35] However stylized, the painting was a record of what he had witnessed.

As a picture of a nameable (albeit unnamed) individual, it was the first portrait he had embarked on since a painting of his father in 1955. In the years in between, he had used real people as models – his friend, Mo McDermott, for example – but he had simplified them to the point of being unrecognizable.[36] The idea of this being *any* Beverly Hills housewife converges with, perhaps conflicts with, the specifics of this woman, this house, this moment in time. *Beverly Hills Housewife* thereby carries to an extreme the two fundamental components of portraiture, as theorized by the German art historian Erwin Panofsky:

> A portrait aims by definition at two essential and, in a sense, contradictory qualities: individuality, or uniqueness; and totality,

> or wholeness. On the one hand it seeks to bring out whatever it is in which the sitter differs from the rest of humanity and would even differ from himself were he portrayed at a different moment or in a different situation; and this is what distinguishes a portrait from an 'ideal' figure or 'type'. On the other hand it seeks to bring out whatever the sitter has in common with the rest of humanity and what remains in him regardless of place and time and this is what distinguishes a portrait from a figure forming part of a genre painting or narrative.[37]

Hockney's version of Freeman has an air of subtle yet profound *difference*. She is a symbol, seemingly, of nothing other than herself. At the same time, the title of the painting beckons us to see her as symbolic in the more usual sense – an instance of a larger pattern. As an archetype. The inscription on the back of the canvas, which is slightly longer than the commonly stated title, reinforces the idea: 'A Beverly Hills housewife. Los Angeles, 1966–7 David Hockney'. In the end, however, it is not so much her status as *a* housewife that connects her with 'the rest of humanity', but the very quality that had appeared to set her apart – that air of aloofness or separation, of being suspended for ever within her own consciousness.

A painting begun slightly later, *Portrait of Nick Wilder* (1966), portrays the art dealer standing in the swimming pool of his apartment building in West Hollywood, up to the chest in water. Like Freeman, he appears at the centre of a larger picture, a small detail but also the focal point. His head and shoulders rise like a portrait bust above a patchwork of swaying blues. He is framed by the nondescript setting: the blank facets of the building, a curtained window, tidy crops of greenery. Hockney's friend Mark Lancaster had taken preparatory photographs of Wilder standing in the pool, and the artist referred closely to these in the creation of the painting. His tendency to schematize is apparent, however: far from being a 'photorealist' depiction, Wilder's face is as simplified and impassive as that of a Modigliani. His surroundings have been stripped of individualizing details.

All the same, the painting marks a turning point. The inexpressive, sun-kissed 'figures' of Hockney's earlier pool paintings have consolidated into a known quantity. 'This was about the first absolutely specific portrait I'd painted for many years, since being a student in Bradford probably', he wrote in 1976.[38] Wilder, unlike Freeman in *Beverly Hills Housewife*, appears solely as himself – not as an epitome, for all that he exhibits a masklike impassivity.

A similar transformation is apparent in the near-contemporaneous painting *Peter Getting Out of Nick's Pool* (1966; also begun shortly after *Beverly Hills Housewife*), in which Peter Schlesinger is seen from behind as he levers himself out of the water. The 'beautiful boy' of earlier pictures has resolved, here, into a real person, although the painting concentrates on Schlesinger's body more than his face (of which we catch only a glimpse in profile). The art historian Paul Melia has gone so far as to argue that the painting is not a portrait all, conveying instead 'a figure as the passive erotic object of the spectator's gaze'.[39] In his insistence on Hockney's gaze and its 'colonizing' force, Melia ignores the intimation, both here and in the subsequent painting *The Room, Tarzana* (1967), of Schlesinger's own incipient desire. Hockney *was* painting a type, the beautiful Californian boy, but he was also making a portrait of a lover. Schlesinger, in these first pictures, is both an erotic object and a sentient subject: a character with his own selfhood and agency. Hockney's early portraits in California in this way conjure paradigms – ideal forms – out of real life, but the reality of the situation is never quite lost.

In antiquity, portrait busts were objects that gravitated between lifelike and generic styles. In the period of the Roman Republic's transition into an autocracy under the first emperor Augustus, busts gradually lost their individualizing, 'naturalistic' details: they became more standardized, more idealized. Hockney's pictures of 1966 seem to express a reverse shift, but the transformation is not absolute – nor is it a simple one-way movement. Even as the generalities of his previous works seem to be resolving into tangible realities (no longer a nameless showering man but the artist's boyfriend), real life appears to be transfiguring *back* into an ideal.

Beverly Hills Housewife was the first of a grand sequence of panoramic-format portraits that Hockney made between 1967 and 1974, each realized with crystalline precision in bright acrylic shades, while steering notably clear of photorealism. Freeman is in some ways the progenitor of Celia Birtwell in *Mr and Mrs Clark and Percy*, in which Birtwell stands with similar theatrical poise. The artifice of these pictures – their staged, static quality – is often overlooked in discussions of their notional realism. The power of *Beverly Hills Housewife* lies in the fact that Freeman is not straightforwardly real. She retains the semblance of a cipher, a formula, a model – perhaps to herself as much as to others.

4

Californianess

'Thank Christ he had decided to come to Los Angeles, where people were really alive and things happened right now as well as in the past.'

ALISON LURIE, *THE NOWHERE CITY*, 1965[1]

Over the time that he was painting *Beverly Hills Housewife*, Hockney led a charmed existence. His relationship with Schlesinger was in its early, blissful phase, never mind the privations of the lodgings on Pico Boulevard, into which Schlesinger moved in early 1967. There were problems, not least the objections of Schlesinger's parents to their son's relationship with an older man. But in the main, these were carefree days, spent either at home – Schlesinger was too young to drink in bars – or in the company of friends. 'Looking back on it,' Hockney wrote in 1976, 'it was certainly the happiest year I spent in California, and it was the worst place we lived in.'[2]

Christopher Isherwood and Don Bachardy were regular dinner companions, as were Jack Larson and Jim Bridges. 'They were the first gay people I met', Schlesinger remembers, 'who lived together as couples, completely openly, like husbands. Now it's normal. Then it wasn't.'[3] Indeed, for all their openness, Larson and Bridges would always attend parties with a pair of female 'dates'. Everyone knew the reality, according to Schlesinger, but a degree of masking was customary.

The first time that Hockney and Schlesinger went to a gay bar together, in 1966, it was a speakeasy where visitors were required to knock on the door before entry to demonstrate that they were not the police.

At that time, according to Bachardy, 'You could still find yourself in a raid and arrested as well.' Isherwood and Bachardy rarely went to gay bars, but did accompany Hockney and the artist Patrick Procktor on one occasion in 1965: 'They were three years younger than I, but I was the only one to get carded', Bachardy recalled.[4] Only two years later, there were many more gay bars – including one that did not serve liquor and so allowed an underage crowd. 'It changed so fast from being hidden – worried about being arrested – to big discos', says Schlesinger.[5]

That sudden shift from secrecy to openness suggested the lifting of a veil rather than the explosion of a new phenomenon. Despite its reputation for sunny hedonism and openness, Los Angeles has always been a city of private lives, as the novelist Raymond Chandler knew; and this may have been part of its allure for Hockney. 'It was physically hard to get to', Schlesinger points out. 'People didn't fly about as they do now. It was isolated, way out there on the West Coast, and the nature of the city is that you can live much more privately – people were in houses, far away from each other.' The British actor Charles Laughton had a house in Santa Monica, Schlesinger points out, where he would take his boyfriends – out of sight of his wife, Elsa Lanchester.[6] 'A lot of people did that. They had a little house down by the beach for their girlfriend or their boyfriend.'

In love with Los Angeles

The literary critic David Fine has described Los Angeles as 'the land dedicated to the proposition of the fresh start'.[7] Although that proposition has always carried an element of wishfulness, the myth of the city as a place for self-reinvention exercised a powerful hold on the young Hockney. It was a myth, moreover, that he succeeded in realizing. For him, the dream of renewal and transformation that LA has so often inspired (and also cruelly denied, as Fine relates in his study of the city's noir-ish incarnations in fiction) became a reality, at least for a time. For Betty Freeman, arriving in 1950, the city's combination of domestic privacy and openness was beguiling. When she was asked, forty years later, if moving from Boston had been a shock, she flatly replied: 'No, I loved it. No shock, only pleasure, which still remains.'[8]

Los Angeles at that time was a radically new city – a place almost without a history, having been conjured out of a wilderness of desert and mountains less than a century earlier. Hollywood had been laid out as a real-estate subdivision only in 1887, with the film industry arriving from the East Coast in the early 1910s. In a short story of 1928, F. Scott Fitzgerald crisply observed: 'Everything in the vicinity – even the March sunlight – was new, fresh, hopeful and thin, as you would expect in a city that had tripled its population in fifteen years.'[9]

In the middle of the twentieth century, more strongly even than now, the entire metropolis seemed like a mirage, teetering on the edge of nature or set to be swallowed up. The cultural theorist Mike Davis has pointed out that Los Angeles has 'the longest wild edge, abruptly juxtaposing tract houses and wildlife habitat, of any major nontropical city': no other cities in the United States '*enfold* wild terrain in the complex fashion of Los Angeles' with the possible exception of tropical Miami, bordering the Everglades.[10] The Santa Monica Mountains taper into the Hollywood Hills; the San Gabriel and San Bernardino Mountains descend into Pasadena. The urban settlement, which is vaster and more all-encompassing than practically any other, is nonetheless encompassed – imperilled – by nature, as the devastating wildfires of January 2025 demonstrated.

The city's enlivening sense of impermanence was a salient characteristic for Christopher Isherwood, who remarked in 1969:

> This country is fundamentally desert country. It's country which has been [...] adapted and planted and settled, but nevertheless it is desert country. [...] As a devotee of Hindu philosophy, it's very much like the Hindu account of the universe as being just a kind of projection or maya. Sometimes the great noble traditional buildings of Europe seem rather too solid in comparison. Here on the coast it's all rather dreamy and strange and kind of merging. It's very characteristic. [...] I adore this area. I think this has a wonderful kind of horror.[11]

Freeman voiced a similar idea of the city as a primordial being in a speech that she delivered at the Los Angeles County Museum of Art in 1990, recalling her first arrival:

> Way back in the dim prehistoric days of 1950 when I moved to the frontier town of Los Angeles from cultured Boston there was no Music Center, no Santa Monica Freeway, no high rises and no Mayor Bradley. The dinosaur bones were still sticking out of the La Brea Tar Pits. The Pasadena Art Museum was in a Chinese pagoda and the LACMA was at the Colosseum in the National History Museum just past the dinosaurs.[12]

For a long time before Freeman moved there, Los Angeles had suffered from a reputation as a cultural backwater. Compared with the grand institutions of New York or the great European capitals, the city was considered provincial and arriviste. The art dealer Irving Blum once dejectedly professed: 'When you talk about the art establishment, it exists in New York and nowhere else.'[13] But for many who came to Los Angeles in the mid-twentieth century, this was mere snobbery – its own LA myth. The fact that the city did *not* share the grand institutions, the accumulated heritage, of New York and other places was precisely what gave it an edge.

For Freeman in the 1950s, as much as for Hockney a decade or more later, the fact that Los Angeles had not been fully recognized was a gift; it allowed them to define the city on their own terms. According to the British composer Thomas Adès, who became a close friend of Freeman's in the 1990s: 'The attraction of the place, instead of New York or Boston, was exactly the fact that people weren't looking over your shoulder and saying, "You have to do it this way." You could do what you wanted.'[14]

From the outset, Freeman loved Los Angeles for its atmosphere of incipience, the pervasive sense of a beginning. The character of the city resonated, in this regard, with her own character – a nature that was open to new sensations and ideas, and open to reinvention. It's a wonderful world out there', Freeman said in 1991 of the Los Angeles cultural scene. 'It's so *open*.'[15] 'She gave me the impression that LA was *the* art place of the world,' Adès adds, 'that every room you walked into would be full of the most amazing artists. She just loved it here. The one word would be freedom. She loved the latitude. She always said very proudly, "You can drive for eighty miles and not leave the city."'

In Hockney's own estimation, *Beverly Hills Housewife* was not 'absolutely specific'. The ghost of an *idea* – of a paradigm – remained. In this regard, however, the painting captured something innate in Freeman. As a person, as a persona, she was possessed of the openness – the sense of brimming potential – that she prized in her adopted city.

Anglo-American

> 'You can't write a story about L.A. that doesn't turn around in the middle or get lost.'
>
> *Eve Babitz,* Slow Days, Fast Company: The World, the Flesh, and L.A., *1977*[16]

In the early summer of 1971, *West*, the weekly colour magazine of the *Los Angeles Times*, announced that the city's art scene was in the throes of 'a London–LA love affair'. British artists, Hockney foremost among them, had become a fixture on La Cienega Boulevard. The new arrivals were conspicuous on the sidewalks and in the galleries for their 'ice cream bellbottoms, striped soccer shoes, Our Gang spectacles, perky suits in Oliver Twist checks and ties that look like bibs made of the Union Jack'.[17]

The breezily satirical tone of the article by William Wilson in *West*, which served as a preface to a longer essay by Reyner Banham (a British architectural historian who became one of the great theorists and eulogists of Los Angeles), is telling. Its description feels closer to a synopsis for a comic movie than a piece of reportage. Whatever 'love affair' was under way was a vibe, one suspects – something in the air rather than an easily measurable phenomenon. The fancy-dress caricature penned by Wilson stood for something harder to grasp.

Seemingly, the embodiment of the Anglo-American cross-pollination was to be found in *Beverly Hills Housewife*, the main illustration to Banham's essay. Four years after the painting's first exhibition, it had become the symbol of a cultural ambience – a record of a moment. Not simply a portrait of a woman at home, but a crystallization of something as elusive as a 'zeitgeist'. Central to that zeitgeist, according to *West* magazine, was an 'infatuation for things Angeleno'. This had afflicted not only Hockney, but 'many of his English colleagues', not to mention American artists such as Ed Ruscha.

That casual mention of Ruscha causes the thesis to unravel somewhat, indicating that Englishness was beside the point. Artists, wherever they came from, were turning their attention to the city in new ways. The same year that Hockney began *Beverly Hills Housewife*, Los Angeles-born artist James Turrell had bought a run-down hotel in Santa Monica, and was busy converting it into a site for experiments in light projection; Turrell's hotel was an installation literally wrought from the fabric of the city. A year earlier, in 1965, the photographer Lee Friedlander (an adoptive New Yorker, although he had studied in LA in the mid-1950s) had captured a Californian sunset spreading above a strip mall, complete with a cut-out advertisement of a smiling couple (*Los Angeles, California*, 1965). The same conjunction of open sky and concrete city can be seen in Latvian-born painter Vija Celmins' photorealist painting *Freeway* (1966), a motorist's eye view of the route to San Diego; here, the human character has been relegated to beyond the bounds of the picture, their presence signalled by the impingement of the car dashboard. In 1967, painter Richard Diebenkorn – one of the leaders of the Bay Area Figurative school in the 1950s – began his 'Ocean Park' series, heavily abstracted landscapes that derived their linear structures from the steel-frame windows of his Santa Monica studio.

In the opinion of Banham, Hockney was 'one of the two or three people in the world who actually know what Los Angeles really looks like [...] his pictures induce the genuine shock of recognition'.[18] But despite the artist's own claim that there were no paintings of LA when he arrived ('people then didn't even know what it looked like'), the era saw a prolific variety of responses among artists – British or otherwise – to the architecture and 'aesthetic' of the city, not to mention its sociopolitical inner workings.[19] It is arguable, then, that the 'love affair' perceived by Wilson was only one manifestation of a broader and deeper engagement with the look of Los Angeles, both by those who came from elsewhere and those who had long dwelt in the place – an engagement, moreover, that balanced the reality of the city against symbolic or mythic possibilities.

In *West* magazine, *Beverly Hills Housewife* was reproduced above seven smaller pictures from Ed Ruscha's photographic series *NINE SWIMMING POOLS AND A BROKEN GLASS* (1968).[20] Ruscha's pictures dispense with style and subject matter in the traditional sense, presenting affectless views of empty pools in undisclosed locations (opposite and fig. 11). The pictures appear vacant of semantic import – as empty, in a sense, as the pools themselves. Each pool is a plain blue sheet, distinguished only by ripples or reflections of trees or lane markings beneath the water. Undisturbed by human activity, they nonetheless convey a sense of the potential for activity, events, life – but all is held in suspense.

Ruscha's pools are at once specific and generic. Each is a particular pool in a particular place, but each conforms to an archetype. He has described Los Angeles as 'the ultimate cardboard cut-out town', a statement that acknowledges the preponderance of the *type* in the city's scenery.[21] The pools, palm trees, freeways, and signage create their own iconography of forms; every real thing has a corresponding archetype – a Platonic 'ideal' Form. Not simply a pool, but a Pool: an idea as well as a mundane reality. In Ruscha's series, the distinguishing details of each pool (a diving board, a round shape versus an oblong one) are subservient, in the end, to their common aspect: each is a shimmering capsule of cobalt blue.

As if to undercut the order and purity of the sequence, the tenth photograph in Ruscha's group sounds a jarring note. A smashed water glass lies on wet, reflective concrete – the image resembling a still from a noir movie. It is a fragment, incongruous and unexplained, with none of the placid self-containment of the preceding images. It shifts us from a dually formalist and documentary register (variations upon the formula of the swimming pool, viewed from above) to a narrative one. A contingent, mildly unsettling reality intrudes.

The capacity for the universal to collapse, in this way, into a specific instant, reflects a deeper aspect of Los Angeles and its manifestations in art, film, and literature. The iconography of the city is mythic, a constellation of Forms; but it is equally a hard – sometimes friable – reality. A water glass that might be smashed.

This doubleness is felt in depictions of the place. In his 1971 psychogeographic ode to the city, *Los Angeles: The Architecture of Four Ecologies*,

Ed Ruscha, *NINE SWIMMING POOLS AND A BROKEN GLASS*, 1968 (detail)

Banham proposed that the vast freeways of LA amounted to their own ecosystem or 'Autopia', 'in its totality [...] a single comprehensible place, a coherent state of mind, a complete way of life'.[22] In Banham's coinage, utopia and dystopia seem to have converged (albeit with more of a stress on the former), and this of course is the point: the concrete reality and the soaring myth are inextricable. One's personal experience, the specifics of an individual life, can be subsumed by the myth, just as the myth can condense into a singularity: the dull fact of smashed glass, say. (It is curious to note that Banham's thesis has itself routinely been accused of mythologizing or romanticizing the city – a charge that first occurred in Peter Plagens' riposte in *Artforum* in 1972, a review headed 'The ecology of evil').[23]

Joan Didion, in her 1970 novel *Play It As It Lays*, evoked this sort of oscillating motion between mythic and real visions. The protagonist, a lonely married woman called Maria, slips free of the painful constraints of her life when she drives alone on the Los Angeles freeways, or even when she remembers doing so: 'She imagined herself driving, conceived audacious lane changes, strategic shifts of gear, the Hollywood to the San Bernardino and straight on out, part Barstow, past Baker, driving straight on into the hard white empty core of the world.'[24] A momentary and annihilating sense of escape is to be found on the city's freeways, even as the character retreats into herself. The commonplace spectacle – a freeway stretching into the distance, the blue void of a swimming pool – has the potential to induce a sense of the infinite, of something immeasurably larger than itself. Then again, it might simply assume its commonplace aspect.

On the pages of *West* magazine, Ruscha's photographs formed a kind of 'predella' of smaller panels beneath the 'altarpiece' of Hockney's painting. The juxtaposition was a reminder of what lies just beyond the space of Hockney's scene: the swimming pool in Freeman's garden. The pool is not readily visible in the painting, although the diagonal lines of light on the leftmost pane of glass might be reflections from the surface of water. The picture hints, in this way, at what lies beyond – whereas Ruscha's photographs are concerned only with what lies within the camera frame. Their pools are ordinary – literal and untransformed;

there is only the faintest glimmer of some larger, stranger possibility. In *Beverly Hills Housewife*, by contrast, a human protagonist has entered the stage. The latent sense of untapped potential in Ruscha's series – of a narrative not yet disclosed – swells into a human drama. The balance between dull reality and infinite possibility is more finely weighted.

In striking such a balance, *Beverly Hills Housewife* conveys the nature of Hockney's embrace of Los Angeles. He was looking, from the moment he arrived, for confirmations of what he had hoped to find. Looking, that is, for concrete instances of the Platonic ideals that had developed in his imagination. A defining characteristic of his work of the 1960s was its progressive movement from generic spectacles – such as that imagined in *A Bigger Splash* (1967, based on a photograph in a pool maintenance magazine) – towards specific people and places. When he first arrived in 1964, he was seduced by what seemed to be the pervasive glamour of the city. By 1965, he was locating that glamour in particular things: buildings, objects, bodies. 'For the first time I began to paint the physical look of the place. It took me a couple of years to do it much more realistically.'[25]

Beverly Hills Housewife, more than any other work from this time, embodies both the glamour and the 'physical look'. It bridges the poles of the generic and the singular, such that they become hard to tell apart. Hockney's myth of Los Angeles, the mental image that he had carried with him since childhood, is still alive in his vision of Betty Freeman at home; but it is coming into contact with (merging with) the reality of an actual woman in her actual house.

In the painting, the pool is an ideal – a concept – but in a different way from Ruscha's photographs. It is a lurking hypothetical rather than a string of clear-cut examples – but a presence nonetheless. Is it simply a coincidence that Freeman seems almost to walk on water? The stillness and uniformity of the idealized swimming pool have extended, transmuted, into the features of the picture at large: the gleaming glass, the oblongs of light and shade out of which the architecture of the house coheres, the simplified forms of her dress and face and blonde hair, as much like reflections as palpable objects.

In his article for *West*, Banham was inclined to see Hockney as representative of a wider phenomenon, that of British artists falling in love with Los Angeles – the 'ultimate American Pop city' – in particular, the new generation of British artists that also included Anthony Donaldson, Richard Smith, Allen Jones, and their 'Pop Art' ilk.[26] But Hockney's 'infatuation' was more complex and changeable than Banham supposed; and it is hard, in retrospect, to see Hockney's response, or those of any of the artists named in the article, as part of a pattern. Each was responding in distinctive terms to a common object: Los Angeles both as an idea and a reality, an *elsewhere* and a *here*.

Even if Banham was overoptimistic in perceiving a resonance between the art scenes of London and LA in the 1960s, the perception itself had a value – as he acknowledged: 'This inference may have been justified or not, but that's how we felt; the action that interested London was in Los Angeles, for there was a deeply felt affinity between the two scenes.' He went on: 'The Pop art generation in London really *needed* Los Angeles both as a fact (the Beach Boys, Kustom cars, etc.) and as a myth.'[27]

The power of *Beverly Hills Housewife* lies precisely in its tilting towards a symbolic, mythic register (as though it really were a picture of Beverly Hills, of all Los Angeles, compressed in a single frame) while insisting, at the same time, on the *particularity* of the woman, the objects, the setting. What seems like a contradiction in the painting – the superficial glamour versus the everyday reality of a person at home – is a confirmation of Los Angeles' own doubleness: its ability to operate between myth and reality.

The pool in the window

Those filiform bars of light on the glass behind Freeman suggest the nearness of the water, just beyond where she is standing. But a closer look at the glass panels behind her, just to the left, reveals something else. Behind one of the panels, amid the slanting light, the room immediately inside the doorway remains visible, although the glass has tinted its orange carpet to a duller brown. At the back of the room, a white sofa can be seen against the grey wall (fig. 9). In the next panel of glass to the left, something has changed. The sofa and the surrounding room have been superseded by an oblong view of trees and ground, with a slither of blue sky at the top. The remainder of the panel is cast in shadow. At a glance, it is easy to assume that this is a vista through the house to a front window, similar to the line of sight immediately behind Freeman to the right – the open doorway that offers a clear view of the orange-carpeted space, through to the dark inner sanctum of the music room and the bright window beyond (fig. 10). But by comparing the final painting to the source photographs, it is clear that this is a reflected image, not a view into the house. It reveals a fragment of the garden beyond the perimeter of the picture, framed on two sides by the overhanging structure of the portico. A similar reflected image occurs in *A Bigger Splash*: the glass doors of the house in the background throw back a monochrome impression of distant buildings, palm trees, and sky.

One catches sight, then, of the space into which Freeman is gazing. And within this reflected space, it is possible to see her swimming pool. The ground of the garden is conveyed by grey, lightly washed canvas. It is a uniform 'blank', apart from where – close to the interface between the ground and the trees – a darker, irregular shape intervenes. Two tiny spurs of grey emerge from the top edge: the twin rails of a pool ladder, similar to the rails that appear in Hockney's later print *Gregory in the Pool E (Paper Pool 4)* (1978). The shaded area signifies the water (again, one of the original Polaroids confirms this), although the pool is practically invisible – little more than a shadow, half-hidden by the repeating blades of sunlight on the glass. Within the hard, bright scenario of *Beverly Hills Housewife*, certain things are closer to phantasms than facts.

A poppy maniac

When the Freemans moved from Boston to Beverly Hills, they crossed from an outpost of 'old Europe' to a place that was barely a city. It had been laid out only in the early twentieth century; before then, the land had been agricultural – a swathe of ranch country at the base of the mountains. Its population boomed after the arrival of the Beverly Hills Hotel in 1912, at a time when Los Angeles was the fastest growing city on Earth.

Hillcrest Road is a slice of Californian picturesque. Bordered by colonnades of towering palm trees, it runs through the 'Flats' of Beverly Hills from north to south – one of a sequence of gently inclining avenues, each planted with a distinct species of trees. Number 703 was the Freemans' second home in the area. Built in 1930, it is open to the road, set back from the sidewalk by lawns, but screened by shrubs and trees planted close to the house, including a large olive tree and a tropical explosion of birds of paradise. It is not fenced and gated, like many homes in the neighbourhood. But nor is it nakedly visible. Broken up by greenery, the colonial-style villa glimmers in and out of view.

The Freemans moved into the house in mid-1959. 'When they first bought it, the interiors were really ugly', according to their daughter Shelley, who was ten at the time. 'It had gold shag carpeting all over it, and green straw wallpaper.'[28] Between 1958 and 1960, Betty and Stanley commissioned extensive renovations, including the creation of a skylight above the hallway and the laying of large black paving stones in the garden. New interiors were mapped out in a sequence of models.

'She had found this interior designer who was truly eccentric', says Shelley. A year older than Freeman, Eszter Haraszty – herself newly arrived in Los Angeles from New York – was renowned for her extravagant floral-patterned interiors. At Hillcrest Road, she created an extraordinary set of bespoke rooms, their garish colour schemes and hallucinatory patterns expressing an affront to the architecture of the house itself – both to the minimalism of the new portico at the rear and to the colonial classicism of the overall property.

Freeman's choice of designer reflected the increasingly maverick nature of her tastes. Haraszty, conspicuous for 'her blond Twiggy-style crop, Gabor-sisters accent, thigh-high skirts and bug-eye sunglasses', in the words of a posthumous *New York Times* profile, had designed stage costumes in Budapest before the Second World War.[29] She emigrated to New York in 1947, where she rose to become head of the textiles department at Knoll, the Bauhaus-inspired manufacturer of modernist furniture, whose founder Hans Knoll notoriously died in a car crash in 1955. At the time that Freeman commissioned her, Haraszty and her husband had recently moved to a hilltop house in Coldwater Canyon, where she grew flowers in vast crops of distinct colours.

The breakfast room or 'dinette', where *Beverly Hills Housewife* later hung, was dominated by a table in the shape of a poppy (fig. 12) and a deep-pile yellow rug embroidered with floral motifs. The floor was adorned with white and blue tiles in a schematic pattern of poppies. Haraszty admitted to being 'a poppy maniac', and gave the term 'needle-painting' to her self-taught method of embroidery.[30] By contrast, the adjoining living room, or music room, was dark, subdued, and formal. A fireplace was removed to create an uninterrupted wall of thirty-five feet. It was here that Freeman hung her monumental work by Clyfford Still, *1955-H*:

> We panelled the entire living-room in navy blue velvet, floor to ceiling. I only know that in the back of my mind must have been the great painting itself. For this is where it is, in the center of this long darkly glowing wall.[31]

The theatricality and detail of Haraszty's designs, bordering at times on the psychedelic, seemed to express an emergent aspect of Freeman's temperament: the radical experimenter inside the sober persona of the Beverly Hills housewife. One senses, above all, that it was Haraszty's determined individualism, her lack of self-regard, that appealed to Freeman. The designer once claimed: 'I abhor giving advice or acting as an arbiter of taste. Those roles are directly opposed to what I believe in the most: the freedom to express personally one's own creative self.'[32]

Haraszty was a perfectionist when it came to that act of expression. The transformation of Hillcrest Road was frustratingly slow, with Freeman confiding to Still in November 1959: 'I live out here now in a house that has no furniture except for rugs, beds, and 12 dining room chairs spread through the house. But I do have up the paintings I love.'[33] As Shelley recalls: 'My mother had such problems with Eszter, but in the end she loved the design. It was just so outrageous at that time.' Stanley took no part in the renovations, leaving the oversight entirely to his wife: 'He was mystified, that's all I can say. Mystified by everything that she did.' Indeed, the metamorphosis of the house from a traditional 1930s Beverly Hills residence into an avant-garde showpiece prefigured Freeman's own metamorphosis from a middle-class wife into an impresario of new music – that is, away from the identity that

Stanley had recognized and approved of. In turning the house into an artistic spectacle, not to mention a repository for her growing art collection, she was setting the stage for her own transformation.

Until Freeman's death, the decor of the house remained largely unchanged. Everyone who went to Hillcrest Road in her lifetime remembers the velvet walls, the tiled floors, the flower-shaped table. 'It was very fifties', says Julia Bloomfield. 'And *so* Betty.'[34]

Apertures

The scene in *Beverly Hills Housewife* is strikingly flat. The space of Freeman's garden is shallower, more compressed, than in most of Hockney's realist scenes of the 1960s and 1970s. Objects seem barely to cast a shadow.

But the flatness is not absolute. Behind the woman in pink, all the way through the interior of the house, is a window (fig. 10). The distant aperture invests the scene with an element of recessive perspective. At the same time, the reflection in the glass to the left affords a view of grass and shrubbery and sky – projecting outwards, into the garden. The multiple bars of light that score the window panes – anticipating the ciphers of Roy Lichtenstein's 'Mirror' series – themselves refer to the world beyond the scene itself: the shimmering pool, the sunny day.

Within the contained environment of the painting are hints at what lies outside. A similar effect is discernible in a number of the double portraits that Hockney made in subsequent years. In *Mr and Mrs Clark and Percy*, the husband and wife are seen either side of an open window that admits a hazy view of the terrace of houses opposite (the scene is the Clarks' flat in London's Notting Hill, although much of it was executed in Hockney's nearby Powis Terrace studio). *Henry Geldzahler and Christopher Scott* (1969) shows Geldzahler seated on a plush pink divan while his lover stands to one side, like a mundane anoraked version of the angel in Fra Angelico's *The Annunciation* (fig. 7).[35] In the background, a window offers a view of the assorted high-rises of Seventh Avenue, blanched by sunlight. The glinting lenses of Geldzahler's spectacles and the blades of light on the glass table meanwhile imply a separate light source somewhere beyond the picture.

Speaking in 1969 about the creation of this painting, and in particular the light glancing off the table, Hockney remarked: 'If I were Jan van Eyck, I'd put my whole picture in that little reflection.' The statement reinforces the idea that the tiniest reflection might function as a microcosm. The reflection is an index, in other words, of something larger – the limitless reality outside the picture.

Hockney had long been intrigued by the composition of Van Eyck's *Arnolfini Portrait* (1434), in the National Gallery, London, which portrays a merchant and his wife at home in Bruges (fig. 13). The scene might be the occasion of their wedding, or possibly just a portrait of the couple at some other time. The man, Giovanni Arnolfini, is raising his hand, maybe in greeting. In the background, in the circular mirror between the two figures, is an image of what he sees: two men entering the room, one of them raising his hand in turn. Most likely it is the artist himself, accompanied by a servant, saluting his patrons.

Van Eyck's painting has been described as the first modern oil painting, precisely because it is a picture of a real, secular room, filled with everyday things – a bed, a chandelier, oranges on a sideboard (Bruges was a centre of global trade). Everything has been carefully arranged within a believable, perspectival space. The mirror, which the art historian Meyer Schapiro called 'the beautiful, luminous, polished eye' of the *Arnolfini Portrait*, condenses and extends that space; it flips the perspective around so that we are looking back out, through the 'fourth wall' of the painting, permitted to see what Arnolfini and his wife are seeing.[36] In the process, the mirror subtly destabilizes the sense of time in the painting, as if this were not simply a frozen moment but a parallel reality that we could enter into, something still happening.

In *Beverly Hills Housewife*, the bright oblong of the front window betokens external reality. So do the bars of light on the glass behind Freeman: like Geldzahler's glinting spectacles, they indicate a separate, nearer source of light. These details function in the same way as the mirror in the background of Van Eyck's painting, the specular 'eye' that breaches the space of the picture, in whose surface we see the external world of the painter, even our own world.

'The entire picture is looking out at a scene for which it is itself a scene.'

Michel Foucault, The Order of Things, *1971*[37]

Like Van Eyck, Hockney uses apertures and reflections to imply the extension of space and time beyond the setting of the picture. In doing so, he reminds us of his own presence: the space outside the picture is also the space of the artist. In Diego Velázquez's *Las Meninas* (1656), those separate dimensions – the fiction of the painted scene and the 'real world' – are collapsed together (fig. 14). The painter inserts himself into the image, to one side of the Infanta Margarita Teresa, daughter of King Philip IV of Spain and Mariana of Austria, and her entourage of maids, dwarfs, and dozy mastiff. Velázquez is imagined having paused work on a canvas that intrudes to one side of the scene, its painted surface invisible. Behind the group of figures, at the distant end of the room (one of the chambers of the royal palace in Madrid) is an open doorway leading to a corridor and staircase. A figure can be glimpsed beyond, holding back a curtain and pausing, as though unsure of whether to enter.

Alongside the open door, amid the murky pictures that cover the back wall, is a shimmering image. It is a mirror, as many critics have argued, that reflects the two figures whom the painter is looking out at. As Michel Foucault observed in his foundational analysis of the painting, the image in the mirror is of a space entirely outside the representational scene; there is not even the kind of inversion of the painted scene (the architecture of the room, say) that one finds in the *Arnolfini Portrait*: 'It is reflecting nothing, in fact, of all that is there in the same space as itself: neither the painter with his back to it, nor the figures in the centre of the room.'[38]

In *Las Meninas*, the door at the back of the room marks the threshold of the depicted space. There is a second, subtler portal at the right of the picture (probably a recessed window), revealed by the light that floods through it. Together, the doorway and window imply an 'outside'. The mirror, on the other hand, abruptly breaks into that exterior realm or, as Foucault puts it, 'by means of a violent, instantaneous movement, a movement of pure surprise, [the mirror] leaps out from the picture in order to reach that which is observed yet invisible in front of it'.[39]

A similar effect is produced by the dark, embedded reflection of Freeman's pool in the pane of glass. It too constitutes 'a movement of pure surprise', whereby we are permitted to see what the woman in the painting sees.

In distinct ways, the open door and the shining mirror of *Las Meninas* project the viewer into the fictive world of the painting, as though we were the object of those numerous alert gazes. By the same token, they seem to extend that world outward, so that the illusional space of the Spanish court and the external, real space of the viewer appear contiguous. The artist is physically, almost swaggeringly present, looking out at us – or more immediately, at the two figures that he is painting: the king and queen of Spain, Philip IV and Mariana, who are generally accepted to be the two figures reflected in the mirror.

In *Beverly Hills Housewife*, by contrast, Hockney is not literally visible, neither a physical presence in the scene nor a reflected image. (In a later, unfinished portrait of his parents from 1975, he does appear, *Arnolfini*-style, in a mirror between the two main figures.) The image of the artist is metaphorical – displaced. We discern him in those glints of light, reminders of the luminous day of which the picture is only a capsule, a fragment. More obliquely, he is present in the figure of Freeman. She is both his subject and his covert reflection. And just as in the *Arnolfini Portrait* and *Las Meninas*, there is an accompanying sense of the extension of time. It is a painting of a single instant, and yet its intimations of the world beyond are intimations, too, of the future and the past.

A heap of stones

At the far right of the picture is a stack of green shapes: two boulders separated by a darker, squarer block (fig. 15). The totemlike arrangement stands apart from Freeman, a little way across the terrace. But it is not quite the anomaly that it seems. Together with Freeman herself and the palm tree that rises like a toy from the grass, it makes up a triad of vertical forms in the right-hand panel of the painting.

Lama (1961), a sculpture by William Turnbull (fig. 16), was one of the only works of art in Freeman's collection not by an American – and the only one that Hockney depicted in his painting. In its own mute way, the sculpture is another reflection of the zeitgeist. Turnbull's organic-minimalist constructions had become *de rigueur* in the mid-1960s among Californian collectors of contemporary art. 'There is an intelligent and discriminating American audience for vanguard British art', opined the author of an article on Turnbull in the Spring 1966 issue of *Art in America*; 'in the West, where New York is distrusted, London is not'.[40]

'There was a lot of sculpture by Bill Turnbull', Hockney remembered; 'Somebody'd been and sold them there a few years before.'[41] In *California Art Collector*, the imaginary patroness sits facing one of Turnbull's piles of stones in a tranquil face-off. A heftier sculpture features in *American Collectors (Fred and Marcia Weisman)*, where the top-heavy stack resembles a primeval altar between husband and wife, as though a Greek tragedy were playing out in modern lineaments – the Weismans as Aegisthus and Clytemnestra, the warring king and queen of Mycenae (fig. 17). Turnbull's works were modishly modern, but they also had the aura of archaic relics.

Freeman had bought her Turnbull piece from Felix Landau, an Austrian émigré who was one of the trendsetter art dealers of La Cienega Boulevard in the 1960s, and from whom she would later buy *Beverly Hills Housewife*. In the spring of 1966, just months before she was photographed by Hockney, her sculpture had featured in an exhibition by Turnbull at the Pavilion Gallery, Balboa (forerunner to the Orange County Museum), an event that sealed his reputation in America. A review in *Artforum* lamented the artist's tendency to replicate the styles of eminent Americans – especially in his paintings, which suggested studies of Barnett Newman – but also praised Turnbull the sculptor for his 'simplified totemic images in beautifully worked wood, stone and bronze [...] One has the feeling that he is deeply committed to the organic, the calm, in addition to the tense and industrial.'[42]

A generation older than Hockney, the Dundee-born Turnbull – who always preferred to be known as Bill – was in some ways an entirely different artist and man: retiring, phlegmatic, and heterosexual. But there were also curious parallels between them. Turnbull looked beyond the confines of Britain and Europe from an early age, as Hockney would. Both were possessed of a wanderlust that came to inform their art.

In 1938, aged seventeen, Turnbull joined the Royal Air Force; he subsequently served as a pilot in India and Ceylon. ('How the hell do you get as far away from a trench as possible?' he said towards the end of his life. 'An aeroplane seemed a good idea.'[43]) In 1947, his first year at the Slade School of Fine Art, he travelled to Italy, where he favoured the 'primitive' muralists of the fourteenth century, as well as ancient art – much as Hockney would a decade or so later. The following year, disillusioned with the traditionalism of the Slade, Turnbull moved to Paris, where he lost no time in acquainting himself with the leading lights of the art scene – again anticipating Hockney's rapid forging of connections when he came to Los Angeles. He befriended Alberto Giacometti, 'an extremely affable man', and Constantin Brancusi, who proved harder to charm. After hearing that Brancusi would not let anyone into his studio unless they were 'of note', Turnbull showed up unannounced: 'I put my foot in the door and he couldn't close it. And I just kept saying to him that I admired his work so much. [...] Why couldn't he just let me in for a little while?' Brancusi relented, admitting Turnbull to his studio before leaving the room to allow a solitary experience of the sculptures. 'He came back after about fifteen, twenty minutes and said, "You have to go now."'[44]

The stacked, columnar sculptures of Brancusi made an indelible impact. Later, in the early 1960s, Turnbull travelled to Japan and Cambodia, where he was struck by the totemic statues of religious sites. The result of these various experiences was a series of works in bronze, stone, and carved wood, predicated on ideas of equilibrium and gravity. *Lama* consists of two cast bronze sections – roughly akin to an orb and a cylinder – separated by a 'trunk' of dark rosewood. Vertical in stature and apparently abstract in signification, it might be regarded as a counterpoint to the painting within which it features: condensed, abstract, and three-dimensional versus Hockney's lucid pictorial style and cinematic breadth of field.

Fundamentally, however, the sculpture suggests an analogy to Hockney's scene. In the stacked forms, we might discern a compressed, diagrammatic equivalent to the painting's stratified structure: the turf, the portico, the sky. More specifically, the sculpture acts as a gently comedic double of Freeman. The pile of bronze and wood mirrors the 'pillar' of the standing woman. Titled after the Tibetan for 'high mother', it is one of a larger group of stacked compositions that 'stand substitute' (in Turnbull's words) for the human figure – not outright abstractions, then, but corporeal assemblages in stone, wood, and bronze. 'I wanted to make a head-object as autonomous as a football', Turnbull said in relation to one of his totems.[45] Indeed, he regarded the works as 'person objects', a status that Freeman ironically assumes in the *mise en scène* of *Beverly Hills Housewife*: the glamorous woman envisioned as just another beautiful object.[46]

A similar transference is at work in *American Collectors (Fred and Marcia Weisman)*, where Turnbull's 1958 sculpture *Hero II*, a three-part stack of stone and bronze, forms the radial point. It divides the figures of husband and wife as they stand at perpendicular angles to one another, but at the same time connects them in the manner of a pivot. Along with the totem pole that rises in the background, the sculpture seems like a mirror image of the wealthy Americans – a figure for their own stiffness discernible in Marcia's gritted teeth and in Fred's clenched fist.

In *Beverly Hills Housewife*, Turnbull's heap of stones and bronze can again be regarded as a proxy body, almost as a secondary character. There is a sense, even, that *Lama* and every other object in the painting is an extension of the human character at the centre. In this regard, *Lama* helps to imbue the static scenery of the picture with a stealthy animism, as though Freeman had been multiply incarnated: everything we see is, at some level, *her*.

Working against this, the implied equivalence between sculpture and woman underscores Freeman's stillness – her statuesque bearing. As much as her aliveness radiates out to the objects around her, their stasis rebounds back on her. She is a human interloper in a still-life scene; and the stillness of the objects can be seen to enact its own influence, rendering her an object, stopping her in her tracks. Seen this way, she shares the inanimate quality of everything else in the painting. She is like a piece of antique statuary transposed into twentieth-century guise, her dress reminiscent of the *peploi* worn by the maidens of Greco-Roman sculpture (in the preparatory studies, it has the same vertical pleats, the same appearance of being belted at the waist). Like Pygmalion's statue at the point of coming alive, she teeters on the brink of animation. She is real, but seems like a work of art.

10 David Hockney, *Beverly Hills Housewife*, 1966–7 (detail)

11 Ed Ruscha, *NINE SWIMMING POOLS AND A BROKEN GLASS*, 1968

12 *Beverly Hills Housewife* hanging at 703 Hillcrest Road, with Freeman's table in the form of a white poppy

13 Jan van Eyck, *The Arnolfini Portrait*, 1434.
Oil on panel, 82.2 × 60 cm ($32\frac{1}{4}$ × $23\frac{5}{8}$ in.)

14 Diego Velázquez, *Las Meninas*, 1656.
Oil on canvas, 318 × 276 cm (125¼ × 108⅝ in.)

15 David Hockney, *Beverly Hills Housewife*, 1966–7 (detail)

16 William Turnbull, *Lama*, 1961.
Bronze and rosewood, height 154.9 cm (61 in)

17 David Hockney, *American Collectors (Fred and Marcia Weisman)*, 1968.
Acrylic on canvas, 213.4 × 304.8 cm (84 × 120 in.)

Both *Lama* and *Beverly Hills Housewife*, for all their differences, are historically allusive. They act as foils for one another, but they are alike in intimating the forms and structures of older art, of remote times. Other sculptures in Turnbull's series had titles such as *Agamemnon*, *Oedipus*, and *Janus*, names that not only underline the objects' anthropomorphism, but situate them in a far-flung, semi-mythic past. For all their differences of attitude and form, Hockney's and Turnbull's works are alike in being *of their time* and apart from it, answering what the painter Francis Bacon once called the 'long call from Antiquity'.[47]

Light and space

> 'Seeing is a very sensuous act – there's a sweet deliciousness to feeling yourself see something.'
>
> *James Turrell, 2013*[48]

Prior to 1966, Hockney's last portrait of a nameable subject had been a painting of his father made in Bradford in 1955. Here, the older Hockney sits in a bare, dimly lit room, dressed in a black frock coat, striped tie, and flannel trousers. All we see apart from the seated man is the edge of a framed picture, leant against the wall behind him. There is little to link this early work with Hockney's later portrayal of Betty Freeman, except perhaps the subject's averted gaze, which suggests a turning away from the artist – from the entire outward world – into the depths of self.

Eleven years on, with Bradford having given way to Beverly Hills, the 'portrait' was only one component of a larger design. *Beverly Hills Housewife* is a painting of a situation, a place and time, rather than simply an individual. It is a painting, too, about the essentials of artistic composition – a painting about space and volume, sculpture and architecture. The human figure is the pivot around which Hockney balances a number of elements: grass and sky, transparent glass and opaque render, 'inside' and 'outside'. Lurking inside the representational scheme, moreover, are the same concerns that had been preoccupying modernist painters for several decades: the flatness of the painted surface, the minimization of the illusion of spatial depth. The interlocking planes of the walls and windows carry a faint but telling echo of De Stijl modernism; the near-blank facets of the house are small, semi-serious moments of colour-field abstraction. Hockney is interested, in one and the same moment, in the structure of things (the way space is carved up and modulated) and the structuring of a painted surface. To this extent, the painting is determinedly physical – self-consciously modernist.

At another level, the painting is a study in the *immaterial* – in what has been termed the 'light [...] peculiar to Los Angeles, a shimmer which is at times oddly airless'.[49] Seen like this, the physical objects in the picture come to appear almost incidental: the true subject is the feel of the place, its tense tranquillity, the razor's edge between something happening and nothing happening.

For all that it depicts an individual figure in a sealed space, *Beverly Hills Housewife* offers clues to the artistic atmosphere of the time, to the broader cultural climate. It is not as solipsistic as it first appears. Even if Freeman is daydreaming, the painting itself is attuned – cognizant of the moment from which it has sprung. The William Turnbull sculpture makes an explicit nod to one of Hockney's artistic contemporaries, although it is impossible to know what Hockney made of Turnbull's works beyond recognizing them as faddishly popular: *Beverly Hills Housewife* is the third California picture in which a Turnbull totem constitutes a shibboleth of good taste.

The painting contains implicit allusions, besides, to artistic trends in the 1960s. The sharp delineation of colour and light, down to the blades of reflected light that strike the windows, invokes (however inadvertently) the 'finish fetish' aesthetic of West Coast abstraction, then at its height. Spanning sculpture and painting, this ultrafine style was a riposte to the gestural 'mess' and perceived emotional turbulence of Abstract Expressionism. It found its epitome in the auto-lacquer paintings of Billy Al Bengston – metal surfaces sprayed with car paint in slick chevron patterns. Bengston had been a motorcycle racer until he broke his back in an accident in 1964; he had begun to incorporate the attributes of motorcycles, surfing, and 'Kustom cars' into his work of the early 1960s, channelling what Tom Wolfe had described as the 'freedom, style, sex, power, motion, color' of customized automobiles.[50] 'None of us were going to paint like de Kooning — we just weren't', Bengston later said of his peer group. 'You don't punch with the heavyweight champion. So you invent the new fight.'[51]

Finish fetish applied also to the 'industrial chic' sculptures of that group, for instance the glass boxes of Larry Bell, who had originally come to Los Angeles in the hope of becoming a Disney animator. While working part-time in a framing shop, he learned how to cut glass. His earliest boxes, dating from 1963, featured scratching in their surfaces, as well as superimposed segments of Formica or mirrored glass (arrayed in stripes, checkers, and ellipses) – elements that complicated the transparency of the object. As well as instancing an exquisite sharpness of form, Bell's glass cubes had a way of projecting beyond themselves, encompassing the space of the viewer within their reflections; as the gallerist Nicholas Wilder observed:

> You walk up to this cube that's forty inches of smoked glass, and you start inspecting it. And then what happens? In the refractions of the light, the object is extended several meters out – and you. Your body is physically inside the parameters of the object you're viewing.[52]

Ironically, then, finish fetish could end up blurring the boundaries of an art object: hard edges and pristine surfaces did not necessarily seal off the sculpture as a thing in itself. 'The thing I liked about glass', Bell later mused, 'was that it reflected light, it transmitted light, and it absorbed light all at the same time […] and it had a shelf life of something like three million years.'[53] The same effect can be traced in the planar surfaces of *Beverly Hills Housewife*, alternating between transparency, opacity, and glinting reflection. It is a painting in which the 'hard edge', a term used to describe the so-called 'LA look' of the 1960s, achieves a kind of primacy.[54] The glass panels of Freeman's house – disclosing and deflecting by turns – are not a long way from Bell's glass enclosures, if one looks past their domestic function, their pictorial role.

The uniform light of *Beverly Hills Housewife* – an even, anodyne brightness that conditions everything in the picture (flattening forms into near-silhouettes) – itself connotes a larger artistic phenomenon. At the time that Hockney was painting the picture, light had become a primary concern of conceptual artists, a medium as well as a motif. Bell was one of the instigators, along with Turrell and Robert Irwin, of what became known as 'Light and Space', a broad church of artists in southern California who combined the precepts of East Coast minimalism with a pronounced interest in metaphysical and perceptual phenomena. In 1966, just down the street from Sam Francis's studio in Santa Monica, Turrell had purchased the disused Mendota Hotel, in which he engineered his first light projections, the *Mendota Stoppages*. For these, he sealed over the interiors with plaster to create a sort of cavernous mausoleum, before carving holes in the exterior walls to create incidental light projections from sources outside – car beams, stoplights, moonlight, sunlight. 'The Mendota *was* a Plato's cave,' Turrell said, with typical grandiloquence. 'These rooms were camera-like spaces that apprehended the light.'[55]

Beverly Hills Housewife, like Turrell's light show in Santa Monica, equivocates between inside and outside, the minute and the macrocosmic. It is the pictorial counterpart to a work such as Peter Alexander's *Small Cloud Box* (1966), a five-inch resin cube whose translucent material seems to enclose a cloudscape over an open plain.[56] A student of Hockney's friend Oliver Andrews, Alexander had begun to experiment with industrial resins at UCLA in the mid-1960s; it was the same substance that he had used to glaze and repair his surfboards. He regarded the resulting cast 'boxes' – made by injecting hot wax into layers of uncured resin – as 'little rooms, places to go'.[57] *Small Cloud Box*, along with other early works, suggests a desire to capture the light and immensity of southern California (its sea, sky, and spreading terrain) in finite, geometric form.

As a piece of West Coast conceptualism, *Small Cloud Box* is of a different order from Hockney's painting. But a comparison of the two helps to underline the pictorial quality of Alexander's work (its resonance, say, with Dutch landscape painting), as well as the metaphysical component of Hockney's. There is a *multum in parvum* principle to *Small Cloud Box*, in its evocation of boundless space, that has a direct equivalent in *Beverly Hills Housewife*. In Freeman's garden, light is at once diffuse and compartmentalized – framed by a sheet of glass, concentrated into linear glints. Space is tightly sequestered and at the same time, if one glances up at the sky, boundless. That neat strip of blue, like the window at the front of the house, beckons a leap into the void.

In subversively classical terms, then, *Beverly Hills Housewife* aligned itself with the practices of contemporaneous artists. The bourgeois 'episode' of the painting is also, in its own deceptive way, a paean to light and space. The sunlit scene embodies the sense of possibility that Freeman and Hockney had both perceived in Los Angeles' 'openness' when they first came. ('Light and space, I think, were very, very strong attractions', said Hockney in 1984. 'The light after all is very intense in California, compared to England, compared to London. But also I think it's the sense of space. It's a horizontal city.')[58] Despite appearing to eschew artistic fashion, as a representational painting in the heyday of conceptualism, *Beverly Hills Housewife* was peculiarly of its moment; and while the alignment may have been subliminal, it is hard to believe that Hockney was unaware of the innovations of his American contemporaries. His 'early Renaissance' lucidity of depiction – seen in the hard, bright forms of the picture – had a distinctly 1960s resonance.

On the scene

In style as well as content, *Beverly Hills Housewife* points to Hockney's awareness of contemporary trends in Los Angeles, for all that he was never part of a home-grown movement. He was a regular, all the same, on the art scene, which in those days was still in its relative infancy. Whenever John Kasmin was in town, the two of them would frequent 'Gallery Row' on La Cienega Boulevard, the three-block stretch along which the leading commercial galleries were located – the 'Los Angeles Left Bank', in the words of critic Jules Langsner in 1956.[59] Starting in 1962, the Monday-night Art Walk was a weekly fixture. The galleries stayed open until 10 p.m., and visitors would stroll from venue to venue. These occasions brought out a parade of hippies, society ladies, and art students in *Easy Rider* gear – or, as *Los Angeles Times* columnist Art Seidenbaum put it, 'the braided and bearded and bare-toed' flocked together with 'those dressed for dignity'.[60]

As well as mixing with artists, Hockney and Kasmin fraternized with dealers including Rolf Nelson, Irving Blum, and Douglas Christmas, the owner of Ace Gallery. (Christmas was also a long-time friend of Betty Freeman's, despite his reputation for 'creative accounting'; as her friend Julia Bloomfield puts it: 'Doug was the sleaziest art dealer you could imagine, but totally and utterly charming – a lovely man.'[61])

The two heavyweight Los Angeles galleries of the day were Felix Landau and Ferus, the former specializing in modernist and surrealist art, the latter representing the vanguard of the contemporary scene. Ferus had been established in March 1957 by Walter Hopps and Edward Kienholz, originally occupying the back area of an antique store on La Cienega. The gallery was born partly of Kienholz's frustration at being rejected by existing dealers, and the realization that in order to be shown anywhere he would have to found a gallery of his own; he had first established the Now Gallery, a precursor to Ferus, in a disused theatre on La Cienega in 1956. Hopps had also run a gallery of his own – Syndell Studio, located in a makeshift building in Brentwood constructed from pier pilings and telephone poles covered with tar (its name bizarrely purloined from that of a Midwestern farmer, Maurice Sindel, who had been killed on an Ohio highway in 1953).[62] Out of these unlikely forebears, Ferus soon became a beacon of the avant-garde, although it did not flourish commercially: Hopps was ill suited to selling, and Kienholz wanted to concentrate on his art (he would paint in the gallery's back room, with a mirror to alert him to visitors coming in).

After Irving Blum bought Kienholz's share in the business in 1958, the gallery moved across the road and grew more professionalized, dropping the cooperative model on which it had previously run. As the artist Ed Moses later put it:

> Irving Blum came into town from New York. [...] Chico [Hopps] interviewed him to be the director. And Irving said:

> 'First we've gotta get rid of about thirty people.' There were all these funny guys that were in there, with beards and babies and wives with babies. [...] Irving got rid of all these guys.[63]

Blum was an urbane salesman who had worked at Knoll International Showroom in New York ('my first "brush" with modernism'); he funded the Ferus purchase partly through the proceeds from work on a porn film, *The Immoral Mr Teas* (1959), for which he had contributed the story and voiceover (the tale of a timorous man who is able to look through people's clothes with X-ray glasses, it was one of the first softcore movies to gain a widespread theatrical release).[64]

'I decided that to do a gallery in California might be a provocative thing at that time', Blum recounted in 1977. He was drawn by the pioneering spirit in Los Angeles, so unlike the hierarchy and competitiveness of the East Coast – a spirit that was embodied by the Pasadena Art Museum (where Marcel Duchamp held his first American retrospective in 1963) and the relocation of *Artforum* magazine from San Francisco to LA (in offices above Ferus) in 1962.

Under Blum, Ferus broadened its scope – giving Andy Warhol his first show in the city, for example, which featured a sequence of 'Campbell's Soup' prints – although as Blum admitted: 'We operated squarely behind the eight ball. Really in the red year after year after year.'[65] The Warhol exhibition, for one, seems to have been ahead of the curve: an outraged dealer across the street arranged a stack of actual Campbell's soup cans in his window, with a sign that read: 'Get the real thing for only 29 cents a can.'

The coterie around Ferus, which included Billy Al Bengston, Larry Bell, Ed Moses, Craig Kauffman, Robert Irwin, and Ed Ruscha, became known as the LA Cool School. Bengston had first come to the attention of Hopps and Kienholz when he strode into the gallery and declared: 'I'm going to be the world's greatest artist. [...] I'll take you all to lunch.'[66] His chutzpah was representative. 'You could hang with this group if you were serious about what you did', said Bell in 2007. 'You had to take a lot of shit and be ready to dish it right back. But it was all based on a profound love and sense of humor.'[67]

The group dominated the art scene of the 1960s, just as it has dominated subsequent histories of art in Los Angeles. The architect Frank Gehry, a sometime member of the in-crowd, recalled: 'There was certainly a group of figures who were welded together. They hung out together, dated each other's girlfriends. It was a communal scene, and every night we met somewhere.'[68]

It was an avowedly macho scene, heavily invested in motorcycle and 'hot-rod' car culture; in this regard, it was a foil for Hockney's gay sensibility and subject matter, whatever subtler continuities may have existed. (By his own admission, Hockney remained somewhat apart from the local fraternity: 'I'm a little antisocial here. I don't go out that much.'[69]) Bengston and peers such as Moses, Irwin, and Ken Price styled themselves 'The Studs' – also the title of an exhibition staged by the four at Ferus in 1964. Judy Chicago, one of Bengston's students, was the only woman admitted into their milieu:

> They spent most of their time talking about cars, motorcycles, women and their 'joints'. I knew nothing about cars or motorcycles, couldn't really join in on their discussions about women, and obviously didn't have a 'joint'. They made a lot of cracks about my being a woman and repeatedly stated that women couldn't be artists. I was determined to convince them [...] that I was serious.[70]

A regular haunt was Barney's Beanery on Santa Monica Boulevard, a greasy-spoon diner and drinking den that had attracted artists and musicians since the 1920s. It was the scene of some of Jim Morrison's greatest alcoholic excesses – he once relieved himself on the bar; later, Janis Joplin would go drinking there on the last night of her life. Peter Schlesinger remembers the place as 'very heterosexual – very homophobic. One knew not to go.'[71] Notoriously, a sign hung behind the bar that read 'Fagots Stay Out' (sic); it was only finally removed in 1985, under pressure from gay-rights activists. Hockney nonetheless seems to have dropped in from time to time, taking his London friend Patrick Procktor in 1965. A willowy Englishman with a refined, theatrical air, Procktor was as different from the regular habitués as could be imagined. Perhaps Hockney shared the insouciance of James Dean, the bisexual screen idol who made a point of eating at Barney's in the 1950s – and wryly remarked of the sign: 'That's why I like coming here.'[72]

In one of his best-known works, Kienholz recreated the hangout in a walk-in installation, *The Beanery* (1965), complete with soiled floors, electric lights, battered furniture, and an odour of bacon fat and urine, along with a soundtrack of babbling voices. 'I don't know if it's art,' he said at the time, 'and I don't give a damn.'[73] The plaster bodies and recycled mannequins in Kienholz's 'adult dollhouse' have stopped clocks in place of heads, in reference to the 'surrealist time' experienced by the boozy escapists in the bar, versus the 'real time' of the outside world. One of the female figures wears a mink stole that Kienholz had acquired for $600. When the work was exhibited at the Dwan Gallery, New York, Procktor declared: 'It's more like Barney's than the real thing.'[74]

Kienholz's *Beanery* is a crucible of Los Angeles life entirely unlike that of *Beverly Hills Housewife* – a nightmarish Dionysian enclosure; and yet Kienholz's conception of two modes of time running in parallel (the surreal time of the sealed interior, the real time of the world beyond) applies with uncanny force to the image of Betty Freeman in her garden. There, too, a different dimension of time has come into play.

'That little window of time and space and place in the '60s in Los Angeles was unique', according to Hockney's LA dealer Peter Goulds.

> That Beverly Hills circle: it was a world that Irving Blum, for one, wanted to enter. He was, after all, a furniture salesman who happened to have a loud voice and would sometimes rent a Bentley to attract his clients. Ferus was his portal to that world.[75]

Freeman was emblematic of that Beverly Hills world. She shied away from first-night openings, and Blum's Bentley would probably have made little impression; still, she bought art from Ferus and other local galleries, including Felix Landau and Nicholas Wilder.[76] A particular friend was Virginia Dwan, one of the few women dealers in an avowedly male scene. The heir to a Minnesotan mining dynasty, Dwan had established her gallery in Westwood in 1959, and had been instrumental in bringing

East Coast heavyweights such as Franz Kline and Philip Guston to Los Angeles. Like Freeman, she was conscious of her position as a woman striving to cultivate an independent identity. She would later state:

> In that era, there weren't many precedents for women to own galleries, and feminism was only just beginning in the US. I wasn't so concerned about that, but I do remember telling my husband that he shouldn't go to parties or dinners with me, as people always assumed it was his gallery.[77]

While she avoided gallery openings, Freeman was a regular at other art-world events. In the spring of 1964, she attended a lecture by the formidable modernist critic Clement Greenberg on the occasion of his exhibition at the Los Angeles County Museum of Art, *Post Painterly Abstraction*. 'Greenberg's visit & lecture were an awful dud', she confided in a letter to Sam Francis. 'He couldn't get off the subject of collectors & was only irritating.' She took a sceptical view of the show itself, in which Francis's work had been included: 'I did think your 3 paintings looked out of place there. [...] It amazed me how yours seemed so separate & not belonging in a group show whose purpose seemed to be light & color.'[78]

Four years later, she reported having been to a party given for Claes Oldenburg by Stanley and Elyse Grinstein, collectors whose gatherings had acquired a legendary status. 'All the pals were there like fraternity brothers with their inside games – Billy Al, Kenny, etc. [...] Billy Al thinks that a year & a half is nothing, that even seven years are needed for something good to cook.'[79] Oldenburg's massive sculpture of a typewriter eraser would later stand on Freeman's back terrace at Hillcrest Road (fig. 22).

The Oscar Wilde of LA art

Few people were more synonymous with the Los Angeles art scene, in those years, than Nicholas Wilder – a person of significance for both Freeman and Hockney. After opening his gallery in a storefront space on La Cienega Boulevard in April 1965, he had fast achieved success through a combination of prominent East Coast artists (Barnett Newman, Helen Frankenthaler, Cy Twombly) and LA discoveries such as Bruce Nauman, to whom he gave a one-man show in 1966. His horn-rimmed glasses and dapper, blazered attire belied an instinctive mode of working; he had seen one of Nauman's fibreglass sculptures at the artist's studio at the University of California, Davis, and been unable to get it out of his head. Nothing sold from that first show, but Wilder's fixation was prescient.

Wilder had been born into an upper middle-class family in Rochester, New York, and retained the outward gentility of an affluent Wasp. His father had helped to invent Kodachrome film, which had led to a prevailing – false – impression that Nicholas was the heir to vast wealth. He had become interested in art as an undergraduate at Amherst College, Massachusetts (he majored in political science), while working as a guard and projectionist in the art department. When Marcel Duchamp came to give a lecture, Wilder ended up giving the artist a tour of the collection. Determined to live in California, he then embarked on a law degree at Stanford (partly to satisfy his family's expectations) before switching to art history. 'My education literally started as a slide projectionist – not as a student', he remembered. This origin story accounts for his enduring belief in art as visual communication, not a platform for critical posturing: 'The art historians and the critics today do not like art and, therefore, try to make it into something other than it is through philosophy or their theories or codes. They don't interest me very much.'[80]

Free-spirited and fast living, Wilder was 'the Oscar Wilde of LA art', in the words of an admirer.[81] Interviewed in 1988, when he was dying from Aids, he semi-joked: 'I think being an art dealer is an occupation for disenfranchised people. I never knew a good one who wasn't a woman, Jewish or gay.'[82] In the 1960s, his gallery was at the centre of a happening scene. 'It was a kind of a golden age. There were about six galleries and 30 artists that counted.' Hockney soon became one of that select group: 'Nick was about my age, quite intellectual, and I instantly got on with him. He was only just opening his gallery, where he showed very radical artists for the time. He was a bit disorganized, but very driven.'[83]

In 1966, Wilder was living in a small apartment in Larrabee Street, off Sunset Boulevard. It was here that Hockney stayed when he returned to Los Angeles early that year; by the summer, Peter Schlesinger was also a regular guest, when he was not staying at Hockney's Pico Boulevard studio. Another painting that Hockney completed in 1966, aside from *Sunbather* and *Portrait of Nick Wilder*, was *Peter Getting Out of Nick's Pool*, which depicted his lover drawing himself out of spangled water, seen from behind. Here again, the human character is the picture's focal point, albeit that Schlesinger is turned away – exposed yet remote. *Beverly Hills Housewife*, which he had begun first, took the longest to complete.

Wilder's apartment – the outside of which appears in both the 1966 paintings as a plain modernist block – was at odds with his preppy deportment. The living room contained almost nothing apart from a yellow John McCracken sculpture on which Wilder used to hang his underwear. There were no beds, only mattresses on the floor. 'You would have gone to bed and some boy would come in and get on the mattress with you', according to Hockney. 'You never knew what was going to happen any night.'[84]

When Schlesinger's parents discovered that he was living with Hockney, they insisted on their son seeing a psychiatrist. It was Wilder who helped to allay their fears. Schlesinger recalled how his father went to see Wilder, who told him: 'Your son is with a very wonderful person and would you rather have him hustling on Santa Monica Boulevard?'[85] Ironically, Wilder and Hockney both had some familiarity with that hustling scene. Among the last of Hockney's panoramic realist paintings of the 1960s and 1970s was *Santa Monica Boulevard* (1978–80) a vast drive-by scene of the strip of West Hollywood known as Boys Town, to where Wilder's gallery had moved in 1970. Amid the twenty-foot frieze, a young man leans against a lamp post, attired in tight denim shorts, white socks, and a visor, one hand planted on his hip in languid provocation.

While longer-lived than many, Wilder's gallery symbolized a finite moment. He left Los Angeles in 1979, finding himself overtaken by an ever-more corporate and thrusting gallery model; he subsequently pursued a career as an artist. 'I wasn't a very good businessman', he admitted. 'I got into all kinds of trouble. As the name of the game changed in the 1970s, that's when I decided I wanted to get out because it became a viable business situation, which isn't why I got into it.'[86]

Nicholas Wilder and David Hockney at Chateau Marmont, Los Angeles, March 1976, photograph by Betty Freeman

A photograph taken by Freeman in March 1976 shows Wilder and Hockney on the roof of the Chateau Marmont on Sunset Boulevard.[87] Hockney was back in Los Angeles that spring to make a series of lithographs, and had been staying in the penthouse of the hotel with his boyfriend, Gregory Evans, a young designer who had formerly been Wilder's partner.[88] Wilder was acting as Hockney's dealer by this time, but in the photograph, he is momentarily Hockney's subject: the artist leans back to take a snapshot, and Wilder strikes a casual, vaguely diffident pose.

The photograph is black-and-white, but the softness of the light suggests the time of day known in Los Angeles as the 'magic hour', the phase of twilight during which the sky fades to a faint lilac or a rose gold. The lights of Hollywood twinkle behind Wilder as he stands – cigarette in hand, arms folded – and gazes upwards, past Hockney. In the background, blurred but recognizable, is the French Revival landmark of the Voltaire Apartments, a stylistic double of the Chateau.

Wilder has some of the insouciance, here, of Freeman in *Beverly Hills Housewife*. Once again, we see a person on the threshold of middle age, gazing serenely into space, surrounded by a picturesque fragment of Los Angeles (the metropolis, in this case, rather than one of its houses). There is even a miniature palm tree in the left foreground. Hockney's presence in the photograph literalizes – renders explicit – his unseen presence in *Beverly Hills Housewife*, where we sense him observing, recording, capturing, from just beyond the limits of the picture.

In a way, it was true

In the summer of 1967, Hockney taught for three days a week at Berkeley, on the east shore of San Francisco Bay. He was given a studio and put up at a hotel on nearby Telegraph Avenue. Unlike the summer class he had taught at UCLA the previous year, the graduate students at Berkeley showed talent and promise: among them were the painters Mary Heilmann and Alan Turner.

San Francisco was in the throes of the 'Summer of Love' – the Bay Area, a hive of countercultural energy. Hockney was aware of the political climate, but little affected by it. His feelings might have chimed with those of Kenneth Clark, the art historian and former director of the National Gallery, who remarked of student protestors in Paris in 1968 that they were 'impatient to change the world, vivid in hope, although what precisely they hope for or believe in, I don't know'.[89] Hockney attended a few anti-Vietnam meetings, but he was primarily concerned with his own painting, his teaching, and his romance with Schlesinger. His works of the 1960s were a pointed demonstration of his indifference to political causes – an indifference that would persist throughout his career.

In *Beverly Hills Housewife*, Freeman's garden is a little strip of Eden, a private idyll within the larger sanctuary of Beverly Hills. The band of green that runs along the base of the painting is an impassable boundary, enclosing her within the space of the picture and underlining the fact that a spectator can never access her world, only look on.

Inevitably, the painting invokes a certain formulaic conception of Los Angeles as a paradise on Earth, one that can be found in the first travel books on southern California. For the journalist Charles Nordhoff, who published a popular guide to the region in 1873, Los Angeles was a magical idyll akin to the Land of the Lotus Eaters, the beautiful isle that the Greek hero Odysseus visited – a place so beguiling that its inhabitants never wanted to leave: 'Below us, as we looked off a hill-top, lay the suburbs of Los Angeles, green with the deep green of orange-groves, and golden to the nearer view with their abundant fruit.'[90]

Later commentators have characterized such visions of Los Angeles as no more than marketing hype ('promotional "myths" that were designed by chambers of commerce', as the critic Norman Klein has stated), or even as racist propaganda.[91] Mike Davis, the pre-eminent 1980s theorist of the city, referred in withering terms to the 'writers, antiquarians, and publicists [...] who at the turn of the century created a comprehensive fiction of Southern California as the promised land of a millenarian Anglo-Saxon racial odyssey'.[92] Such judgments suffer from their own whiff of mythmaking, in their effort to ascribe a firm intention – and a sinister, programmatic ideology – to what was in large part a collective daydream, however wishful or distortive.[93]

Descriptions of the city have long been said to suffer from a tendency towards 'boosterism', waxing indulgently lyrical, or conversely to revel in the dystopian underbelly of urban life, as seen in classic noir fiction or a neo-noir movie such as *Blade Runner* (1982).[94] Even Nordhoff, in his early hymn to the beauty of southern California, acknowledged that 'the Puebla de Los Angeles – the town of the angels – is not, in its present state, a very angelic place. It is irregularly built, the older part having but one principal street.'[95] Hockney's early preconceptions of the place – and perhaps his later experiences – had oscillated between the poles of sunlit beauty and abject underlife, highlighting the possibility that the two visions might in fact coexist: the myth and the anti-myth did not have to be mutually exclusive.[96]

As Hockney found, the 'lotus-land' of Los Angeles' earliest celebrants was no fable. The myth was real. The fiction (or parts of it) had become a self-fulfilling prophecy. Other realities existed, of course, but what he perceived in their midst was an earthly paradise. Even the fantasy scenarios of *Physique Pictorial* could be glimpsed from certain angles, in certain lights, never mind the tawdry facts of Bob Mizer's AMG studio:

'The photographs portrayed what a certain life was like in Los Angeles; and, in a way, it was true. They were accurate if you looked for it.'[97]

That perception of 'a certain life' – as opposed to anything more earnestly documentary – soon formed the essence of his work. But it would be wrong to suggest that his art offers nothing more than a blithe, boosterist view of sunshine and tranquillity. In *Beverly Hills Housewife*, the sunniness is offset by an unresolved tension, the faintest hint of a darker undercurrent. Here and in *American Collectors*, the possibility of melancholy, even of a noir narrative subtext, lingers within the Californian light.

Californianess

There is a fundamental contrariness to the painting. Its brightness seems to admit the potential for darkness. In its very apartness from 'real life', the staged scene harbours a paradoxical sense of the real. The objects have a hard, unarguable precision about them. Freeman is a matter-of-fact presence, for all that her stilled, erect figure channels distant times and places. Working against – or in tandem with – the impression of a sealed, inaccessible world is a semblance of life unfolding by the second, as though Freeman inhabited a dimension of time very close to, or identical with, our own. Except for certain period details – her dress, say, or the blockish white furniture in the room behind her – we could be in the present.

All of Hockney's California paintings have this quality about them: a sense of being both of their time and occurring in the present tense, both recorded and live. His pictures of the 1960s, along with those of Ed Ruscha, have often been regarded as synonymous with Los Angeles – as perennial expressions of the city's look and feel. In 1971, for instance, *A Bigger Splash* featured as the cover image of Reyner Banham's study of the city, *Los Angeles: The Architecture of Four Ecologies*, the ruptured pool becoming a microcosm of a time and place.

According to this view, paintings such as *Beverly Hills Housewife* and *A Bigger Splash* are paintings of Los Angeles – among the first serious attempts, even, to give artistic representation to the city. Admittedly, Hockney himself encouraged such a reading. Speaking in 1975, he remarked:

> In Los Angeles, there were no ghosts; there were no paintings of Los Angeles. People then didn't even know what it looked like. And when I was there, they were still finishing up some of the big freeways. I remember seeing, within the first week, a ramp of freeway going into the air, and I suddenly thought: 'My God, this place needs its Piranesi, Los Angeles could have a Piranesi, so here I am![98]

Freeman herself would go on to acclaim him as a great American painter: 'I'm very opinionated and very biased, but I think David is a great artist. He's the great artist of Los Angeles, the great artist of America, and the great artist of the twenty-first century.'[99] The art historian Howard N. Fox mirrors her sentiments when he compares Hockney's images with the landscape paintings of California Impressionists such as Granville Redmond and Guy Rose, positing that Hockney's art 'shares the outlander's fascination with the region's quality of light, lush natural settings, and ineluctable sense of place — of "Californianess"'.[100]

The universalizing tendency of *Beverly Hills Housewife* – its apparent gravitation towards the type, the ideal – encourages interpretations of this variety, but, of course, the gravitation is offset by a contrary movement. For all its tilting towards the universal, the painting offers a single view of a single life at a moment in time. Unlike *A Bigger Splash*, where particulars have disappeared in a burst of water (we will never know who has jumped beneath the surface), the scene in *Beverly Hills Housewife* refuses to flatten altogether into an emblem.

Those segments of bright space in the background of the picture, reducing the outside world to sky and greenery, leave other realities suspended. The window visible through the house gives only a fragmentary glimpse of the sunny suburb beyond. *Beverly Hills Housewife* makes no claim to capture the city at large, never mind Hockney's profession of himself as the Piranesi of Los Angeles. If anything, it evokes the hermetic, luxuriously enclosed nature of the suburban enclave. Beverly Hills had been created in the early twentieth century as a sequestered neighbourhood, away from the fray of the metropolis. Writing in 1968, Banham identified the incorporated city as an 'enclave of unrelieved middle class single-family dwellings', representative of a larger tendency, throughout Los Angeles, towards 'total indifference to the needs of all communities except one's own [...] the ugly backside of that free-swinging libertarian ethic that makes so much of Angeleno life irresistibly attractive'.[101]

Other realities existed, though, as Hockney would have admitted. Only a year before he painted *Beverly Hills Housewife*, in August 1965, the Watts neighbourhood in the south of the city had seen the worst racially aggravated rioting in the history of Los Angeles: six days of disorder following the arrest by the LAPD of a young Black man named Marquette Frye. Pitched battles between residents and the police and National Guard resulted in thirty-four deaths. The riots laid bare the racialized strife that had run in parallel, always, to the city's reputation as a utopia. For Joan Didion, the unrest had been a symptom of the city's inherent volatility – its vulnerability to conflagration: 'What struck the imagination most indelibly were the fires. For days one could drive the Harbor Freeway and see the city on fire, just as we had always known it would be in the end.'[102]

In *Beverly Hills Housewife*, as with all of Hockney's Californian paintings, the chaos that had erupted in Watts – the harsh realities of Los Angeles in the mid-1960s – remained invisible. (In much the same way, social and religious discords of the seventeenth century made little impression on Johannes Vermeer's scenes of Dutch bourgeois life.) Like the mythic visions of Los Angeles entertained by early settlers, Hockney's vision was partial, necessarily incomplete. Other artistic accounts of life in LA were being made, including those of the city's 'underground'. Those accounts have since been recuperated by art history, with a deepening emphasis in recent decades on non-canonical artists – whether the Chicano movement of the 1960s and 1970s (with its dimension of civil-rights activism) or the Black Arts Movement that was born of the Rebellion. But Hockney's desire, then and for ever after, was to bear witness to the pleasures of his senses and imagination – to record what he saw and felt.

This impulse, redolent of a fin-de-siècle cast of mind, has laid him bare to charges of superficiality. According to Mike Davis, the booming city of the 1980s found itself able to 'afford to buy the international celebrity architects, painters, and designers – [Richard] Meier, [Michael] Graves, Hockney, and so on – capable of giving cultural prestige and a happy 'Pop' veneer to the emergence of the world city'.[103] In its throwaway cynicism, the comment betrays a familiar view of Hockney's Californian paintings as lacking in depth of meaning – not to mention an overeasy equation of his work with 'Pop', a label that the artist has always resisted. In a similar vein, the art historian Paul Melia has stated – with a ring of disapproval – 'Not only is the Californian sky in Hockney's paintings devoid of '*noir*' clouds, but the social realities of the city are negated so that it serves as a *spectacle*.'[104]

But to omit is not to negate. Observations of this kind not only discount the possibility of perceiving an undercurrent in a painting such as *Beverly Hills Housewife* (the merest touch, say, of noir unease), but overlook the fact that Hockney's art bore no obligation to probe 'social realities'. They equally overlook the possibility that meaning might reside in so-called veneers, that 'spectacle' does not have to be inimical to truth. As *Beverly Hills Housewife* suggests, an entire identity might be constructed (or played with, or dismantled) through visible things, through alluring surfaces. If nothing else, Hockney's paintings are an eloquent reminder that surfaces may be all we have.

Beverly Hills Housewife is not a work of social realism in the vein of the grand nineteenth-century novel, far less an earnest act of reportage (Melia does at least concede that Hockney 'never claimed to be a social realist').[105] More than many of his 1960s generation, Hockney had no interest in subordinating his art to polemic. His personality was characterized by a roaming curiosity, pleasure-seeking, and a fascination with the quandaries of image-making. It resisted being moulded by political principles. Even his abrasively sexualized paintings of the early 1960s, which unapologetically professed his gayness, were personal testaments rather than statements of political intent.[106] Far from being a symbol of Los Angeles, *Beverly Hills Housewife* symbolized Hockney's desire to communicate the haphazard truth of his life at that time, without commentary or critique. Californianess was a cast of mind: a way of seeing rather than a state of affairs.

Another reality

'He works in the farthest-out region. Like Charlie Parker, you always know Kienholz is there.'

Anonymous Los Angeles artist, 1966[107]

A very different vision of American life can be found in the art of Edward Kienholz. Hockney knew Kienholz, albeit vaguely, through their shared patronage of Barney's Beanery. Kienholz was a far more devoted customer than Hockney, admittedly, and moved in a distinct circle; he later reminisced: 'In the Beanery, we talked about who was a good fuck and where we were going to get six dollars so we could buy gas for a car to go to, you know, the Valley and get drunk.'[108] Kienholz's life-size dioramas might be regarded as the diametric opposite of Hockney's Californian paintings, not to mention a stark departure from the finish fetish and sun-'n'-surf influences of other Los Angeles contemporaries. Abjectly physical – composed out of the literal detritus of city life – and pulsing with sex and violence, they betoken an engagement with the gritty reality of LA (especially its fringe, freakish elements) of the kind that Hockney's paintings conspicuously lack.

For all their differences, though, Kienholz and Hockney shared a fascination with the capacity for everyday reality to transmute into a mythic version of itself. Neither artist was a sober documentarian. It was the very interplay of reality and myth in their works that made both so in tune with their adopted city.

Kienholz was a character straight out of the American West, plain-speaking and physically imposing. (He once attacked the desk of an airline employee who refused to reimburse him for a Tiffany lamp that had shattered in transit.) He had grown up in the 1930s on a farm in Washington state, a rugged and isolated existence that moulded his artistic ethos in place of any formal training. From an early age, he was schooled in the economics of the Depression – instilled with a sense of the value of throwaway things. On the farm, broken machinery was typically adapted rather than thrown out; his father ('an angry disciplinarian') would repair wrecked cars in his spare time. Kienholz's mother was a Presbyterian whom he characterized as 'religious and in a way fanatic'; as a boy, he used to help her to fashion crèches for the Sunday school where she taught – the first tableaux he made.[109]

After working various odd jobs in his twenties, including stints as a used-car dealer and an attendant in psychiatric hospitals, Kienholz moved to Los Angeles in 1953 – in part because 'it was so big that you could be completely anonymous. [...] And I liked that. I liked that feeling. It wasn't a small town; people looking over your shoulder.' The amount of junk that the city produced was another attraction: 'If you're living on the edge of the economy like that, all the waste filters through your awareness and you take what you want.'[110]

Kienholz abandoned painting in the late 1950s (around the same time he was running Ferus with Walter Hopps) in favour of three-dimensional assemblages; these ranged from box-sized, quasi-ceremonial assortments of items to extravagant scenarios incorporating cast bodies and found objects, often slathered in varnish or sprayed with fibreglass compound. By the middle of the following decade, he had become recognized as a foundational figure in the city's art scene – a status that was cemented by his solo exhibition at the Los Angeles County Museum of Art in the spring of 1966, a year after the institution had reopened in a new building. The exhibition became a cause célèbre, with Kienholz's installation *Back Seat Dodge '38* (1964) in particular igniting a moral panic.[111]

The work consists of a 1938 Dodge car marooned on artificial grass (fig. 18); Kienholz had scoured the city for precisely the model that his father had owned. The car has been surreally truncated: its front hood and fenders seem to have collapsed together with its back compartment, the driver's section having entirely disappeared. Strewn over the interior of the car and the surrounding turf are beer bottles. On the back seat, visible through the car door, a couple are imagined having clumsy sex – their bodies rendered out of cast plaster and semi-exposed chicken wire, their heads appearing to merge at the far end of the seat. The girl's plaster legs are splayed. On top of her, the boy sprawls as a transparent wire 'ghost', his erection nonetheless unmistakeable. One of his hands has been given solid form; fumblingly, it creeps beneath the girl's skirt, between her legs. A beer bottle is clamped in one of her hands.

The installation was based on an episode from the artist's youth, when he had picked up a girl at a dance and taken her for a drive. The car evokes a generalized wartime 'yesteryear'; the interior scene homes in – as though by a telescoping motion – on the feverish night-time episode. Kienholz wanted it to appear as though viewers had stumbled upon the vehicle: the back door was left closed, but could be opened by anyone wishing to see more clearly inside. According to Lyn Kienholz, his wife at that time, 'Ed always said that the best installation [of the work] was one that would make it look like you came upon this couple at night, up at a necker's spot on Mulholland Drive.'[112]

It was just the kind of thing that prompted Marcel Duchamp to say of Kienholz: 'Marvelously vulgar artist. Marvelously vulgar artist. I like that work.'[113] Others were less approving. When *Back Seat Dodge '38* appeared at the Los Angeles County Museum of Art, the County Board of Supervisors tried to have it and another work removed on grounds of obscenity. (The other piece was *Roxys*, 1961, a brothel scene featuring decapitated female sex automata constructed out of furniture and animal bones.) Supervisor Warren M. Dorn wrote to Edward Carter – one of the museum's founders – lamenting that it was 'tragic to me that with so much talent available we should have this kind of depressing, nauseating and revolting expression displayed in our magnificent new public facility'.[114] At a news conference, Dorn professed – apparently without irony – 'My wife Phyllis knows art. I know pornography.'[115] The Board voted to close down the show, but the museum refused to remove the works from display.

Back Seat Dodge '38 is the type of work, one suspects, that Betty Freeman would have looked askance at, even if she had not shared Dorn's censorious opinion. In a letter to Sam Francis in 1965, she expressed a negative view of Kienholz's work – a view that she had conveyed with characteristic candour to LACMA's curator of contemporary art, Maurice Tuchman, who would organize the offending exhibition the following year:

> I said in my opinion K. was an obsessive putter-together of details, that the finished piece ended up for me like folk-art and that the men who remodel cars into hotrods are more imaginative.

> Well, since then he hasn't spoken to me, as he thinks K. the greatest artist on the west coast.[116]

At one level, Kienholz's installation is the antithesis of *Beverly Hills Housewife*. In place of the well-heeled lady in her manicured home, the work re-enacts a drunken fumble in the back of a car. What the two works share, however, is an intent focus on an individual, intimate moment – whether that intimacy is of a sexual or psychological variety. Both are depictions of private lives, and in both instances, there is a sense of profound self-containment about the subject: it exists apart from the viewer – we might be no more than trespassers.

Neither work, moreover, has an overt social or moral 'message'. Where *Beverly Hills Housewife* offers a vision of an American home – semi-anonymous, faintly idealized – *Back Seat Dodge '38* offers a tawdry picture of the other principal 'living space' of America, the car. Their respective strips of turf (Kienholz's patch of fake grass, the pristine lawn of Hockney's picture) both signal everyday domesticity, however different the ambience. Kienholz's and Hockney's episodes are closer together than they appear; they might even belong to the same story. It is not too fanciful to suppose that, however removed Kienholz's scene might have been from Freeman's own existence, it was not so far from those of her teenage children.

Both works are imbued, too, with a powerful sense of the artist's subjectivity. We see each scene not as an objective fact, but as the artist experienced it or remembered it. More than many artists, Hockney and Kienholz magnified the effect of their own impressions (to the extent that Hockney's most 'photorealist' creations are in fact highly choreographed – products of the imagination rather than the camera). For all its verisimilitude and graphic detail, *Backseat Dodge '38* is not simply a frozen moment, just as *Beverly Hills Housewife* is not a photograph transferred into paint. As Tuchman observed at the time of the LACMA exhibition, 'Kienholz does not so much recreate the past as frankly proffer a view of the past by the present.'[117] The assemblage embodies the artist's nostalgic view of the long-ago event; even its resinous coating seems like a patina of age, as if to remind us that this is a remnant of an old story, a dredged-up relic. An element of mythmaking is discernible in the plaster and wire figures of the lovers (or tusslers), which suggest spectres as much as physical bodies. The car is attenuated – literally compressed – in a way that could make sense only in a dream, with the result that all our focus is trained onto the back seat. It is as if the car had no function (no autonomous existence) beyond the urgent activity inside.

As with *Beverly Hills Housewife*, however, the intensity of the vision is countered by a disarming blankness. Indeed, in Kienholz's installation the blanks are far more explicit: we do not see the faces of the two characters, nor any other distinguishing features besides certain tantalizing details (a watch worn by the boy is engraved 'To Our Son Harold', the girl has a gold pin bearing the name 'Mildred'). There is an imprecision about their gestures and motions. As much as Kienholz seems to be willing us to interpret the scene as a real event, the elements of redaction work against any sense of the one-off, the immediate. It is revealing that in his retelling, the adolescent experience became an admixture of hard detail and forgetfulness: 'We got some beer and pulled off in the tules someplace and [I] did intimate and erotic things all over her and we sat there and drank beer and had a nice time.... And I couldn't remember her name later.'[18] And so the restaging is partly anonymized. The individual moment has metamorphosed into something akin to a myth. Like that of *Beverly Hills Housewife*, the episode is not quite singular, nor is it flatly generic.

Beverly Hills Housewife and *Backseat Dodge '38*, for all their contrasts, throw light on one another as elliptical records of their 1960s moment. Both are rooted in the artists' experiences, and yet both are imbued with the universality of a trope. Each has the quality of a film still, at once representing a fractional moment and intimating a larger story whose shape cannot be discerned. One of the most striking resemblances between the works is this aspect of the cinematic. *Back Seat Dodge '38* is like a film set, or a fragment of one, down to the soundtrack of pre-war music that plays through a radio. To describe the work is to set a cinematic scene. *Beverly Hills Housewife* likewise implies a set or a stage. Freeman standing in her garden is a scene *in medias res*: the wide-angle shot that precedes the close-up. The subject is witnessed in a moment of pregnant pause, as though something momentous were about to occur – or already had.

A later installation by Kienholz, *Five Car Stud* (1969–72), has a more explicitly social dimension.[119] In this, four cars and a pickup truck, their headlamps blazing, form the setting of a lynching. Six effigies of white men are shown exacting revenge on a Black man – the backstory is that they have discovered him drinking with a white woman. At the centre, one of the assailants crouches to castrate the victim, attending to the task with horrible delicacy. A child waits in one of the cars. In another, a woman – maybe the Black man's lover – bends over to vomit.

To say that *Five Car Stud* is nothing more than a critique of racial prejudice and violence in the 1960s – for all that it certainly *is* that – would be to traduce its power and complexity. Like *Backseat Dodge '38*, the work transcends its historical moment. As well as exposing the horrors of what was happening in America at the time, the tableau conveys the commonplace nature of evil now and always. It lays bare the mundane or even comedic appearance that violence can assume. In willing us to interpret the grotesque scene as a mere puppet show, it exposes our own propensity to dismiss or trivialize acts of violence.

By comparing Hockney's *Beverly Hills Housewife* with the creations of an artist, Kienholz, who was supposedly a committed social critic, one begins to sense the limitation of questing for topicality in either artist's work. Kienholz's sordid and grisly installations, as much as Hockney's Californian paintings, resist being analysed purely in terms of their social relevance. The social element in each artist's work is part of a larger aesthetic and narrative framework. In the end, it is reductive to interpret Kienholz's work in terms of its 'politics', just as is reductive to dismiss Hockney's scenes as apolitical. Kienholz himself was wary of any such attempt. He once declared: 'Politics are an abdication of our responsibility – it's about getting people to do our thinking for us!'[120]

Beverly Hills Housewife, like all of Hockney's Californian paintings, evinces a bright, slender shard of its 1960s era. Any attempt to read it as a social document, or to decry it for what it overlooks, denies its autonomy as a work of art. What connects it with works as ostensibly different as those of Kienholz is its double ability to embody and to transcend the specific moment – to overlay that moment with a sense of the mythic, in such a way as to mirror LA's own unending traversal of reality and myth.

5

The Dreamer

There is something not quite real about her. Her stillness, her straightness, her marionette impassivity – all these work against the supposed naturalism of the picture. The artifice of Betty Freeman's stance is that of early Renaissance art, of a saint or a Madonna in a fresco by Giotto. It can still be glimpsed, as we have seen, in those three figures in the foreground of Piero della Francesca's *Flagellation of Christ* (fig. 8), the unidentified men who stand and wait as Christ is scourged behind them. The artifice comes close to breaking into naturalism, but stops short, as though the entire picture were possessed of tension that had failed to dispel.

Semi-Egyptian

In her tautness and simplicity, Hockney's woman in pink has older antecedents, not only the archaic statuary of Greece, but the Egyptian art that had entranced him as a boy. In the early 1960s, he began to incorporate references to Pharaonic Egypt into his work. *A Grand Procession of Dignitaries in the Semi-Egyptian Style* (1961) depicts three boxy figures in a line, descending an incline like puppets in a theatre – their shapes set against a near-plain ground of raw canvas and beige primer.[1] A line of tassels hangs along the top of the image, suggesting the fringe of a raised curtain. A bulbous storm cloud descends like a piece of stage scenery. The leftmost character has the upturned head of a bird, suggesting a variant upon Horus, the falcon-headed Egyptian god. The central one wears the classic headdress or *nemes* of the pharaohs. The third wears a top hat.

A Grand Procession of Dignitaries in the Semi-Egyptian Style was Hockney's most ambitious work to date. In its sheer scale and pageantry, it was a departure from the more concentrated, interiorized pictures of preceding years. For a while, it hung in the hallway of John Kasmin's house in Earl's Court – an incongruously modest setting. 'I thought, what am I going to do with it?' Kasmin later recalled. 'I finally sold it for £150 to a man who swore to me that it was going to his children's primary school. […] Then at auction 25 years later it's sold for $2.2m to some mad lady in America.'[2]

It was a fittingly eccentric, extravagant legacy for a painting in which personal and ancient mythologies mingled and combined. One inspiration had been Constantine Cavafy's poem 'Waiting for the Barbarians' (written in 1898): Hockney had only recently discovered the work of the Alexandrian Greek writer. The poem imagines the leaders of a city state awaiting the arrival of an invading force:

> Why did our emperor get up so early,
> and why is he sitting enthroned at the city's main gate,
> in state, wearing the crown?
> Because the barbarians are coming today
> and the emperor's waiting to receive their leader.[3]

Something of the purposeless ceremony imagined by Cavafy found its way into Hockney's toylike masquerade. More specifically, the picture translated the schematic figures of Egyptian wall painting into a graphic-expressionistic style – endowing those figures with an ironic, cumbersome three-dimensionality: 'The Egyptian style of painting of course is flat, and since I was breaking the rules of flatness, it was the semi-Egyptian style.'[4] Hockney perceived a tussle, then, between the artifice of ancient Egyptian art and the incipient 'realism' of three-dimensionally rendered bodies. In this respect, *A Grand Procession of Dignitaries* reflected his movement away from a semi-abstract mode of depiction to one that invoked the classical realism of the high Renaissance.

Not that those two modes were always in opposition. On a visit to the Pergamon Museum in Berlin in 1962, Hockney happened to catch sight of his friend Jeff Goodman standing next to a painted Egyptian sculpture of a woman, and it amused him that they looked so 'united'.[5] The painting that resulted, *Man in a Museum (or You're in the Wrong Movie)* (1962), shows a man in modern dress turned away from a hulking effigy that seems to peer at him with interest. The human figure is oddly stiff, while the painted deity swims into life – its arms flexing, its dress flowing off the edge of the picture. There is an impression of time frames colliding, as though disparate eras – not to mention dissimilar styles – had fallen together. A similar impression occurs in subsequent paintings such as *The Second Marriage* (1963), where a man in sunglasses and a lounge suit sits alongside an idol-like bride.

Hockney visited Egypt for the first time in October 1963, on a trip funded by the *Sunday Times*. He was as captivated by the everyday life of the place as by its ancient remains. In crayon and ink, he depicted subjects including a packet of matches on a café table, men in traditional Egyptian dress walking past a shuttered garage (a mural of Gamal Abdel Nasser on the wall behind), exhibits in a vitrine at the Cairo Museum, and an intersection of roads seen from the Nile Hilton hotel. In one sketch, a pair of Arab youths are seen walking down a street in Luxor; one is dressed in a green-striped galabeya, the other (a ghostly outline in pencil) places his hand on his friend's shoulder. Even as Hockney confirmed his love of ancient Egypt, his attention was turning to the 'here and now'.

His portrayal of Freeman has none of the symbolism or the masquerade of *A Grand Procession*. It is an everyday domestic scene. And yet, in her unassuming way, Freeman exhibits a trace of the semi-Egyptian style. Flattened by the pastel-pink blankness of her dress, and by the simple blocks of tone that denote her head and arms, she is like a figure from an ancient wall painting. A reincarnation, even, of the pink-robed goddess Isis, as painted in the thirteenth century BC on the tomb of Queen Nefertari – or perhaps the queen herself, whom Isis is leading by the hand (fig. 21).[6] Nefertari was the wife of the pharaoh Ramesses II, but in this painting her husband is absent, underlining the queen's independent stature.

'The housewife's head is recognizably her own,' wrote Henry Geldzahler of *Beverly Hills Housewife* in 1976, 'and yet its individuality is deliberately downplayed; like her house and her objects she exists in a windless clarity emphasized by her frozen stance, wooden dress and three-quarters profile, Egyptian and expressionless.'[7] Freeman is not expressionless, however, so much as deadpan – wryly solemn. That is, her expressionlessness is a piece of performance, an act. So too her Egyptian 'rigidity'.

Freeman is no mere cipher: turned half towards us, half away, she breaks out of her hieroglyphic flatness into three dimensions.

Two totally different people

Certain objects in *Beverly Hills Housewife* are symbolic of Betty Freeman. The recliner designed by Le Corbusier is one (fig. 19), the stacked form of the Turnbull sculpture another. It is possible to glimpse the emergence of a new, independent self in these objects. Like the attributes in a portrait by Hans Holbein, they frame and reflect the human subject. The recliner, however, is not simply an expression of her taste. The skin of a zebra that Stanley Freeman had shot covers the sleek, articulated framework. In this one object, their opposing selves appear reconciled. And yet the recliner is empty. Betty stands before it as though another person ought to be there – as though the picture should have been a double portrait.

Stanley was an outward-bound, masculine man. Big-game hunting was his great love, whether in the wastelands of Wyoming or on safari in Africa. Other recreations included golf at the Hillcrest Country Club and all-day waterskiing at their Bass Lake retreat. Clyfford Still once joked: 'I doubt [...] that I will ever challenge Mr Freeman to a race on water skis.'[8]

It is perhaps no accident that Betty seems indifferent, in the painting, to the antelope's head on the wall – another of Stanley's spoils (fig. 9). She did not share her husband's enthusiasm for hunting; indeed, her view was closer to that of her composer friend Harry Partch, who once remarked – looking back at his boyhood – 'I'd seen my father and sister hunt animals, and I had this terrible feeling about anyone who has hunted any animal.'[9] One of Stanley's most prized trophies, not shown in the painting, was a stool fashioned from an elephant's foot. In later years, Betty would regularly admit that she had known that her marriage was a mistake from the first day. She wanted to divorce at once, but her father advised against it. Every year, she would ask him again for permission. But by the time her third child had arrived, it was too late.[10]

'They were essentially two totally different people', says their daughter Shelley. 'If you were to take any category, they wouldn't match. My mother loved exotic, weird food, and my father liked hamburgers.

My mother loved modern art and music. My father liked forties band music.'[11] When they first moved to Los Angeles, the marriage was defined by their four children, but by the late 1950s, the children were away at boarding school for much of the time. 'Please forgive me if I write you too often', Betty wrote to Still in 1961. 'Stanley says that perhaps it might bore you. But it makes me happy to write to you. Maybe it is that so few people understand and I do sometimes feel a bit lonely.'[12]

Stanley was frequently away from home. By this time, he had left Witco – having found Robert Wishnick impossible to work for – and was running his own business, which involved the manufacturing of magnetic tapes for consumer and industrial markets. There is a revealing tale – told by Partch's long-term assistant – that Betty was appalled to discover that her husband had sanctioned the cutting up of video tape to reuse as audio tape, a notorious rogue practice in Hollywood.[13] They espoused different value systems. If she looked askance at his business dealings, then he was frequently baffled or even antagonized by what she did. Shelley speculates that her father had left Hillcrest Road by the time Hockney visited in 1966, or at least was away for much of the time. If he had been there, she conjectures, 'He would have freaked out.' For a period, they slept in separate bedrooms – a decision that Betty ascribed (at least to her children) to her husband's tendency to snore.

Once the children had grown up, there was little to hold the marriage together. By February 1968, Betty was confiding to Sam Francis:

> I am trying very hard to work out what is important and what is not. There are a lot of things I would like to change in my life but I don't know if I am strong enough. There are things I admit to myself but not to anyone else. [...] So much of what I do doesn't mean anything to me anymore, nor do many of my relationships.'[14]

It came to light that Stanley had been having an affair with his secretary. He did not want to leave Betty. In fact, he would have been content to maintain the affair and the marriage in parallel. It was precisely the kind of humdrum compromise – tragic with a small 't' –

that Betty was no longer prepared to entertain. The impetus to divorce was finally too strong to ignore. 'The decision was totally my mother's', Shelley states. 'Because although my father certainly wasn't happy, he was against divorce.'[15]

But he had no choice in the matter. The Freemans' divorce was finalized in 1971, twenty-eight years after they had married. Stanley moved to a rented apartment in a high-rise block in the mid-Wilshire area. (The property was owned by an older lady, and decorated accordingly.) Betty stayed at Hillcrest Road, ownership of which had transferred to her. 'She mostly didn't care what my father did afterward', says Shelley. 'She was totally on her own, and very happy.' *Beverly Hills Housewife* captures her in this new phase of independence. 'I don't know if you would call her an *alone* person', cautions her friend, art historian Pamela Kort. 'I would prefer to use the word aloof. I think she was essentially always someone of an independent spirit.'[16]

Seen in relation to the large-scale double portraits that Hockney made afterwards, *Beverly Hills Housewife* is all the more striking for *not* being of a couple. The other half of the marriage is an absent presence, a vestige glimpsed only in his dead trophies. All the same, the painting shares characteristics with those later portraits: the panoramic shape, the setting of a figure in a domestic interior. Freeman's posture might be interpreted as a conflation of those of Celia Birtwell in *Mr and Mrs Clark and Percy* (1971) and Christopher Scott in *Henry Geldzahler and Christopher Scott* (1969). If the double portraits sometimes convey the air of an annunciation – the earthly recipient, the heaven-sent messenger – then Freeman collapses together the two roles. She is her own divine visitation.

Writing about Hockney's double portraits, Edmund White remarked: 'In every couple, there is a disturbing element.'[17] In *Beverly Hills Housewife*, the disturbance does not lie in the dynamic between two subjects. The tension or discord has been interiorized. We see Freeman on a threshold between the identities of wife and independent woman. For all her new-found contentment, one suspects that she was capable of feeling at odds with herself. In *Beverly Hills Housewife*, her past and future selves are locked in a mute, deceptively tranquil stand-off.

A real, real, real love affair

No one quite remembers how Franco Assetto entered Freeman's life. A lean, shaven-headed Italian, he had moved to Los Angeles in the early 1970s, although she first met him while travelling in Italy. Some time later, he paid her a visit in Beverly Hills. 'He was a character', says Pamela Kort. 'I recall Betty saying that he rang her door one day, and was standing there with a gigantic box of candy.'[18]

Assetto was close to seventy, a decade older than Freeman, but seemed years younger. He was an artist of some note, having begun to make sculpture in his native Turin in the 1950s. An early series had consisted of bronze casts of loaves of bread, anticipating the lowly subject matter of *arte povera* by more than a decade. Over the years, he had turned his hand to surrealist drawings, relief paintings, ceramic vessels, designs for fountains, and sculptures of candy in moulded plastic.

His personality had been forged in the Second World War, when he served as a captain in the Bersaglieri, a light infantry unit famed for its feathered hats and its flamboyant, sprinting parades through the streets of Turin. 'Franco was still enamoured of the bersagliere when I went with him and Betty to Turin', says Kort. 'There we saw these soldiers running through the town, and he had once been one of them – it seemed to me that he still thought he *was* one.'

At first, Freeman was not impressed. She found him loud and overbearing. But gradually her view shifted. He wrote her love letters and made drawings for her. She came to see through his bombastic exterior. In contrast to the retiring, conservative Stanley, Assetto was an extrovert who drew Betty out of herself. Throughout her first marriage, and to some extent her life, she held back from expressions of emotion. 'She wasn't made to be a mother', according to her daughter Shelley. 'She didn't know how to be. In the 1950s and '60s, people didn't know what they know now. She was very closed – never expressed what she felt, ever.'[19] Family life had been imbued with restraint. (Her eldest son, Robbie, grew up idealizing James Dean – the paradigm of the moody, subdued outcast.) In Assetto's company, she blossomed. They married on Christmas Day of 1979, at the Candlelight Wedding Chapel, Las Vegas. A photograph from the occasion shows them posing beneath the chapel's

Betty Freeman and Franco Assetto at their wedding reception, Las Vegas, Christmas Day 1979

giant red sign, Betty grinning as she lifts one leg off the ground in fairy-tale-princess style. 'Franco literally changed my mother', says Shelley.

For six months of the year, the couple would stay in Turin. Italy was their hinterland, far away from the frenetic socializing of Los Angeles. 'The life here is so full of beautiful small things,' Freeman wrote in a postcard to Sam Francis in the 1980s, 'such as fresh flowers every day, special red wine made by the electrician, long dinners till midnight with friends.'[20] For the remainder of the year, they would be based at Hillcrest Road, where Assetto took over the garage as his studio.

Betty Freeman at home, with Roy Lichtenstein's *Frolic* behind her, photographed for the *New York Times*, 1997

Vivacious and extrovert, Assetto's personality was the antithesis of Freeman's. One of his eccentricities was to crow like a rooster first thing in the morning. Freeman used to say that her Roy Lichtenstein painting *Frolic* (1977) made her think of the pair of them because of its resemblance to a rooster chasing a hen. 'Franco was tasteless, but she adored him', according to her friend, the theatre director and playwright Robert Wilson. 'It was a real, real, real love affair. They had different tastes, different aesthetics. But that was what was wonderful about Betty. It didn't interfere with her love for him.'[21]

In the summer of 1981, the American composer Virgil Thomson stayed with the couple in Corfu. By then in his mid-eighties, he was one of the titans of modern classical music, albeit not the easiest house guest. During his stay, he created 'musical portraits' of each of his hosts. These off-the-cuff compositions, made in the style of a drawing from life, were part of a long-running series (amounting to some two hundred pieces) that he had first begun in the 1920s, with portraits of Pablo Picasso and Gertrude Stein.[22] Thomson sat down Freeman and Assetto in turn, and began writing from the top of a sheet of music paper. An hour or so later, the portrait was done.

Thomson's two scores were a neat encapsulation of the couple's divergent natures. Hers was entitled 'Dead Pan: Mrs Betty Freeman'. Both the title and the music alluded to her cool, unflappable demeanour. 'The piece itself is innocuous,' she wrote afterwards, 'a simple dance-like round in the 18th century style. But the title was terrible and I didn't like it at all.'[23] The other portrait, 'Franco Assetto: Drawing Virgil Thomson', was created while Assetto sketched Thomson, each man capturing the other in a sort of silent duel. It is striding and energetic – marked *vigoroso* on the score.

The Corfu trip was a notorious instance of how Freeman's patience – and Assetto's easy-going charm – might be tested. 'Either I overestimated my ability to handle old men of genius or I underestimated Virgil Thomson's talent to be difficult', she wrote afterwards.[24] Upon being invited to stay at the villa, Thomson had initially declined – then abruptly announced that he would stay for five weeks. Freeman had holidayed with him before, travelling through Tuscany together, but by this time he had grown slovenly and disinhibited. He complained continually about the food that Assetto made, declaring on one occasion: 'You're not feeding me well. I'm getting tired of tomatoes and eggplant.'[25] At mealtimes, he ate with his fingers, belched freely, and picked at his teeth – never mind Freeman's placement of a box of toothpicks beside his plate. Unable to bear it any longer, Assetto burst out: 'Tell me, Virgil, were you born on a farm?'[26]

Thomson departed with a show of ingratitude that Freeman recorded in an unpublished memoir on the composer:

> I was hoping that Virgil would say he had enjoyed being at Corfu or at least had had a good rest or even would say he felt well. No, no such word much less a word of thanks in any form. Only, 'Well, goodbye, I'll see you in New York.' And off he went, lugging his own heavy-as-rocks ancient Louis Vuitton wood luggage, a little round figure, strong-willed and tough as a cock.[27]

Assetto, too, had been appalled by Thomson's rudeness. There were no more holiday invitations, but Freeman continued to support Thomson until his death in 1989.

CHAPTER FIVE

The dreamer who remained

> '[T]hey who have done kindnesses feel friendship and love for those to whom they have done them, even though *they* neither are, nor can by possibility hereafter be, in a position to serve their benefactors.'
>
> *Aristotle,* Nicomachean Ethics, *fourth century* BC[28]

For six weeks in 1964, Harry Partch lived in his car in the Mojave Desert, a vast wasteland of south-eastern California. The self-taught composer, by this time in his sixties, had considered himself as a hobo since the years of the Great Depression, when he had lived a wandering existence in the western states. His lapses into homelessness were the consequence of genuine poverty, but also a reflection of his peculiar temperament, marked by a refusal to assimilate to social norms. Partch was a tragi-comic character, the kind of person it was easy to dismiss as insane Or a dreamer who appeared to be on an endless, impossible quest.[29] 'Hobos', he once said, 'are extraordinarily individualist people.'[30]

As the year progressed, Partch's fortunes began to reverse. He had been invited to give a concert (or 'lecture demonstration') at the Pasadena Art Museum in November, as part of a new programme of performances entitled 'Encounters'.[31] On 12 October, he drove up from Del Mar, a coastal suburb of San Diego where he was living temporarily, to Pasadena for a meeting with the organizers.[32] Among them was Betty Freeman.

Freeman had only recently joined the board for 'Encounters' at the invitation of her friend Walter Hopps, who at thirty-two was the youngest museum director in the United States.[33] Since his appointment as a curator at the museum in 1962, Hopps had instigated a series of radical exhibitions that sought to counter Pasadena's reputation as 'rich, reactionary and retired'.[34] His first show was a retrospective for the German Dadaist Kurt Schwitters; the following year, he organized the first American retrospective of Marcel Duchamp, hailed as 'the mystery man of modern art' by the *Los Angeles Times*.[35]

For Freeman, the Pasadena Art Museum was a liberating environment. She later recalled: 'I realized – this is the time I gave up the piano – that I enjoyed much more working with people than practicing.'[36]

She was involved with 'Encounters' for nine years, helping to stage three concerts a season in an attic space at the museum. Forty years on, she recollected how nervous the board of directors had been that the performance artist La Monte Young would attract pot smokers to the museum ('Which didn't happen but there were police checking').[37] It was through 'Encounters' that she met a number of the pioneering composers whom she would continue to support, among them John Cage, who came to perform in January 1965. She quickly become devoted to Cage, helping him to secure other engagements in California, including one in Santa Barbara: 'The lady from Santa Barbara called me one day in hysterics. She had just received the list of equipment Cage needed for his concert. And she said, "What is this: a composition for 'Eleven Mixing Bowls'!"'[38]

In the autumn of 1964, Freeman had never met Partch – probably never heard of him. Within recondite musical circles, he was known mostly for his complex scores and his construction of eccentric, unwieldy instruments bearing names such as Zymo-Xyl, Boo, Gubagubi en Chrychord, and Chromelodeon. He had created his own notational system: an octave divided into forty-three intervals, rather than the typical twelve, which he considered 'the basic ingredient [of the] chaos' of Western music.[39]

They were an unlikely pair: the well-heeled patron and the dishevelled quixotic. But at that first meeting in October 1964, they hit it off. Afterwards, they went for lunch at a nearby drugstore, where Partch busied himself with trying to smoke his pipe through a piece of gauze (he had heard that this reduced the risk of cancer). Freeman found him 'utterly charming'.[40] She invited him to stay with her and Stanley that night in Beverly Hills in order to attend a Schoenberg concert, but he declined.

Partch was a lifelong outsider. Born in Oakland, California, in 1901, he had grown up in the frontier territories of Arizona and New Mexico. His parents had been Presbyterian missionaries, but had renounced their faith after ten years in China. Partch's own life had been a long, determined renunciation of institutions of all kinds. In 1922, he dropped out of the University of Southern California's School of Music; he dispensed with the twelve-tone model soon after. In his twenties, he worked variously as a theatre usher (conducting a brief affair with the actor Ramon Novarro), piano teacher, and sailor. Partch travelled to London and Europe in 1934 on a grant, during which time he met W. B. Yeats, whose play *Oedipus Rex* (first staged in 1926) he wanted to set to music. At the poet's house in Dublin, Partch gave a recitation to Yeats on his own Adapted Viola, a viola with a long cello's neck ('I sensed guarded worry and disbelief', he noted afterwards in his journal, 'when I went ahead with my exposition, chanting the words of the 137th Psalm').[41] He persuaded Yeats to intone choruses from the play, despite the poet's lack of confidence in his own musical ability. Yeats's words upon their parting remained with Partch for years afterwards: 'You are one of those young men with ideas, the development of which it is impossible to foretell, just as I was thirty years ago.'[42] After returning to the United States, however, Partch lived an itinerant 'hobo' life for close to a decade. He continued to compose music and build instruments during his wilderness years. If he needed money, he picked fruit or washed dishes.

That day in Pasadena, Freeman may have detected an unlikely kindred spirit: Partch's iconoclasm was an extreme version of her own quiet, still-nascent individualism. His many fascinations included the occult and mythology, as reflected in theatrical works such as *Oedipus* (1951, the final realization of his adaptation of Yeats) and the dance satire *The Bewitched* (1955); Freeman herself was intrigued by transcendentalism, as a follower of the Russian philosopher and mystic George Ivanovich Gurdjieff.

Freeman was struck more than anything by the lack of support available to a 'solitary genius' like Partch.[43] Such figures operated outside the university sector. Whatever following they commanded was usually among artists and choreographers rather than fellow musicians. The National Endowment for the Arts would not be established until the following year, and so radical composers were compelled to fend for themselves, or – in Partch's case – subside into penury. Among his contemporaries, the problem was commonplace. When Freeman visited John Cage in New York in 1965, she was so appalled by his state of living (in a dank basement apartment) that she began paying him an annual grant of $5,000; this continued until Cage's death in 1992.[44]

Her support for Partch was just as determined, if more casual at first. ‘I couldn’t not help him and just let him die’, she later said. ‘It was that serious.’[45] She began to cover expenses here and there, sensing that he might refuse a direct handout. After he had moved to a rented house in the Los Angeles neighbourhood of Van Nuys in November 1964, she became a regular visitor, sometimes bringing him a bowl of pears or a little whisky. The house was an improvement on many of Partch’s former situations, but cramped – its main advantage being a sunlit backyard where he liked to sit beneath a tree, drinking mint juleps. But he could not stand the yuppie community in Van Nuys.[46]

Freeman took it upon herself to establish Partch in a permanent studio; a suitable location was soon found in a disused laundromat in Venice. Writing in thanks, he admitted that the gesture was ‘momentous to me in a way you probably cannot realize [...] For over twenty years I have been the strangest kind of hobo – a hobo with over two tons of “weird” instruments to take, wherever.’[47] From this time on, Freeman acted as his guide and protector. She regularly drove him around (he had no car of his own), claiming that this was how she became acquainted with the LA freeways. In early 1965, not long into their friendship, she accompanied him to Cage’s concert for ‘Encounters’ at the Pasadena Art Museum. In a letter written the following day, Partch gave a dry distillation of the performance:

> [Cage] ate cut carrots with an amplified knife for ten minutes, put them through an amplified juicer for another ten, then drank the juice through an amplified gullet for another ten (while smoking an unamplified cigarette). Knife, juicer and gullet all came out the same – static. When he began to cut carrots again I left.[48]

Partch was not the only sceptic in the audience that night. Freeman’s friend (and eventual bête noire) Clyfford Still had also turned up; it would have been hard to imagine a character less suited to the occasion. So far as music went, Still was a staunch classicist – a man who loves great pianists, like [Sviatoslav] Richter and Beethoven and all’, according to Walter Hopps. Recollecting the episode, Hopps stated: ‘Cage takes a Waring blender, all contact-miked for heavy amplification and

Harry Partch on the set for the film *The Dreamer That Remains*,
San Diego, 1972, photograph by Betty Freeman

blends carrots and cucumbers and so on. Mr Still is just sitting there, grim as death.'[49]

In the years that followed, Freeman assisted Partch in buying a house in Encinitas, a beach town north of San Diego, and footed the bills for the continual hefting around of his instruments. The ultimate testament to her devotion was her decision to finance a television documentary about him in the early 1970s. She found a director in Stephen Pouliot, a master's student at the film school of the University of Southern California. Upon meeting Pouliot, Partch was besotted – describing the younger man as a living Apollo.[50] Although his feelings were not reciprocated, he later credited Pouliot as the inspiration behind the music that he composed for the film, which he planned to subtitle 'My Love Song for Stephen' (eventually settling for 'A Study in Loving').[51]

The resulting film is part documentary, part performance. It features Partch's extended composition (for a chorus and fifteen instruments) of the same title – drawn from a poem by Roy Campbell, 'The Flaming

Terrapin': 'I am the dreamer that remains, / The Man, clear-cut against the last horizon!' Screened on television in the United States and abroad, the film did more than anything to shore up Partch's legacy, ensuring that his maverick career had a lasting record.

Freeman went on to underwrite Partch's final work, *Delusion of the Fury* (1971), a musical stage play inspired by Japanese Noh drama and written for twenty-five custom-made instruments. 'The concept of this work inheres in the presence of the instruments on stage', Partch explained in the notes accompanying the recording, 'the movements of musicians and chorus, the sounds they produce, the actuality of actors, of singers, of mimes, of lights'.[52] Freeman was instrumental in making this elaborate concept a reality, lobbying the music department at UCLA for two years – unsuccessfully – to stage the play before convincing the theatre department to do so. The premiere took place at the UCLA Playhouse in January 1969.[53] 'After that,' according to Partch's long-time assistant, Danlee Mitchell, '[he] moved back to San Diego again and died [...] with all the instruments in his house.'[54]

Partch's meeting with Freeman in 1964 had proved life-changing for both of them. 'Betty adored him', in Mitchell's words, 'because she was dealing with a very unique, creative, intelligent person.' In this regard, her relationship with Partch was a forerunner to that with Hockney. She loved him for his foibles, but above all she revered him for his determined pursuit of a singular path. Not that Partch was a readily loveable figure. In his final years, as he succumbed to alcoholism, he grew ever more stubborn and cantankerous. 'I shall curse you', he once wrote to Cage – furious at his contribution to a volume of artists' statements having been abridged without his permission. 'You have been cursed before but never by me.'[55] For John Adams, it was a sign of Freeman's strength of character that she had been willing to put up with Partch's 'crusty, abusive' behaviour for a decade until his death.[56]

She was still married to Stanley during much of her friendship with Partch. But as early as 1966, when Hockney painted her, the marriage was in its end phase, and she was contemplating a new life. This arose through her involvement with Partch and other composers who came to the Pasadena Art Museum – with the kinds of outsider characters who might have seemed, on the face of it, to belong to a different universe from hers. Freeman saw through Partch's gruff and shabby facade to the humorous visionary within. His contrarianism, which set him apart from society, was precisely what endeared him to her. 'He was a very spiritual man', she observed. 'But he was so against the church, and so against organizations. [...] There are not so many, who tell you to think for yourself.'[57] Each of them, though, combined a resolute individualism with a respect for cultural tradition – Partch himself insisting in 1967 that 'meaningfulness must have roots'.[58]

Their friendship was characterized by a mutual regard, each seeing something precious in the other. 'Partch did not think of himself as an innovator – but I do', Freeman commented in 1988. 'Partch affected my life by showing me the possibility to think in ways other than those the world takes for granted.'[59] Upon completing the music for *The Dreamer That Remains*, Partch inscribed a copy of the score: 'For my Beloved Producer Betty – that unfailing source of Strength, Belief, Regeneration – and with the hope that it is worthy'.[60] The dedication is aptly reminiscent of those made by the great Roman poets of the first century BC to their patron Maecenas, a benefactor and socialite who lived on the Esquiline Hill.[61]

In late 2008, a few months before her death, Freeman recounted of her old friend: 'The strangest thing about him was his system of talking in voices, which scared me at first. It was always Partch in another room, such as a bathroom, with another man shouting in a completely other voice. And the door would open and Partch would walk out with nobody else there.'[62] The idea of a single imagination encompassing multiple selves – of Partch as a many-voiced creature – seems to have struck her at the inception of their friendship. In the mid-1960s, she was undergoing an identity crisis of her own, struggling to reconcile the conflicting parts of herself. The title of her documentary on Partch, *The Dreamer That Remains*, seems appropriate to the character that ultimately emerged from this mid-life struggle. As she channelled her energies into music and musicians, the Beverly Hills housewife carved out a new persona. Through her support of Partch and those who followed, she proved that her sometimes dreamy demeanour belied a staying power.

6

A Modern Medici

'Betty was a true patron. Vivacious, engaging, totally interested in art and the music world.'

ED RUSCHA, 2025[1]

In *Beverly Hills Housewife*, the door to the house is open. Behind the woman in pink is a tunnelling vista through her home. The music room appears as an oblong of dark space, a flattened interval between the scene in the garden and the front window in the distance. It is a sign, however elliptical, of Betty Freeman's recent past – the musical career that she had lately abandoned. In the music room was the piano that she had ceased to play, always perfectly tuned as though she might return to it. But the dark gap of the room can also be seen, in retrospect, as a glimpse of the future.

A prison bailout

'Although I was musically trained as a classical pianist, it was through marijuana and the art world in 1961 that I became a music patron', Freeman said in 1992.[2] That year, the artist and composer La Monte Young had been arrested and jailed for drug possession while driving in Connecticut. Young's friends attempted to raise funds for a lawyer, and they approached Freeman – a known art collector – for help.

She had never met Young, and had perhaps only heard of him in a roundabout way – 'I must have known La Monte through his writings', she later surmised – but all the same, she sent a cheque. 'I thought it was terribly undignified putting a composer in jail for marijuana. […]

So I put in $100, which was a substantial donation in 1961. La Monte sent me a record of his, and I loved it.'[3] Young's record was the first piece of modern classical music she had listened to with interest. She was struck by how inventive and forceful it was, without being obnoxiously 'noisy'.[4]

Her random act of charity had brought her into contact with a character, and an entire world, that would come to fascinate her. Then in his late twenties, Young was a blazing light in New York's downtown art scene, notable for his wild appearance (goatee, straggling hair, beret, and cape) as much as his jazz-inspired music. His performances had a quasi-theatrical dimension – requiring participants to build a fire, for example, or to release a batch of butterflies. Young had arrived in New York from California in the early 1960s, soon becoming one of the main instigators of the Fluxus art movement.[5] He organized concerts at Yoko Ono's loft in SoHo, one of which consisted of him pacing the floor obsessively. His own apartment (shared with artist Marian Zazeela) became a beatnik hangout. According to Billy Name, the photographer and Warhol acolyte:

> Young was the best connection in New York. [...] When you went over to La Monte and Marian's place, you were there for a minimum of seven hours – probably end up to be two or three days. It was a pad with everything on the floor and beads and great hashish and street people coming and scoring, and this droning music going on.[6]

By the time he came to Freeman's attention, Young was known for starkly minimalist scores such as *Trio for Strings* (1955), a composition dominated by single notes sustained for long phases. *Composition 1960 #7* consisted solely of a B and an F sharp, with the instruction that the two notes 'be held for a long time'. In another performance, Young played the same piano chord 1,698 times.[7] Freeman was an early – and enduring – enthusiast. She would later recall: 'He's somebody who can play one note for four hours, but it's what he does with this one note.'[8]

It was this kind of music that Hockney recollected hearing at Hillcrest Road in the early years of his friendship with Freeman. In the mid-1960s, she had begun to hold the occasional performance at home – 'after-dinner punishment/entertainment', as she termed it.[9] 'She used to have these kind of concerts on Sunday afternoons,' Hockney related:

> And sometimes they were almost like a joke. I remember one afternoon when there was this event which consisted of just one note going on and on and on for hours and hours and people just sat there and didn't really know what to do. I think it was by Harry Partch, a composer she had rescued off the streets. Then Betty would come in and say, 'You can talk if you want to.'[10]

Peter Schlesinger has a similar memory. He and Hockney attended a concert given by La Monte Young at Hillcrest Road, a minimalist rendition that felt like a feat of endurance. Around the same time, Freeman took him and Hockney to a Frank Zappa concert – another indication of her evolving taste. 'It was more modern than we were. David didn't like any contemporary music, unless the singer was cute.'[11]

Seen in retrospect, Freeman's assistance to Young in 1961 was an early sign of her partiality for outsiders – a partiality that drew her soon after to Partch. Like Partch, Young was an example of the difficulties faced by those on the fringes of the art and music scenes – the lack of either financial assistance or institutional recognition. In later years, she would observe that several of the composers she supported had been recognized more fully by the art world than the music world; indeed, she saw little distinction between their work and other forms of avant-garde art.[12]

In January 1968, Young performed four concerts at the Pasadena Art Museum. Freeman's original gift to him, which had ignited her interest in new music, had come full circle – with Young becoming part of the 'Encounters' programme that she had helped to organize since 1964. 'They were an unexpected huge success', she wrote to Sam Francis, after the concerts; 'overflowing crowds every night, mostly kids. I went twice & find it extraordinary & very exciting, not at all dream-state. He makes sound an actual thing you feel in your stomach & also see.' For some of the cognoscenti of Los Angeles, however, Young's experimental music was a step too far. Freeman added that 'Martha [Jackson] walked out, poor thing', and that Marcia Weisman likewise fled the concert 'as it bothered her physically'.[13]

A decade on from her donation to Young, Freeman introduced herself to music critic John Rockwell, whose articles she had read in the

Los Angeles Times. He invited her to his house in Laurel Canyon, a tiny property crammed with books and records, and played her the music of Steve Reich and Philip Glass. As she subsequently recalled, she experienced an 'instant excitement'. In her eyes, Reich and Glass came to represent the 'fathers' of new music, along with Young and Terry Riley. She had long been a devotee of Partch by this time, but her musical horizons were expanding.[14]

In *Beverly Hills Housewife*, Hockney created a mid-twentieth-century answer to Pablo Picasso's 1905–6 portrait of Gertrude Stein. Each painting presents a stylized image of a real woman, known to the artist. There is an oft-repeated legend that when Picasso was told that the picture did not look like Stein, he answered: 'No matter; it will.'[15] For decades afterwards, Stein held court in her Paris apartment from a chair directly beneath her likeness; the image not only mirrored but actively shaped her real-life persona, almost to the extent that life imitated art. Stein later stated that Picasso 'gave me the picture and I was I still am satisfied with my portrait, for me, it is I, and it is the only reproduction of me which is always I, for me'.[16] In similar vein, Hockney's painting of Freeman hung in her house for forty years as a paradigm that she might emulate or depart from as the mood took her.

Picasso's portrait, like Hockney's, was a celebration of a formidable female patron of the arts – a picture that enshrined its subject in the cultural imagination of its era. At the time that Picasso painted her, Stein had already begun hosting weekly salons, populated by a shifting assortment of European and American artists and writers. Ernest Hemingway recounted in his memoir *A Moveable Feast* (1964): 'It was easy to get into the habit of stopping in at 27 rue de Fleurus late in the afternoon for the warmth and the great pictures and the conversation.' He remembered the main studio as being 'like one of the best rooms in the finest museum except there was a big fireplace and it was warm and comfortable and they gave you good things to eat and tea and natural distilled liqueurs made from purple plums, yellow plums or wild raspberries'.[17]

Six decades later, Freeman was beginning to gather a milieu of her own, with Hillcrest Road as its centre of gravity. The music room is no more than a passing detail in *Beverly Hills Housewife*, a featureless space punctured by the distant aperture of a window. In 1981, that space would transform into a stage.

Salotto

> 'People do listen when you speak. This is something that money cannot buy, and you have it. I feel very lucky to have been a part of your circle of influence.'
>
> *John Adams, letter to Betty Freeman, 29 April 1987*[18]

Soon after turning sixty, Freeman began to hold regular private concerts in the music room. This marked the beginning of a decade-long programme that would evolve into an institution of the classical-music scene in Los Angeles. 'I remember clearly the epiphany (that's the exact word) that happened to me late one afternoon in 1981 standing in my large sparsely-furnished-but-with-a-Steinway living [room]', she later said.

> It began (the epiphany that is) wishing that someone in Los Angeles would have home performances of contemporary music similar to the mythic Evenings on the Roof of Peter Yates. Simultaneously the realization occurred that that someone was me. I had the space and piano, I had the financial means, I already knew many of the important composers – most of all, I had a passion for contemporary music. Even more I had an Italian artist husband who loved to cook ... and would have a good excuse to stay out of the living room.[19]

She had wanted for some time to create a salon in the style of 'Evenings on the Roof', a long-running series of performances of modernist classical music that had originated at the Silver Lake home of music critic Peter Yates and his pianist wife Frances Mullen Yates in the late 1930s (the earliest seasons taking place from 1939 to 1941 in a rooftop music studio).[20] The compositions of Arnold Schoenberg and Igor Stravinsky featured prominently at these occasions, with Stravinsky himself attending. When Freeman first moved to Los Angeles in 1950, she knew nothing of the concerts, but later became a regular attendee (the venue had changed by this time to the Wilshire Ebell Theatre).

The catalyst for her own venture was her introduction to Alan Rich, a seasoned music critic and broadcaster – then aged fifty-seven – who had arrived in Los Angeles two years earlier.[21] Rich was known in the music world as a man who was unequivocal in his tastes, and wilfully idiosyncratic ('He revered Mozart and Schubert and was very dismissive of Brahms', according to his *New York Times* obituary in 2010).[22] Urbane and articulate, he could also be acidly critical of performances that did not meet his standards. But when he first came to Hillcrest Road for a screening of Freeman's film about Harry Partch, *The Dreamer That Remains*, she warmed to him – and sensed an opportunity.

Five or six times a year, between November and April, Hillcrest Road became the setting of a private concert or 'musicale'.[23] Invitation cards carried the words 'music room' in an Art Deco typeface, alongside a pictogram of a grand piano. Rich's and Freeman's names appeared beneath, together with the address. From the beginning, the tone of the occasion was discreet, simple, and serious.

The recitals took place in the dark chamber that is visible in the background of *Beverly Hills Housewife*. The walls were clad in black velvet on three sides, with white velvet and drapes on the fourth wall. Spotlit paintings shone out like illuminated glass: a horizontal Sam Francis on the main wall (where Clyfford Still's *1955-H* had previously hung – since loaned to the Los Angeles County Museum of Art), and two Roy Lichtensteins flanking the entranceway. A shagpile rug with a geometric design stretched the length of the room, bordered by polished black wood.

At first, the gatherings comprised a small circle of friends. Before long they numbered as many as a hundred a time. The music room, which measured thirty-three by twenty-eight feet, could barely accommodate the guests: people would spill out into the lobby, up the stairs, and into the breakfast room, where they sat on a ledge beneath *Beverly Hills Housewife*. The Music Room, as the concerts were always billed, had become – in Rich's words – 'the Los Angeles Sunday afternoon hot ticket'.[24]

18 Edward Kienholz, *Back Seat Dodge '38*, 1964.
Paint, fibreglass, and flock, 1938 Dodge, recorded music and player, chicken wire, beer bottles, artificial grass, and cast plaster figures, 167.6 × 304.8 × 396.2 cm (66 × 120 × 156 in.)

19 David Hockney, *Beverly Hills Housewife*, 1966–7 (detail)

20 John Kasmin, *David Hockney and Betty Freeman at Freeman's house*, 1967

21 Queen Nefertari being led by Isis, *c.* 1279–1213 BC, watercolour copy by Charles K. Wilkinson. Tempera on paper, 70 × 46 cm (27⁹⁄₁₆ × 18⅛ in.)

22 Claes Oldenburg, *Typewriter eraser*, 1976.
Painted aluminium, stainless steel, ferroconcrete, and bronze,
227.3 × 203.2 × 177.8 cm (89½ × 80 × 70 in.)

23 The landing at 703 Hillcrest Road, showing Ronald Davis's *Blue Tray* (1966) and Dan Flavin's *'monument' for V. Tatlin* (1967)

24 and 25 Stills from Jack Hazan's film *A Bigger Splash*, 1973, filmed at Betty Freeman's house

26 and 27 David Hockney with Betty Freeman (top) and his mother Laura Hockney (bottom) at Betty Freeman's house, 1981

'At the concert', a photograph by Leigh Weiner taken in November 1986 for a feature on Betty Freeman's musicales in *Connoisseur* magazine, February 1987

A photograph from November 1986 gives a glimpse of one such event (above). Listeners sit on rows of folding chairs. A group of women in the foreground are turned out in chic 1980s style (patterned dresses, prominent earrings, bouffant hair), facing the action that lies just beyond the field of view. A man in a jacket and tie stands behind, beneath Roy Lichtenstein's *Frolic*, his hand cupping to his chin as he listens. Close by, another man reads along to a giant score. Through the doorway, in the adjoining bar room, other guests gaze on, spotlit against the dark night beyond. The mood is solemn, attentive, immersed.

'You were seeing people you knew well, and at other moments there were people you'd never met – and who it was a thrill to meet', recounts theatre director Peter Sellars, who became a close friend of Freeman's after moving to Los Angeles in 1987. 'It was a gorgeous alchemy. There was a hint of that old Diaghilev glamour in the air: Stravinsky *and* Cocteau *and* Picasso are coming over this afternoon.[25] Freeman and Rich devised the programme, with a typical concert featuring two composers: an established talent and a debutant. By pairing a younger artist with an older one, they generated an 'encounter' that carried a subtle echo of the painting in the next room: the woman in the prime of middle age observed by the young, aspiring artist. But just as in *Beverly Hills Housewife*, any apparent hierarchy tended to be levelled or upended. 'Often they are each other's antithesis', as Freeman put it.[26] Each performance framed the other.

At the beginning of a performance, Freeman would ring a set of wind chimes to summon attention. Rich would then act as master of ceremonies, introducing the evening's participants and occasionally throwing in a sardonic reference to a negative review of their work. There would follow a recital or tape recording or sometimes simply a talk.

The opening of a poem written by the art historian Melinda Wortz after a musicale in 1986 captures the charged air of the occasion:

> I arrive late – 5:30
> Morton Feldman's 'metaphorical fugue' performed on the piano
> by Robert Krupnick
> has already begun
> I make not a sound, and sit in the foyer, on a bench
> beside the Dan Flavin.
> Feldman's piece is subtle but not severe
> reverberations massing in layers ...[27]

The musicales were a showcase for the new, the maverick, and the offbeat. Participants included characters as eclectic as Pierre Boulez (the French composer renowned for his 'remorseless logic'), American minimalists John Adams and Philip Glass, the prodigious Finnish composer and conductor Esa-Pekka Salonen (who would become music director of the Los Angeles Philharmonic in 1992), and countless virtual unknowns.[28] One of the more outré performances featured a belly dancer in accompaniment to a taped composition by Janis Mattox, a young composer from San Francisco.

'The atmosphere was so special,' recalls Freeman's friend Julia Bloomfield, 'because we were all hearing music that had never been played before, and it was being played by the composers themselves – or else a group or soloist had been brought in to play it.'[29] Hockney's London dealer John Kasmin was one of various attendees who took a more irreverent view:

> I didn't know about the music at first. But then we got invited, and David said, 'We better go to Betty's concert. There's going to be a recital. It might be a bit dull, Kas – it might be hard work. But we'll stand and make a get-away.' Well, it went on for four hours! A musician playing one note.[30]

From early on, a regular at the musicales was Ernest Fleischmann, the executive director of the Los Angeles Philharmonic, conspicuous for his swept-back silver hair and patrician good looks. Then in his late fifties, the Frankfurt-born Fleischmann was a force in the city's music scene – he was once described by an employee as an 'egocentric, completely unprincipled and yet incredibly brilliant monomaniac in music'.[31] He and Freeman enjoyed a close but fractious bond. 'They were like brother and sister', says Thomas Adès. 'They adored each other, but they both had very strong personalities. He was a really high-brow modernist, the kind they don't make any more. He would have known Schoenberg and Stravinsky well.' Their tastes diverged widely at times: Fleischmann, she once pointed out, 'loves Mahler, [whereas] I find him just an emotional bore because it just doesn't touch me, this kind of emotional overflow'.[32] Their fallings-out, as Sellars remembers, were endless: 'But it was a sign of the depth of their friendship that they could hate each other for months at a time and then be the closest people in the world.'[33]

Another fixture at the musicales was Marta Feuchtwanger. The German émigré, whose home in the Pacific Palisades had long been an oasis for Jewish intellectuals, would sit – in Rich's words – 'enthroned like some legendary dowager in a robe of black and gold, not merely a presence but a Presence'.[34] One night in February 1983, when snow-storms in New York had detained numerous guests and disrupted the programme, Feuchtwanger gave a talk on her life – chronicling her escape from an internment camp in occupied France, and her efforts to rescue her novelist husband, Lion, from a separate camp (he had eventually been chauffeured out, disguised as a woman, by a US vice-consul).[35] A photograph from the occasion shows the ninety-two-year-old Feuchtwanger seated on the sofa, swaddled in furs, eyeing Freeman's camera with the sombre gaze of a woman who has seen it all.

Despite the number of big personalities clustering together, the mood – in Sellars' recollection – was relaxed and egalitarian:

> There was a frisson, and an implied challenge in the air, but part of it was just California: laid back, relaxed, chilled out. That was the enjoyment of it, in that Beverly Hills interior where the furniture was so relaxing: everything tended towards the horizontal. It wasn't about pedestals or hierarchies. It was just that we were all hanging out together.[36]

During the performances, Freeman's husband would be in the kitchen, cooking the evening's pasta dish – his baritone cursing audible throughout the house. The dish would vary from a standard carbonara to a

Marta Feuchtwanger at Betty Freeman's home, 13 February 1983, photograph by Betty Freeman

baroque concoction of Franco Assetto's own creation, for instance his 'Pasta with 9½ Ps': pasta, pomodori (tomatoes), prosciutto, parmigiano, pecorino, piselli (peas), prezzemolo (parsley), peperoncini (peppers), panna (cream) and 'paragus (the 'half' referred to in the name of the dish).[37] The pasta became an institution in its own right, as indispensable as the music. Or, for some, the main attraction. In the autumn of 1985, Freeman summarily removed a number of regulars from the mailing list after they failed to contribute to a fundraiser for a music festival that she was championing; she referred to them afterwards as 'the people who came to eat the pasta'.[38] At dinner time, attendees were invited to inscribe Freeman's visitors' book, a leather-bound tome inscribed 'Il Salotto Musicale', brimming with pasted-in photographs, musical scores, invitation cards, and other ephemera. Assetto had coined the term – *salotto* ('big salon') being his word for the musicales.

True to Freeman's nature, the concerts were stylish and discreet affairs, even as they swelled in size. Writing in 2005, Rich reminisced:

> Betty's home, on a Beverly Hills side street, was not one of those legendary show-biz mansions, and there were times when the crowds pushed fairly hard against the walls of her home. I saw her go pale only once, however; that was when Lou Harrison showed up with twenty members of his gamelan, exotic Indonesian percussion instruments and all. We deployed them up and down the staircase leading up to the bedrooms, and their sounds filled the house most gloriously.[39]

Hockney attended many of the musicales. He was also one of only a handful of contributors to the programme who was not a musician. In April 1982, he gave a talk about his set and costume designs for the Metropolitan Opera's Stravinsky triple bill, which had opened the previous year, directed by John Dexter. Photographs from the occasion show Hockney and Rich relaxing after the performance with a Budweiser; Hockney is dressed in a fleecy white jacket and striped tie, and smokes a pipe.[40] In a tape recording from earlier in the evening, he can be heard showing slides to a rapt audience, explaining how he had conceived a separate design for each of the three components:

the ballet *The Rite of Spring (Le Sacre du Printemps)*, the opera *The Nightingale (Le Rossignol)*, and the opera-oratorio *Oedipus Rex*. 'I thought it looked best from the balcony', he reflects, in relation to the final Sophoclean chapter. 'John Dexter thinks [it's because] I spent my youth in the cheap seats in the theatre.'[41]

That same year, Robert Wilson was another non-musician contributor. Like Hockney, he gave a slide presentation, looking back at *Einstein on the Beach* (1976), the opera he had written with Philip Glass. Unlike Hockney, he risked causing a scene. 'Bob was diagramming the opera's structure on white pads while projecting slides of scenes onto my Sam Francis painting,' Freeman recounted, referring to the fifteen-foot-wide canvas that hung on the long wall. 'Bob was just so carried away he got mixed up. He headed to the painting with his felt-tipped marker, intending to draw. I thought to myself, "Should I be a good hostess and leave him alone?" I shouted at him to stop just in time. Later he was very apologetic.'[42]

The marvellous element

By this time, Freeman's attention had shifted away from art towards music, but she continued to collect. 'She wasn't ever really finished acquiring works of art', according to Pamela Kort. 'If something caught her eye and she had a space for it, she just might snap it up.'[43] Throughout the era of the musicales, the house was filled with Freeman's extensive collection. The dark walls of the music room had always been intended to offset the pictures that hung there. *Japan Line* (1957) by Sam Francis, the work that Robert Wilson nearly defaced, consisted of a near-uniform field of white interrupted by a plunging ochre zigzag, streaming drips over the white. Roy Lichtenstein's *Still Life with Cash Box* (1976, one of a series of *Office Still Lifes*) presented a schematic view of an empty cash box and a lamp on a desk – all diagonal lines and hollowed-out white space. This mundane subject was countered by Lichtenstein's *Frolic*, a more up-tempo – if similarly stylized – vision of a nude blonde bombshell cavorting on the shore, based doubly on Picasso's *Bather with Beach Ball* (1932) and Lichtenstein's own *Girl with Ball* (1961). Flanking the main doorway, the two Lichtensteins might be felt to embody the divergent

Robert Freeman, David Hockney, and Alan Rich, at Betty Freeman's home on the occasion of the 6th Musicale, 9 April 1982, photograph by Betty Freeman

character of the concerts themselves, or indeed of their patron: at once sober and exuberant, formal and experimental. Freeman's art was an inextricable part of her home and a defining aspect of her life. 'It wasn't a house that was meant for show', says Stephanie Barron, senior curator at LACMA. 'She lived with the art, it was integrated into the home: it wasn't precious. Things were floor-to-ceiling, edge-to-edge.'[44]

Beside the grand piano in the music room was an installation by Doug Wheeler, one of the pioneers of the 'Light and Space' movement of the 1960s and 1970s. Wheeler's *Untitled* (1968), installed on the wall, consisted of an oblong of Plexiglas, dim at the centre and illuminated around the edges by stark white neon. Similar in size to a hefty Baroque painting, the piece inverted the light and shade of the room itself, as recorded in *Beverly Hills Housewife*.

While expressing the eclecticism of Freeman's taste – from Ab Ex-style abstraction to sleek Pop imagery to transcendental minimalism – these works of art were all of a piece with the decor of the room itself. Black on three of its walls (but white on the fourth, northern side where the piano stood, with a pale curtain drawn across most of it), the space was a paragon of mid-twentieth-century cool. White button-back couches stood around the walls, trimmed with electric pink.

Around the house, numerous works bore witness to Freeman's friendships, whether with Hockney, Francis, Dan Flavin, or Joseph Cornell. Flavin's neon-tube sculptures burned like secular shrines in various rooms. Thomas Adès has a memory, from one of his stays at the house, of the Peruvian maid Bertha coming into the dinette at breakfast time and switching on a Flavin, with a fizz of electricity as the neon tubes came alight. In a guest bedroom, an eight-foot corner piece by Flavin was suspended above the bed, a triangular matrix of red neon tubes with one bar jutting from the corner of the room like an arrow strung in a bow.[45] Adès remembers the experience of lying beneath the glowing crossbars on the bed, which itself protruded from the corner of the room at a forty-five-degree angle. 'You turned them on and they were blood red – a fizzing red cross over your head. Betty said: "Yes, honey, that's a very violent piece. It's about Vietnam."'[46]

Freeman's dream was to have a Flavin sculpture in every room, and the artist himself came to Hillcrest Road to install a number of works. The staggered shape of *'monument' for V. Tatlin* (1967) blazed out from the top of the staircase, a tower of light that mimicked the spiralling form of Vladimir Tatlin's unrealized Soviet megastructure *Monument to the Third International* (1919–20) – an ironic riff on the revolutionary ideal (fig. 23). Freeman had acquired the work in 1984 from Leo Castelli, writing to Flavin the following January: 'It is, as you know, one of the Tatlins and looks like a shining diamond in my entry hall [...] Light itself is certainly the marvelous element and I feel sorry for anyone who lives with those heavy, ugly things made of bronze or iron or metal.'[47]

There was a tongue-in-cheek aspect to the remark, or simply a blithe forgetfulness, because Freeman herself lived with numerous heftier sculptures. And yet the most monumental object in her collection had a self-deprecating air. Outside on the terrace was a giant sculpture by Claes Oldenburg, *Typewriter eraser* (1976; fig. 22), which she had acquired in 1978. Wrought from painted metal and concrete this replicated an everyday office accessory – a hard rubber disk with a brush attached, for scrubbing out printed text – on the scale of a grand equestrian statue: a 'fine anti-heroic subject', as the artist had termed it.[48] Like Duchamp's early readymades, the sculpture reframes a commonplace object as a work of art, in this case by means of artful fabrication and incongruous enlargement. The eraser is balanced on its curved edge, as though it had just struck the ground from a height, the bristles of its brush trailing into the air like plumage. In a different way from Flavin's neon tubes, functional matter achieves a strangely transcendental impact.

In later years, the dress that Freeman had worn for *Beverly Hills Housewife* assumed the status of a work of art. She hung it in a closet beside the front door. According to Adès: 'It was just there, so that if anyone asked, she could say: "This is the dress."'[49]

A modern-day Medici

> 'I like being a patron. Mozart could have used a patron like me in the last years of his life.'
>
> *Betty Freeman, 1987*[50]

When Freeman stood to introduce Marta Feuchtwanger as a speaker that evening in February 1983, she referred to her veteran attendee with affectionate humour as 'the youngest member of our salon', going on to add: 'I think in her person she carries the history of who she is and where she's been and what she's seen. It's just something that emanates from her.'[51] Much the same could be said of Freeman herself. Despite her refusal to be caught up in the past, she was the living sum of her experiences – as a musician, as a wife, as a collector, as a patron. In her laconic way, she emanated the history of where she had been and what she had experienced.

Freeman's position as a powerful female benefactor – and the host of an influential salon – had illustrious precedents, from Gertrude Stein in Paris to Isabella Stewart Gardner and Florine Stettheimer in the United States.[52] In particular, the musicales revived a format that had been popular in the haute-bourgeoisie echelons of pre-war Vienna and Berlin, and that threaded back to the salon culture of the nineteenth century.[53] As Peter Sellars puts it: 'The avant-garde has always been about people working with their friends – parties and gossip and who's with whom and who's *not* with whom. You could *taste* that in Betty's gatherings.'[54]

Even as she ran the show, however, Freeman remained largely on the margins. In the estimation of the music critic Anthony Tommasini: 'Freeman was not a pusher. She was a facilitator. She brought people together and let things happen naturally.'[55] 'There was no one in the world like her', according to Pamela Kort, who began attending the musicales in 1983. 'She was so serious about the music. Even if I sometimes found what I was listening to somewhat boring, I didn't dare nod off.'[56]

At once intimate and formal, the musical salon was the perfect environment for her character. As Hockney had perceived when he first met her, that character was double – charismatic yet self-effacing. Freeman rarely wanted to be the centre of attention, and would busy herself by taking photographs. According to Stephanie Barron, the camera became a means of expressing – but also overcoming – her innate reserve. 'She used the camera as a way of getting close to people and having that intimacy. The camera was an important thing in terms of her relationships with people in the music world.'[57]

In the end, though, the concerts were an embodiment – always – of Freeman's sensibility. That sensibility, like so much about her character, was rarefied and offbeat. Her tastes were eclectic, spanning the high minimalism of Philip Glass, the Balinese-inspired melodies of Lou Harrison, the recondite modernism of the British composer Harrison Birtwistle, and the superhuman rhythms of the Mexican-American Conlon Nancarrow. Beyond classical music, she was partial to rock and roll and country, although she could not stand jazz, describing it as 'very mechanical [...] One instrument plays, everybody applauds, the next instrument plays, everybody applauds, etc., then they all play together, then it finishes.'[58]

She recognized that classical music, especially the experimental variety, was a niche art form – not likely to garner mass appeal – but seems to have believed in it all the more strongly for this reason. 'I am going back to the old-fashioned idea that music is esoteric, that classical music is esoteric', she once said. 'In the seventeenth and eighteenth centuries, it was written only for the court and the church a little earlier. [...] I like the idea of music just being for a few people. I don't dislike small audiences, I am not even shocked if there is.'[59] (On the other hand, she remarked with approval that European music festivals enjoyed a mass following compared with those in America, where classical music remained a largely white, middle-class luxury.[60])

Even those closest to her were mystified by her choices. Franco Assetto cared little for experimental music. Verdi and Rossini were more his thing – the sort of 'lyrical' music that Freeman could not stand. On the few occasions that she took him to concerts, he fell asleep. Her son Robbie, who served drinks and food at the musicales, was similarly ambivalent: he had performed in a rock band while living in Paris in the 1970s, and exuded the air of a young rebel among the connoisseurs at Hillcrest Road.

In 1998, a profile of Freeman in the *New York Times* described her as 'a modern-day Medici'.[61] In the same piece, her old friend John Adams likened her to a grand figure from the French or Russian court. There were hints, in these comparisons, of the somewhat antiquated nature of her patronage – its happily haphazard nature. (The article noted that her gifts might vary from a few hundred dollars to tens of thousands, with the total sum 'in the high six figures'.[62]) A similar view was expressed by Russell Platt of the *New Yorker* in the aftermath of Freeman's death:

> Freeman struck me not as a contemporary phenomenon but rather as something touchingly antique: the last iteration of the eighteenth-century music patron. She was our Prince Esterhazy, and L.A. was her Grand Palais where, instead of Haydn, the likes of Boulez, Birtwistle, and Lachenmann displayed their wares.[63]

She was adamant that her support was not in any sense charity, and had nothing to do with liking a given composer or wanting their friendship. The gender of a composer had no bearing on her view of their music: 'I don't like to think of "women composers". I just like to think of "composers".'[64]

What united her approach was the sustained devotion she showed to the artists in whom she believed, a devotion that was reciprocated in many cases. John Cage entitled a set of violin studies the *Freeman Etudes* (1977–80); John Adams dedicated his 1987 opera *Nixon in China* to her. She was candid, at the same time, about what she disliked – including in situations where her support had produced a work that she did not rate. 'Art was the only thing that mattered to her, in a way', according to Thomas Adès.

Composers were frequently invited to stay at Hillcrest Road. Adès remembers how, when he first came to Los Angeles in 1996, she insisted on collecting him from the airport. 'She set out that first day to seduce me with the city. We had lunch at Yamashiro, which I realize now was so that I would see that view and be duly wowed.'[65] She drove him in her open-topped, cream-beige Cadillac past the Hollywood Bowl, where a sign had been placed to announce his performance with the Los Angeles Philharmonic New Music Group. She stopped the car in the middle of Cahuenga Boulevard to take his photograph.

In time, her tastes inclined more towards European music. She was a great admirer of Benjamin Britten, and regretted that the United States did not boast composers of the same calibre.[66] 'I find European composers are much stronger, more original, more dynamic, more adventurous, more courageous', she remarked in 1991. 'The Americans, with a few exceptions, set out to amuse, to entertain, to do something in the rock style.'[67] Her shift in taste left many aghast. She received 'hate letters', as she termed them, from several American composers she had previously assisted.[68]

'She was really supporting the hardcore old European avant-garde, which was super-unfashionable', says Peter Sellars, who directed the original production of Adams' *Nixon in China*. 'It wasn't done for fashion. It was done because she felt the disruptive, deep power of radical ideas.'[69] For many, though, the tenacity and seriousness of her approach were an enduring inspiration. 'She was *out there* and at the forefront of things,' says Adès, 'and that actually pushed taste along for other people. Other patrons would look at Betty and think: that's how I want to do it.'[70]

Throughout the 1990s, Freeman went every year to the Salzburg Festival – supporting the director, Gerard Mortier, in his promotion of contemporary music. (In the eyes of her former American protégés, Mortier was a Svengali who had stolen her away.) Pamela Kort, who accompanied her on many of these trips, relates that they stayed on one occasion at the Hotel Goldener Hirsch, the opulent medieval inn that Patricia Highsmith had featured in her 1970 novel *Ripley Underground* as one of the haunts of the urbane murderer Tom Ripley – a place imbued with 'the pleasant scent of furniture wax, the aroma of comfort and tranquillity'. Not that the trappings of luxury were ever of great interest. 'It wasn't about being seen at the opera in a nice dress', Kort maintains. 'She was there solely because of the music; she went to every rehearsal she was permitted to attend.'[71]

Robert Wilson had his own memory of going to Salzburg with Freeman, and in particular of her refusal to be photographed in the same room as a portrait of Mozart. 'They were taking a bunch of photographs of Betty and Gerard and myself. The photographer had set up his camera, when Betty screamed: "No! There's no way I'll be photographed in this room. There's a painting of *Mozart*."'[72] The picture perhaps symbolized a heritage that Freeman felt she had cast off long ago. As she once professed:

> I can't understand why anyone would want to hear the same Mozart and the same Beethoven and the same Brahms over and over when there is such great music being written today, [music] that is far more relevant and meaningful. [...] We are not living in either the Classical or Romantic age, but instead [the age] of new discoveries and challenges.[73]

But as Wilson pointed out, Freeman's objection to being photographed with the Mozart portrait may have had a simpler explanation. It was the same painting that appeared on many of the chocolates sold in Salzburg – the kitsch association may simply have been too much to bear.

7

Betty

Don't look back

'I'm not a person of nostalgia', David Hockney once admitted. 'I just live for now.'[1] He and Betty Freeman were kindred spirits. In her later years, she refused to dwell much on the past. When Hockney painted her in 1966, she was in the gradual process of emancipating herself from her own history – the life of Mrs Stanley Freeman. As time went on, she looked back less and less. Writing to Clyfford Still while staying in Florence in 1968, she gave a hint of her inclination to 'live in the present', watching the world rather than gazing into the past or future. 'I'm going to stay here for another month and am discovering many nice things – such as a better relationship with my daughter, Claudia, and sitting for hours in restaurants or cafés like the Italians do and just watch other people without having to rush off.'

Paradoxically, that apprehension of the life around her was also a recognition of the *presence* of the past in the antique European city: 'It's very quiet here – the whole city lives only in its past. [...] I went to Spoleto for the music festival and stayed in a farm house that Galileo lived in. And after the last performance Ezra Pound drifted into the café like a ghost.'[2] The past was experienced, then, as a revenant in the present – as an aspect of real-time experience, never simply a relic.

Similarly, *Beverly Hills Housewife* brims with art-historical revenants – ghosts of one kind or another – while conveying Freeman's peculiar ability to inhabit the present moment. In Hockney's scene, she is like a character in a story or a play – but a character whose narrative has evaporated. The space around her offers little clue as to 'before' or 'after'. The antelope's head, a memento of Stanley, feels incidental; it is an attribute that has ceased to define her. In later life, she spoke strikingly

Betty Freeman, 1975

little of the years of her marriage. 'That was really not the topic', according to Peter Sellars. 'I think that's one of the reasons why, for Betty, it had to be new. She was looking forwards, not backwards.'[3]

'I've always been interested in the new', Freeman professed in 1998. 'I like contemporary painting, clothing, furniture, architecture. So of course I like contemporary music. Old music is fine. But I like complexity, challenge, ambiguity, abstraction.'[4] The past, to the extent that she paid heed to it, was a living thing, something to be experienced anew in the present. Art was always defined by the moment of looking, or music by the act of listening. Her deep affinity for music may have related to the capacity for a melody to fill one's mind, extinguishing memory or anticipation – an idea that had been advanced by the musicologist Victor Zuckerkandl, her teacher at Wellesley: 'It is even a condition of hearing melody that the tone present at the moment should fill consciousness *entirely*, that *nothing* should be remembered, nothing except it or beside it be present in consciousness.'[5]

There was a spiritual dimension, too, to Freeman's outlook. As her friend Pamela Kort explains: 'At some point she became very interested in [George Ivanovich] Gurdjieff. She had sometimes gone to Ojai, and not just to hear the Festival, but also because of its spiritual community, at least that's what I recall Betty telling me. I know that even during her last months of life she remained interested in Gurdjieff's ideas.'[6] Gurdjieff (1866–1949), a Russian philosopher and mystic, had acquired a sizeable following in the United States through his advocacy of an expanded form of consciousness beyond thinking or perception. He argued that most human beings live in a state of hypnotic 'waking sleep', exercising only one portion of their consciousness, and that transcendence of the sleeping state can be achieved through conscious 'work'. Such work was Freeman's modus operandi in the years that she evolved from a Beverly Hills housewife into a leading patron of music. Its success depended on not looking back.

'The past comes screaming back at me,' Harry Partch once wrote to her, 'and I just want to go away and hide.'[7] Freeman herself seems to have been impervious to such hauntings. Whether this state came naturally to her, or required its own determined effort, is impossible to know. But living for the moment – refusing to be burdened by the past – became a precept, almost an article of faith. As Kort recalls: 'She was very practical. She did what she did because that was the best she could do. And she believed that we're all doing the best we can do. That was one of her mottoes.'[8]

Even in her appearance, she seemed to inhabit a perennial present – her hair unfailingly blonde, her smile (at least in photographs) as bright as it had ever been. According to Robert Wilson, her father had once told her how much he loved her hair blonde. She later bought a convertible car that matched its colour. She always took care to look the part. 'The clothes she wore were always brilliant colours – brilliant pinks, just like *that* dress', according to her friend Julia Bloomfield. 'She had a beautiful figure, lovely legs, so she always wore high-heeled shoes and skirts. She knew she cut a figure.'[9] Freeman had never stopped looking like the woman in Hockney's painting. Her ability to live in her own singular dimension of time gave her the appearance, sometimes, of a daydreamer. When Thomas Adès's composition *America: A Prophecy* (1999) was performed in Los Angeles, Freeman announced to a friend: 'Tom wrote this piece for the millennium. I can't remember *which* millennium.' Adès laughs to remember: 'When she said things like that, you just accepted this world in which we could all be a thousand years old. Betty somehow seemed like a broader sky.'[10]

Conservative instincts

In 1973, Hockney's life became the subject of a film, *A Bigger Splash*. It was directed by Jack Hazan, a British-Syrian filmmaker who had come across the artist's work in the catalogue for his 1970 retrospective at the Whitechapel Art Gallery. Hazan had been struck in particular by his double portraits, which suggested to him the possibility of interweaving real life and artifice.

A Bigger Splash amounted to a two-hour dramatization of Hockney's life, part documentary and part fiction. Shot in a languorous and episodic style, the film focused on his break-up with Schlesinger (Hockney was aghast when he first saw it, and tried to block the distribution). Across a chain of discontinuous moments, the artist is seen chatting with friends, travelling to New York, and working at home in London. The action is interspliced with fantastical interludes: four nude youths swimming in a sunlit pool; Peter Schlesinger wandering naked in the gardens of Los Angeles houses like the roaming protagonist of John Cheever's story 'The Swimmer' (1964).

In one sequence (figs 24 and 25), Schlesinger comes across Betty Freeman in her garden at Hillcrest Road. In cinematic form, Hazan restages the *mise en scène* that Hockney originally constructed when he took the Polaroids for *Beverly Hills Housewife*: Freeman, now in her early fifties, wanders beneath the portico in her pink dress. She dusts the antelopes' heads, although this feels closer to a whim of the moment than a chore. Schlesinger then appears like a figment of a dream. He is naked. He swims in the pool, then climbs out and wanders close to the house. Freeman watches him sedately. She doesn't bat an eyelid.

Like the painting on which it is modelled, the interlude in *A Bigger Splash* casts Freeman in a stock role, that of the respectable housewife. It also implies the existence of other kinds of lives, other realities, beyond her own – and the fascination that they held for her. The nude youth is like a visitor from another universe. While underlining what was conventional about Freeman, the episode signals her nonconformism. Few actual Beverly Hills housewives would regard such a visitation with the same wry nonchalance.

In the film, both Schlesinger and Freeman are playing themselves. They are simultaneously 'in character' and undisguised. 'For about two years, Jack was making the film, and he repeatedly asked me to be in it,' Schlesinger recounts. 'I said I didn't want anything to do with it. He took me to lunch in Soho. I was a movie fanatic, so I said, "I'll do it if you treat it like a real movie and pay me. I'll play a part. It's not me."'[11]

Freeman herself seems to have shifted between performed and spontaneous versions of herself, not only in *A Bigger Splash* but throughout her life. Casting her as a type, even a cliché, Hockney's portrait became a sort of doppelgänger – something that was simultaneously her and not her. She played with the persona of the Beverly Hills housewife – assumed it and cast it off by turns. Indeed, this very inclination to play with her identity, however subtly or unconsciously, became a defining characteristic. Shortly before filming began on *The Dreamer That Remains*, she wrote to Harry Partch, urging him not to tidy up his cluttered home before the cameras arrived:

> I do not want to pretty up the workshop – it must stay exactly as it is when you use it. You are not a neat housekeeper and it would be false to have an immaculate workshop as if you were a Mondrian. You are you and that is what we want in the film, not a Beverly Hills housekeeper like me.[12]

Here, perhaps because the situation demanded it, she appropriated the persona of the 'neat housekeeper'. At other times, she held that persona at an ironic remove. In Hazan's film, she is playing a part, but the performance is not a mere pretence. Something of *her* persists in the dusting of the objects, in the dreamy surveying of her domestic world, in the very way she moves.

The making of *The Dreamer That Remains* involved a separate incident that laid bare her innate conventionality. Partch had drafted a libretto – intended for the soundtrack of the film – that gave an allegorical account of his sexual awakening as a child. This featured quixotic details such as a booklet found inside a Victorian harmonium with the punning title 'How to take good care of your organ', and a discussion of 'How to put on your pantyhose' (to which a chorus retorted 'A quick jerk!'). For the second half of the composition, Partch planned to feature a new percussion instrument that he called the Rhythmic Pudenda. A culminating chorus would resound with the word 'Bullshit'. Thirteen years later, Freeman recalled that she found the libretto 'so horrible, with its references to organs and panty hose and jerks that I had to [...] telephone Partch immediately to revise it. Without argument he agreed. [...] I think that when he wrote this first version there was something very wrong, a perversity in his mind.'[13] The long span of time had done little, it seems, to soften her disapproval.

Partch agreed to remove most of the offending elements, but Freeman did not have it all her way. One line in the libretto made reference to male homosexuality: 'Even in public parks, where a couple of people want to improve the darkness with a little loving ...' – a description that the director, Stephen Pouliot, had staged with two men seated on a park bench. Freeman objected to the line and wanted it removed, but Partch stood firm, telling her that the sequence was 'light-handed satire [...] not *seriously* homosexual'.[14] Her objection is hard to square with the image presented in *A Bigger Splash*, where the spectacle of a naked youth in her garden seems like nothing.

She was contradictory. For all her love of radical art and music, she remained the product of her respectable Jewish upbringing. During her first marriage, she and Stanley had been Republican voters, and in 1962 they donated funds to Richard Nixon's unsuccessful campaign for governor of California. How far her political views shifted in subsequent decades is unclear, but she always maintained a strict outward decorum, objecting to swearing or lewd talk. 'She was proper', Thomas Adès observes. 'But it was part of her elegance – her style.'[15] Robert Wilson remembers the difficulty involved in going for hotdogs together in Hollywood, because of a sex shop across the street from the food stand. 'She didn't want to go because she didn't want to look at that trash. "I said, oh Betty, please – let's just go get a hotdog and close your eyes."'[16]

But Freeman was capable of affronting her own conservative instincts. The mere fact of her friendship with Partch and others like him was a reflection of her ability to transcend the prejudices of her background. In 1972, three years before the decriminalization of gay sex in California, her scruples about onscreen homosexuality were nothing unusual. Far more remarkable was her readiness to fund a project such as *The Dreamer That Remains* in the first place.

One quality that she shared with Hockney, by her own admission, was a determined libertarianism. She sympathized with his crusade against anti-smoking legislation. 'I'm all with David in his antipathy to the laws that restrict human interest and human conduct', she stated in 2005. 'I'm not shocked by anything David says.'[17] She had a sign made for her home, bearing the words 'HERE SMOKING IS ENCOURAGED' in large felt-tip lettering. It was put on prominent display in the hallway whenever Hockney and other smokers visited.

At times, she could be breezily permissive. Her son and daughter-in-law, Robbie and Fanny, have a memory of her taking them to a party at an apartment in Paris where Andy Warhol was present. In a room upstairs, everyone was smoking pot. Freeman chatted away, blithely indifferent to the clouds of pungent smoke filling the room. Another time, when her friend Bill Jones – the dancer and choreographer – staged a nude photo shoot, she went along to spectate.[18] In her bedroom at Hillcrest Road was a giant watercolour of a nude woman with a flower over her crotch. 'It was a huge bedroom', Julia Bloomfield remembers, 'with a *huge* double bed. And there was a painting – a view up a lady's legs. Naked. To live with that, to sleep with that, to wake up to that!'[19] Freeman's conservatism always had its limits.

Unradical feminism

Beverly Hills Housewife is a painting of an awakening. In 1966, Freeman had given up her dream of a piano-playing career. She was reassessing the role that she had fulfilled for the past two decades, that of the middle-class housewife. Her suggestion to Hockney of the painting's title was self-deprecating – a nod to the very persona that she was outgrowing.

Her awakening coincided with a larger, societal one: the second-wave feminism of the 1960s. Los Angeles in that decade fostered a dynamic feminist art movement, embodied by figures such as Betye Saar, Judy Chicago, and Carole Caroompas – their works and sentiments a far cry from the macho posturing of the LA Cool School. Freeman's casting off of her marital role chimed with feminist ideals, and yet Hockney's portrayal of her as a solitary, affluent woman in the sanctuary of her home was at odds with the collectivizing, anti-bourgeois ethos of the movement in that era. Indeed, *Beverly Hills Housewife* is a confirmation, more than anything, of the personal nature of Freeman's awakening. She took little interest in radical or mainstream politics – and might well have shared the view of Joan Didion in relation to the Marxist underpinnings of feminist ideology: '[T]o those of us who remain committed mainly to the exploration of moral distinctions and ambiguities, the feminist analysis may have seemed a particularly narrow and cracked determinism.'[20]

Beverly Hills Housewife is not a picture of feminist renunciation. It reminds us that Freeman continued to be defined by the domestic, bourgeois aspects of her existence – by her social status – but on her own terms, after her own fashion. In 1966, she was reimagining what a Beverly Hills housewife could be, not dispensing with that identity wholesale. As she garnered a new sense of self, she built upon the persona that she had cultivated to date: individualistic and enterprising. She was enacting her own private revolution, exercising a freedom that money facilitated – but that depended more fundamentally on strength of character.

8
Other Selves

'Friends ... that's really the only thread running through my life.'
DAVID HOCKNEY, 2014[1]

In 1978, after years of travelling between London, Paris, and the United States, David Hockney decided to make Los Angeles his primary home. From this time onwards, Betty Freeman used to visit him regularly at his studio in the Hollywood Hills. Hockney was a routine presence at Hillcrest Road, too – attending many of Freeman's musicales throughout the 1980s. She would also host private dinners in the room where *Beverly Hills Housewife* hung. As her friend Julia Bloomfield recounts:

> We would eat at the big round table that was shaped like a flower. Hockney would be opposite me, and Gregory Evans, or somebody else from Hockney's coterie. Jack Larson would sometimes come too. And of course, David couldn't hear a flipping word, so he would talk the whole time – but he was fascinating.[2]

Freeman often took visiting composers to meet Hockney. Thomas Adès recalls being driven to the artist's house for the first time in the mid-1990s. 'That environment where the house is, all that richness of the hills and vegetation – it was as if she'd set out to astonish me. It was dawning on me as we were driving who it might be. But she didn't say. At this time, Hockney was painting small oil portraits of his two dogs, Stanley and Boodgie. 'I felt he was a bit lonely. He was surrounded by pictures of his dachshunds – it was an emblem of how isolated he was.'[3]

The seemingly innocuous subject belied a sense of loss. Aids had claimed many of Hockney's friends (Nicholas Wilder among them), and he was leading an increasingly solitary existence. Freeman would often tell the story of how, when she originally saw the dachshund paintings, she spoke the first words that came into her mind: 'David, you're painting your dogs!' He replied: 'No, I'm not painting my dogs. I'm painting my love of my dogs.'[4]

Hockney had been spending more time at his beach house in Malibu, which he had bought in 1988. Walking on the coastline with his dogs, at what he referred to as 'the edge of the Western World', was a new pleasure.[5] On car journeys through the Santa Monica Mountains, he would play Wagner's operas – blasted out on his stereo – in defiance of his progressing deafness. This led him to create a route called the 'Wagner Drive', a carefully choreographed conjunction of landscape and music that reached its crescendo at the mountain summit. Freeman loved the Wagner Drive. It seems to have been one of the few instances in which she welcomed more traditional classical music, perhaps because it took the form of a piece of performance art.

'Some say that they who are blessed and independent have no need of Friends,' observes the ancient Greek philosopher Aristotle, 'for they already have all that is good, and so, as being independent, want nothing further.' But he goes on to counsel that 'the notion of a friend's office is to be as it were a second Self and procure for a man what he cannot get by himself.'[6] Betty Freeman became, for Hockney, that second self.

In a photograph from November 1987, the two of them sit in front of *Beverly Hills Housewife*, both dressed up for an evening out. Side by side, no longer divided by their roles as artist and subject, they look back at the camera with forbearing smiles. Their position between the painting and the camera produces a curious effect. The frozen moment of the painting is interrupted, literally, by the present occasion of the photograph. 'Now' stands in the way of 'then', in such a way as to underline the gulf between the two.

David Hockney and Betty Freeman at Freeman's home, 15 November 1987

On the other hand, the painting looms indelibly; it literally overhangs the present moment. The episode from 1966 refuses to be expunged. In the very structure of the photograph, there is a potent suggestion of the presence of the past. It is as though Hockney and Freeman's friendship consisted of all the moments leading back to the painting – a countless accumulation. Seen in juxtaposition, the staged encounter of 1966 and the casual togetherness of 1987 are part of a continuum. Like so many of Hockney's own works, the black-and-white photograph (itself now poignantly remote in time) is an image of *now*. Nothing exists, the picture seems to say, beyond this.

In this respect, the photograph produces a similar effect to that of *Beverly Hills Housewife*, where the present-tense occurrence of the woman in the garden seems to dissolve past and future into itself. Both photograph and painting give visual expression to Vladimir Nabokov's notion, conveyed in the form of a mock treatise in his novel *Ada, or Ardor: A Family Chronicle* (1969), that 'The Past [...] is a constant accumulation of images. It can be easily contemplated and listened to, tested and tasted at random, so that it ceases to mean the orderly alternation of linked events that it does in the large theoretical sense.'[7]

Were they going to the opera that night? Is the photograph from before the performance or after? Such details are irretrievable. The picture could represent any number of scenarios. On one level a 'random' snapshot of Hockney and Freeman, it is also a symbol of their friendship. It speaks of that friendship's cumulative and ongoing nature, its dual status as a product of long time and as a live reality.

In the fleeting interval of the photograph being taken, Hockney and Freeman have become their own unrehearsed performance – a spectacle of the passing instant, or what Geoff Dyer has termed the 'ongoing moment' of photography.[8] Their attire speaks of the 1980s: the broad lapels of Hockney's jacket; the flaring sleeves and herringbone pattern of Freeman's dress. Such details situate them in time, as do the signs of ageing on each of their faces. And yet the sharp clarity of the image, concentrated in the reflections of the camera flash on Hockney's glasses, seems equally to deny the passing of time, as though the picture really were an immediate, accessible reality – not forty years in the past.

Those glints of light on the glasses, equivalents to the diagonal white bars depicted in the painting behind him, are their own sort of *punctum*, as Roland Barthes called it – the 'sting, speck, cut, little hole' that detains or disturbs.[9] The thing that pricks the attention.

There are numerous photographs of Hockney and Freeman in front of the painting. Having their picture taken with it seems to have become a ritual over the years. A sequence of after-dinner shots from February 1973 show Hockney tricked out in a cream suit and black bow tie with white polka dots (opposite). Longer-haired than usual, and clutching a thick cigar in some of the pictures, he appears debonair and wild. In December 1981, he and Freeman are captured in a lower-key moment, sitting back to back on the bench seat in front of the painting – Hockney this time rather vacant, by contrast with Freeman's primmed smile (fig. 26). A separate shot from the same occasion shows Hockney with his mother Laura, then aged eighty-one (fig. 27).

Hillcrest Road became the recurring setting not only of Hockney and Freeman's friendship, but of others – with Jack Larson, for example, or Celia Birtwell – that were central to each of their lives. (It was Larson, after all, who had first brought them together.) 'Lots of David's portraits are about togetherness', Birtwell has said. 'Togetherness is two people, and it's always an interesting equation for him, because in a way, we're all alone – but it's nice to be part of something, and part of somebody else.'[10]

Celia

Celia Birtwell, alongside Freeman, was Hockney's enduring female muse – and one of the principal characters in his life. He first met her in 1963, after beginning an affair with Ossie Clark – the flamboyant and hedonistic young fashion designer ('giddy, effeminate, mad and exaggerated', as Hockney later summed him up) with whom Birtwell was in a relationship, despite Clark's open gayness.[11] Birtwell and Clark had first met in Manchester in 1959 through a mutual friend, Mo McDermott, whom Birtwell knew from Salford Art School. In 1961, Birtwell followed McDermott to London, where she shared a flat for a while with the artist Pauline Boty.[12] She resumed her affair with Clark when he came to London the following year, ignoring Boty's warning that 'he could be

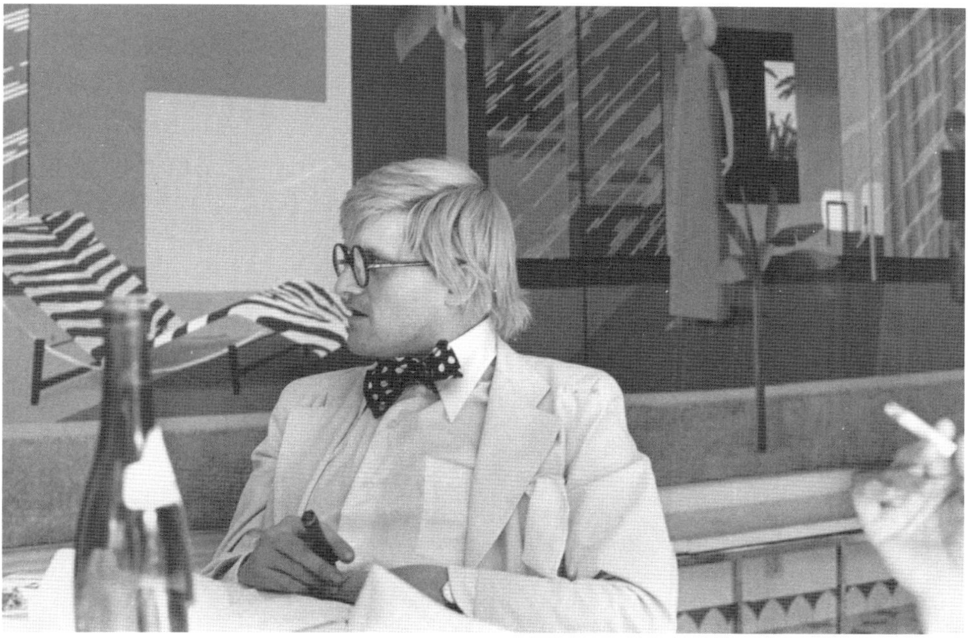

David Hockney at Betty Freeman's house in February 1973,
photographed by Betty Freeman

Celia Birtwell in Betty Freeman's garden at Hillcrest Road (top) and with David Hockney (bottom) in February 1973, photographed by Betty Freeman

a lot of trouble'.[13] She and Clark began living together in a flat above a bicycle shop in Notting Hill – drawn to each other by their shared love of fashion. They married in 1969. Birtwell was initially closer to Peter Schlesinger than to Hockney. Increasingly, though, she sought an escape from Clark's drug-taking and violence in both of their company. 'I soon discovered the great thing about her is that she is very funny', Hockney has said. 'That's what I loved about her.'[14]

A handful of black-and-white photographs taken in February 1973 (the occasion on which Hockney was wearing his dotted bow tie) show Hockney, Birtwell, and Freeman – together and individually – in the garden at Hillcrest Road. The pictures have an ambience of elegant leisure shot through by anticipation. All three are dressed up as though for the opera: Hockney in his pale summer suit; Birtwell in a black Chinese-style jacket and matching dress; Freeman (as if in counterpoint to Birtwell) in a high-necked white blouse embroidered with black geometric patterns. In the monochrome coloration of the pictures, the sunny afternoon has transfigured into steely light and shade.

In one picture, Birtwell is viewed from a distance – across the terrace, amid palm fronds and other foliage, beneath the concrete canopy. She is a mere detail, almost, in the sidelong view of the back terrace. Apparently caught off guard, she is standing in a similar spot to that which Freeman had occupied, seven years earlier, in the Polaroids for *Beverly Hills Housewife*. It is as if, whether accidentally or in humorous half-recognition of the fact, Celia is playing the part of Betty – stepping into the other woman's role.

In another photograph, Hockney and Birtwell are observed close-up – presumably captured by Freeman. He appears solemn; his eyes are faintly challenging behind his black-framed glasses. Birtwell is further forward in the picture, less in focus. With her hair piled up in tumbling ringlets and her eyes circled by mascara, she regards the camera with the same directness as Hockney, but with a flicker of something like equivocation. A touch of uncertainty.

Birtwell, here, is the glamorous – forever slightly ethereal – character of Hockney's paintings and drawings of her, for example *Celia* (1970), a portrait in coloured pencil and crayon. In this, Birtwell – perched on a steel-and-leather chair – meets the artist's gaze, cigarette in hand, her own gaze at once faraway and gently piercing.[15] Her blonde hair spills around her temples in much the same style as in the photographs taken at Hillcrest Road; her blouse is a ruffled, layered, indistinct affair, sketched in pink and grey pencil, although Hockney has observed the striations of the collar and the sleeves with acute accuracy. The drawing is one of a series, made mostly while he was living in Paris in the 1970s, in which he grappled with (sought to dispense with) the naturalism that had characterized his work since the late 1960s, and that *Beverly Hills Housewife* had heralded. In seeking to move beyond this illusionistic, Renaissance-inspired style, he resorted – somewhat paradoxically – to close observational portraits, drawing from life in pencil. The works were inspired in part by Ingres's small-scale drawings of aristocrats, made while the artist was living in Rome between 1806 and 1824 – before Ingres renounced commissioned portraiture as 'a considerable waste of time'.[16]

The photographs taken at Hillcrest Road in 1973 date from a time, then, when Hockney was struggling to break beyond the style that had defined his paintings of the previous decade. He was also struggling to move past his break-up with Peter Schlesinger two years previously. In the years since they had moved to London in 1968, the relationship had slowly fallen apart under the weight of their respective ambitions, and in particular Peter's desire for greater independence. (In December 1971, Hockney had written to Henry Geldzahler: 'I suppose it's all over [...] he won't let me express my love in any way now, and so all I can do is try and forget.'[17]) Perhaps more than ever, his friendships with women such as Birtwell and Freeman were a refuge – shoring up against the uncertainty of his artistic and romantic lives.

A third photograph from that afternoon shows Birtwell and Freeman seated beside the swimming pool. Celia's seriousness is a foil, here, for Betty's smiling ease. The first woman seems pensive and withdrawn, the second, blithely uncomplicated. Birtwell withholds while Freeman reacts. But perhaps the picture can be read another way; because, in the end, it is Birtwell who engages with the camera – with Hockney – while Freeman's eyes are averted. She is semi-distracted, just as she had been in *Beverly Hills Housewife*, still reserving a part of herself.

Celia Birtwell and Betty Freeman by the pool at Hillcrest Road in February 1973

It is easy, of course, to make deductions from a photograph. Yet the two women's expressions in this picture seem to intimate their distinct temperaments. There is the wistful Englishwoman who betrays her vulnerability in a mere look, and there is Betty Freeman: self-contained, turning away, not quite knowable behind her radiant carapace.

Other versions of Betty

David Hockney and Betty Freeman's friendship is traceable in the several portraits he made of her after *Beverly Hills Housewife*. These chart the course of her later life, each capturing her at a precise moment. Like most of Hockney's recurring sitters, Freeman is always recognizable – always at some level *herself* – even if that self was by its nature mercurial.

In 1994, Hockney drew Freeman from life. She sat for him in the studio that adjoins his Los Angeles house – a space that he had created in 1983 on the site of an old paddle tennis court. In the large-scale work on paper (fig. 31), Freeman's likeness materializes out of hatched graphite lines and pockets of shade. The image manages to seem both provisional and exactingly precise. In contrast to other painted or drawn portraits of her, she faces him – returns his gaze.[18]

The drawing is one of numerous works dating from the mid-1990s that reflect Hockney's wider environment, whether social or geographic. It belongs to a group in which he depicted the people he was regularly seeing in Los Angeles. Photographs of him working on the portrait offer glimpses of the paintings that he was making concurrently – a series billed the *Very New Paintings*, whose interlocking shapes were inspired by the Santa Monica Mountains and Pacific Ocean. Another photographic portrait of Freeman – *Betty Freeman, April 21 1991* (fig. 30) – shows her in front of one of these colouristic landscapes (her body amalgamated out of five video stills), her own camera hanging around her neck as she smiles coolly. She had become an indelible part of Hockney's LA life – as much a part of its 'scenery' as the landscapes that he encountered daily.

Freeman kept a print of the 1994 drawing in the corner of her bar room at Hillcrest Road – a cropped view of her head.[19] But she never liked the portrait. This may have been because her usual glint of mirth is absent. There is barely a trace of the irony that flickers, however dimly, in other images of her – forming part of her armour. In *Beverly Hills Housewife*, that quality is likewise hard to discern (one has to return to the Polaroid snapshots to find the hint of a smile), and yet in that instance Freeman appears sealed inside herself, in a sense unreachable. Her averted stance has a distancing effect, protecting her from the artist's scrutiny.

In the 1994 drawing, by contrast, her gaze is direct and defenceless. It is hard to impute a mood or intent to her expression. Her hair sweeps around her temples with its usual grace; the contours of her blouse convey a characteristic elegance. But there is little indication besides of a persona. She looks back at us with the disillusionment – the calm vacancy – of someone whose mask has dropped.

In a second graphite portrait (fig. 32) dating from five years later (made on 20 July 1999) Freeman is in full possession of her image. She is shown seated and looking away, in three-quarter view. Dressed again in a loose-fitting blouse, with her hands folded in her lap, she is simultaneously alert and aloof – a seventy-three-year-old version of the woman in *Beverly Hills Housewife*. Unmistakeably older, she still radiates a deadpan charisma. It is there in her erect bearing – in the concurrent suggestion of inattention and steely awareness.

Hockney's portraits of the 1990s are invested with a sense of the live encounter between artist and subject. He had long dispensed with using photographs as source materials. The encounter was mediated only by the drawing itself, although aided in certain cases by the device of a camera lucida, as in the twelve drawings he made of attendants at the National Gallery in 1999 and 2000.[20] The 1999 drawing of Freeman was not made with a camera lucida, but it shares with those others an impression of close proximity, almost of the subject impinging on the space of the artist. She appears physically present – in touching distance – in a way that she was not in *Beverly Hills Housewife*.

Freeman here has reassumed her public face, stepped into her habitual role. A month before she sat for him, Hockney made a remark about Ingres's portrait drawings that seems appropriate to his portrayal of Freeman: 'They were held up as an ideal in drawing: sensitive, full of character and uncannily accurate about physiognomy.'[21] The effect of the 1999 drawing is similar, certainly, to that found in Ingres's portraits of aristocrats, where character and physiognomy appear inextricably fused. In *Study for the Portrait of Louis-François Bertin (1766–1841)*, for example, the subject stares out with a lack of affectation that seems almost churlish (and yet even here, one senses, Bertin is playing a saturnine part).[22] Ingres's painting of Bertin was the direct model for Picasso's 1905–6 portrait of Gertrude Stein, a picture that became, like *Beverly Hills Housewife*, a kind of archetype of its subject.

Hockney's 1999 drawing of Freeman presents a vision of her, one suspects, that accorded with her view of herself. The smile that so often lurked within her expression is there, if a little fainter than in many photographs. Her character and her physiognomy, her abstract and physical selves, seem to be in interplay. There is a tension – more poignant than before – between the made-up facade (her mascara, her lipstick, her coiffed and coloured hair) and the encroachments of old age. Her face is thinner than in earlier pictures; there are shadows around her eyes and throat. In a group of photographs taken while Hockney was making the portrait, the artist himself seems older in a different way: fuller of face, his hair somewhere between blond and grey. As they advanced in age, he and Freeman remained mirror images of one another.

> 'If Socrates waking and sleeping do not partake of the same consciousness, Socrates waking and sleeping is not the same person.'
>
> John Locke, *An Essay Concerning Human Understanding*, 1689

According to the English Enlightenment philosopher John Locke the 'self' is not a stable entity. It changes constantly – formed from whatever ideas or impressions enter the mind. Over time, even within the span of a day and a night, somebody might be multiple persons.[23] In Hockney's art, people are creatures of the present moment – recognizable as themselves but observed, always, as though for the first time. No subject can be reduced to an essence, and it is for this reason that the artist has returned again and again to the same sitters. They are, in a sense, different people each time he portrays them. His various portraits of Freeman, spanning four decades, are pictures of multiple selves.

And yet, while each picture conveys a distinct version of her, the ghost of *Beverly Hills Housewife* always haunts her image. It is there in all of Hockney's subsequent portrayals, a single constant amid the fluctuations of time and mortality. The vision of the woman in pink became the ideal image (the 'imago', as Sigmund Freud and others put it) that her later selves echoed or emulated or sometimes failed to match. It was the standard against which, knowingly or otherwise, she measured herself – and against which Hockney measured his other depictions of her.

Freeman's multifaceted being – the multiple transient selves that composed her – were the subject of a photographic portrait that Hockney made in 1983 (fig. 29). This montage of colour snapshots is one of an extensive series, begun the previous year, in which he amalgamated numerous photographs of a given subject – taken from differing angles – into a Cubist-style whole. A dozen or so snapshots of Freeman's head and upper body combine into a fractured image. Fragmentarily, she eyes the camera and gesticulates, seeming to have been caught mid-sentence.

The montages, which were originally made using Polaroids, provided a means of dismantling the single-point perspective of photography, which Hockney had long considered a false representation of the actual experience of seeing. Photography, he sardonically observed in 1984, 'is all right if you don't mind looking at the world from the point of view of a paralysed Cyclops – *for a split second*'.[24] The subjects of the montages varied widely. Among the most recognizable is *Pearblossom Hwy. 11–18th April 1986*, which distils a crossroads in the Mojave Desert into a mosaic of conflicting perspectives: juddering road, amplified signage, patchwork sky.[25] There were also depictions of friends – Christopher Isherwood and Don Bachardy, for example, captured in a montage that still hangs in Bachardy's Santa Monica home.[26] Here, the writer and artist are seated in parallel armchairs – just as in Hockney's 1968 painting of them – with Bachardy multiplied into a 'stop-motion' sequence of attitudes. Their friend, the writer Bob Holman, sits facing them. Hockney is the invisible fourth party in the scene, the 'axis' around which the image scatters and clusters like a nebula.

Freeman is her own nebula in the montage that Hockney made of her around the same time: an assortment of looks, gestures, attitudes.

There is a movie-star aura to the component images, amplified by their stark light and shade. Across the stuttering sequence, she seems to turn her head, to flick out a hand, and to draw it back in. And yet, just as in *Beverly Hills Housewife*, her elegance is partly a guise – a veneer. Her facade shines out, but there is a lurking sense of everything that the facade conceals. The fractured viewpoint itself suggests a more complex reality. Freeman used to tell her own photographic subjects not to smile, explaining that 'I prefer that they look inward, rather than out.'[27] In this respect, she was summoning the kind of veiled interiority that characterizes Hockney's portraits of her, where her public and private selves are in perpetual oscillation. In the photographic montage, numerous split-second views condense into a picture that manages to be single and multiple, unified and scattered, all at once.

Freeman's other great friendship in her later years was with Robert Wilson. He often accompanied her to Hockney's studio in the Hollywood Hills; there are various photographs in Freeman's archive of Hockney drawing Wilson from life. 'We used to go almost every time I was in LA to visit David', Wilson recounted shortly before his death in 2025. 'They were fun together. He adored Betty. She was very Californian – laid back, colourful, but also very strict.'[28]

Freeman had first come across Wilson's work in 1974 in Spoleto, when she and Franco Assetto saw a production of his operatic play *A Letter for Queen Victoria* (1974) in a subterranean theatre. 'You could say it was a revolutionary experience for me', she later wrote.[29] They met two years afterwards, in 1976, when Freeman went to photograph Wilson and Philip Glass at work on their new opera, *Einstein on the Beach*. The resulting image shows Wilson lying on the floor with Glass standing over him. 'I had noticed that Wilson often stretched his long Texas frame out on the floor to think', Freeman recorded. 'I guess he was surprised when I asked him to lie down for a photograph because he laughed so hard that Glass came running over and laughed also.'[30]

In later years, Wilson stayed regularly at Hillcrest Road. He designed furniture for the dining room, and a bronze-and-marble chair – the *Betty Freeman Chair* (1986) – for the patio. The pair also created an unpublished book, 'A Faxing Friendship: Robert Wilson + Betty Freeman' (2008), reproducing Wilson's 561 faxes to her over the years. These elaborate hand-drawn designs were miniature works of art. One took the form of a poem in praise of Betty, the letters of her name spreading across the page in spidery capitals, each subtended by a column of stuttering professions:

B [...] IS FOR
IS FOR MY
IS FOR MY BEST
IS FOR MY BEST FRIEND

In her introduction to the book, Freeman made clear that these communications were a sort of emotional live wire, connecting the two of them wherever they happened to be. 'Each fax makes me feel how he is – is he loose and free, or heavily burdened, or swooping up and down in his ever-changing world[?]'[31] Expressing a telepathic connection, the faxes were equivalent – in their playful way – to Hockney's representations of Freeman. His numerous pictures of her are representations, in the end, not merely of her complex and changeable persona, but of his own love and understanding.

Who cares what the name was?

In her unpublished book on the art of Sam Francis, Freeman wrote that the artist's *White Paintings* are 'wrapped in a shroud of remoteness'.[32] In *Beverly Hills Housewife*, there is a hint of Freeman herself being similarly enveloped. It is a portrait of her in a state of separateness – stepping outside the current of her own life. There is a suggestion that she does not quite belong in this place – as though she were experiencing a fleeting sensation of estrangement, a pang of *jamais vu*. She might almost have been drafted in from another scene. Sometimes, in this period, Hockney's figures *were* translated into a painting from a different context, for instance in *Peter Getting Out of Nick's Pool* (1966): the image of Peter Schlesinger pulling himself from the water was actually posed against the back of a car, resulting in a lingering sense of disjunction between figure and setting in the final painting.

Freeman could be standoffish. She had a reserve about her, especially with those she did not know. The musicales, despite the large numbers who attended, were strictly for *her* people. 'You really had to have a strong desire to learn more about new music and be willing to commit to coming regularly to get an invitation and be invited back', according to Pamela Kort.[33] If someone turned up late, Freeman would give them the cold shoulder. And those who fell out with her found it hard to get back into her affections. 'Betty was a bit prickly', admits Stephanie Barron. 'She was opinionated, and I don't think she enjoyed the rigmarole of the art world. She considered herself apart from that.'[34]

While Freeman is remembered with love and loyalty by her children, she was not maternal. Her first daughter Shelley has only one memory from childhood of her mother crying, after an argument with Stanley, and suggests that Betty's emotional reserve was the consequence of her own father's domineering character: 'He was so autocratic, and so strong in her life, that she formed this defence around her in order not to feel pain.'[35] In her later years, Freeman's artist and musician friends occupied a more central place in her affections than her family. 'She was a true Gemini, which means two different people', Shelley remarks. 'To her friends, she was the most charming, generous person. As a mother, she wasn't so great, but that's because she was so involved in what she was doing.'[36] Thomas Adès recalls with amusement an occasion when Freeman's second daughter Claudia brought her young son to Hillcrest Road: 'Betty wouldn't let them in the house because she didn't believe that small children and art were compatible. She just came out and stood on the sidewalk with her grandson. It was magnificent.'[37] What might have seemed callous in another person was simply an expression of her steadfast sense of propriety – and her utter lack of sentimentality.

It was the same quality that Virgil Thomson had observed while staying with Freeman and Franco Assetto in Corfu in 1981. Afterwards, when she told Thomson that she liked the sketchlike musical portrait that he had made of her, but not its title, 'Dead Pan', he protested that it was simply a description of someone who did not show emotion, like Greta Garbo. 'When I read that, I felt that he was teasing rather maliciously and it would be wise to drop the matter right there', Freeman admitted in her private memoir of the holiday.[38] But Thomson clearly took note of her displeasure. When the musical portrait was turned into an orchestral piece for performance in 1982, he retitled it 'A Love Scene' and omitted the identification with Freeman.

Her emotional side was inseparable, perhaps, from the cerebral. Speaking in 1990 of the origins of her tastes, she remarked: 'I discovered that there is a passion of the intellect that attracts me to music as it does to abstract art.' Her 'intellectual passion' for abstract painting found a parallel in her attraction to minimalist music, a genre that she described as 'very regulated, very idea orientated'.[39] The representational style of *Beverly Hills Housewife* might have appeared antithetical to the kind of art that she tended to collect, especially in the 1950s and 1960s. And yet the mood of Hockney's work was itself abstract – indeterminate – in precisely the way that she appreciated. She preferred things to be implicit. After watching the 1990 film *Henry and June* (the story of Anaïs Nin's relationship with Henry Miller and his wife June), she complained about how boring the sex scenes were, precisely because they were so overt.[40] She had a sense of humour, but it was oblique and self-effacing. 'When you told a joke, she would never get it', according to her daughter-in-law Fanny. 'She didn't know she was funny.'[41]

What struck some people as detachment was really a sort of cosmic nonchalance, a freedom from any need to ingratiate herself. She was glamorous, but not in the high-society manner. She could never have been a 'swan' of Truman Capote's – she lacked the vanity. It was this very quality of insensibility that Hockney had perceived in 1966, and conveyed in his painting of her. *Beverly Hills Housewife* invites us to attend to her idiosyncratic glamour. In the painting, as in life, she has an allure that resides not only in her attire and blondeness, but in her apparent indifference – the faculty that some people misread as airiness. Her glamour, such as it was, lay in her insouciance.

For all her reserve, she could be cuttingly direct. Julia Bloomfield, on visits to Hillcrest Road in the 1990s, used to find herself subject to Freeman's critical gaze before she had even set foot in the house:

> Betty would always have the front door open, and would be sitting on the banquette in the entrance hall – right opposite the front door. I'd be walking down the path towards the front door, and immediately

> I could see her assessing what I was wearing, from the moment I'd got out of the car. As soon as I got in the house, she'd comment: 'You look like a farmer, Julia. You can't wear that.'[42]

The architect Frank Gehry, although a good friend, found her intimidating. As Robert Wilson recounted: 'She was good friends with Frank. But for her, Mies van der Rohe was the great classical architect. That "nonsense" of Gehry's made no sense to her.'[43] She was equally damning when it came to certain titans of the music world, among them Leonard Bernstein, whom she dismissed as 'the leading example of what is wrong with American music'.

> I don't think he is a very good conductor, and he is a very mediocre composer except for *West Side Story*. [...] He has been idealized by the concert going ladies, who go to the matinees. He is a lousy conductor, he is embarrassing with all those jumps and motions.[44]

Her judgments on the music that she had commissioned could be just as brutal. The only commission that she regarded as truly worthwhile was Mel Powell's *Duplicates*, a concerto for two pianos, which won a Pulitzer Prize in 1990. She confessed that she regarded the vast majority of modern classical music as academic and trivial.[45] (In old age, she discovered Post-it notes, and would often attach them to CDs that she had been given, annotated with summary judgments such as 'Boring'.) But she had no regrets; the point was that one performance in a hundred might be brilliant.

'So often she would say exactly the wrong thing to somebody – but on purpose', says Peter Sellars. 'She delighted in that. Most people in the music world are extremely intent on currying favour or impressing people or saying the right thing. Betty never had a second for that.'[46] As strong as her opinions were, she never felt the need to impose them. As Adès remembers: 'The rule was: let everyone do what they want. It was very much the way she thought.'[47]

In the end, her reserve – which some people perceived as coolness – was a facet of her uncompromising nature. She did not indulge in niceties for their own sake. Reserve was also a form of self-effacement. She did not want to be the centre of attention, did not need to be. This had been the reason for her original suggestion to Hockney that he title his painting *Beverly Hills Housewife*. Years later, when John Cage telephoned her to say that he was titling the *Freeman Études* after her, she replied: 'Oh, that is very nice, but I have never used my name in connection with anything. I am always a lender in the art world, anonymous, I never used my name. It is arrogant.' Cage managed to persuade her by suggesting that he might instead dedicate the studies to Henry David Thoreau; Freeman retorted: 'No, no, don't take it away, keep it, keep it!'[48]

What mattered, always, was the music that she had helped to bring to fruition – not her own role in the process. In this respect, her personal philosophy was close to that of her old friend Harry Partch:

> I would choose to be anonymous. I'm thinking of those fantastic cave drawings in southern France and northern Spain [...] there's no *author* there, but what a treasure they are! [...] I'm not saying that anything I'm going to do is going to last that long, but who cares what the name was?[49]

An accidental photographer

> 'She now carries a business card that reads, "Betty Freeman, Girl Photographer".'
>
> Connoisseur, *February 1987*[50]

She never meant to become a photographer. For most of her life, she possessed nothing more than a box camera, which she would use to take family photos. It was in the early 1970s, during the filming of *The Dreamer That Remains*, that Freeman found herself making photographic portraits. Harry Partch's assistant, Danlee Mitchell, had asked her to take some production shots: 'I put a camera in Betty's hands. [...] She had no idea what a camera was. So I just said you advance the film with your thumb and you press the button.'[51]

She began taking photographs of composers and conductors, often developing the pictures in a darkroom set up at Franco Assetto's home in Turin. She studied with a number of eminent photographers – Ansel Adams, Cole Weston, and Bill Owens among them. In 1985, she made her public debut with an exhibition at the Otis/Parsons Gallery in Los Angeles. Over time, her repertoire of subjects expanded to dancers, writers, and directors.

The random event of Mitchell handing her that Nikon camera – as incidental, in its way, as La Monte Young's appeal for assistance in 1961 – marked the beginning of a life project. Over the next three decades, Freeman produced a substantial corpus of photographs, the vast majority of them recording the composers and musicians with whom she worked.[52] She developed a theory that the more she liked a composer's work, the better her portrait would be. 'Somebody I am crazy about, like Alfred Brendel, I can get the best photographs', she observed. 'I cannot explain it. It has to do with intensity.'[53]

In this regard, her photographs were pictures of her own appreciation of a person, akin to David Hockney's portrayals of her. Peter Sellars, whose photograph Freeman took on several occasions, goes so far as to suggest that she was a subliminal presence in every one of her pictures. 'You can see in people's eyes that the person they're looking at is her. You are looking at Betty in that photograph.'[54]

The pictures she took of Austrian pianist Brendel in 1988 are among her most powerful. She had tried to take his picture on three former occasions – and never successfully, owing to Brendel's peculiarly unphotogenic appearance (bulging eyes, thick glasses, and buck teeth). But following a performance at the Schoenberg Institute at the University of Southern California, she was able to seize her opportunity. He had just played Schubert's late sonatas to a small audience, and was giving a talk. Freeman captured him in a moment of repose, his expression forlorn. Her six shots of him beside the piano are their own 'man of sorrows' cycle.

'I find that I prefer to photograph older men whose character and history is already written in their faces and gestures; younger men, as well as women, seem more reluctant to expose themselves to the camera', she wrote in the catalogue for her 1990 exhibition *Music People & Others*,

Alfred Brendel, Los Angeles, 1988, photograph by Betty Freeman

which travelled between the Berlin and Cologne Philharmonics that year. 'Unfortunately, the small number of photographs of women is simply because women composers are still rare.'[55]

According to Sellars, once you got to know Betty, you would be allowed upstairs at Hillcrest Road – where pile upon pile of her photographs could be found.

> You got to peer into her universe as a photographer,
> and all the people she was taking these candid pictures of.
> Nobody could say no to her. So you had all these people
> who *always* said no suddenly semi-smiling for a photograph.
> The hardcore European avant-garde would smile for her![56]

Curtain call

In 1991, Franco Assetto died in Turin, the place that had become his and Freeman's second home. He had suffered complications during planned surgery for a hernia, likely related to the anaesthetic, which resulted in a mental breakdown. He had lost most of his memory and required round-the-clock care. Freeman stayed in Italy and tried to take care of him, but returned to Los Angeles once it was clear that the situation was hopeless. 'She just could not deal with it', remembers her daughter Shelley.[57]

After Assetto's death, the musicales changed. Something vital had been lost. The concerts had arisen out of the happiness and confidence that Freeman's marriage to Assetto had brought her. In his absence, it was inevitable that she should decide to stop hosting them. Ten years after starting the 'Salotti Musicali', as she had come to name them, she hosted her fifty-ninth and final concert in October 1991, with the Los Angeles mayor in attendance.[58] She passed the role of host on to a long-time attendee, Judith Rosen, who continued the programme at her own home for another four years.

Over the next two decades, Freeman maintained a frenetic existence. She travelled each year to the Salzburg Festival. In Los Angeles, she continued to commission and sponsor new music, and worked on plans with Alan Rich for a book (never realized) on the musicales. She was constantly engaged. 'She was like a teenager, going out every night,' says her daughter-in-law Fanny. 'I have a memory of coming upstairs to see her doing her make-up before she went out. I didn't know what music she was listening to, but it was unbearable.'[59]

Freeman's final years were a testament to her stoicism. By 2006, she knew that she was terminally ill, but she discussed her situation with few people. She attended to daily matters and sorted through her possessions. 'It was always about the present and the future', observes Julia Bloomfield. 'Always looking forward – that was her personality.'[60]

Hockney retained a central place in her affections. By the early 2000s, he was spending more time away from Los Angeles – particularly in Yorkshire, where he had begun to paint the landscapes of his native county. But they kept writing to one another, and whenever he was back in LA, he came to dinner parties at her home together with old friends such as Gregory Evans, Jack Larson, and the musician Lem Banks. In 2006, Freeman professed:

> He hasn't changed. He's still as generous, he's still as open-hearted, he's still as caring for his friends. His looks changed very little: he still looks always David to me. [...] He's even handsomer now than he was when he was young.[61]

The pink dress, which had hung in her wardrobe since the 1960s, continued to make occasional appearances. She put it on for a German television crew in 2004 (fig. 35), and again the following year when two friends of Hockney's – Gregory Evans and Charlie Scheips – came to record an interview with her.

In February 2006, she suffered an accident at home. As she related in a fax to Hockney: '[I] went to kill a cricket who was jumping around in my bathroom – and carelessly fell backward in the effort & broke my left rib.' She spent two weeks in hospital and was housebound at the time of writing, complaining that she could not attend the concerts of Thomas Adès – then composer-in-residence at the Los Angeles Philharmonic. 'The good news', she added, 'is that the cricket survived and I vowed never to kill one or try again to do so.'[62]

Betty Freeman and David Hockney at Hillcrest Road, 2003, photographed by Philipp Lachenmann

In spite of her injury, she was a regular visitor that summer to the Los Angeles County Museum of Art, where *Beverly Hills Housewife* was on display as part of Hockney's touring retrospective of portraits. At a preview of the exhibition, Larson pushed her around in a wheelchair. One gallery contained the series of gouache portraits that Hockney had made of attendants at the National Gallery, London, in 1999 and 2000, with the assistance of a camera lucida. 'This for me is David's real tour de force', she told a reporter. 'Every one is a human being, and every one is astonishingly individual. They're empathy paintings.'[63]

Freeman downplayed her final illness. She had been diagnosed with pancreatic cancer, but few friends were aware. 'She knew she was dying quite early on, and she didn't want to have any treatment', according to Bloomfield. 'She slowly went through that large house, getting rid of stuff – she was very methodical.'[64]

Adès has a vivid memory of the last time he saw Freeman. She was confined to a wheelchair and being cared for at home by a full-time nurse. Hillcrest Road, for so long an extension of her personality, had become a sort of prison. 'She said, "Here's what's going to happen. You're going to put me in the car, drive me to Venice, and walk me along the boardwalk."

Venice had been one of her favourite places since the 1960s, when the seaside community had been rougher but alive with creativity. Later that day, as Adès wheeled her along the beach, he comprehended the full significance of their voyage out. It had been a final bid for freedom. Freeman would never see the ocean again, let alone most of the city. 'She had a look of complete ecstasy – she was so happy.'[65]

Or as happy as she could be. Adès remembers that she let rip about many of the people who had disappointed her in her life. But it felt more like a catharsis than a rant – the letting go of decades-long resentments.

Three weeks before his own death in July 2025, Robert Wilson recalled his final telephone call with Freeman. 'I was in Frankfurt and I talked to her a couple of days before she died. She said: "Bob, I'm dying. I'll never see you again." I said to her: "Betty, fly with the angels."'[66]

She died at home on 3 January 2009.

Shortly before her death, Freeman had deposited a large archive of papers and photographs at the University of California, San Diego. Her papers contain the numerous faxes that she sent to Hockney over the years. These usually took the form of handwritten letters. On 26 January 1999, on the eve of his major exhibition at the Centre Georges Pompidou, Paris, she wrote:

> I am thinking of you especially for your opening at the Pompidou.
> How well I remember your show at the Musée des Arts Decoratifs
> and can picture how the paintings looked there. So many shows,
> David, and so many paintings always more and more wonderful.
> I'm proud to know you and happy to love you –
> Ever your devoted friend,
> Betty

Three years later, on 19 October 2002, she wrote: 'I just want to let you know that I am missing you and Los Angeles is not the same without you here. This is not just words, I really do think to you and the big hole your absence has made.' And again, on his birthday, 9 July, the following summer: 'I think of you, love you, miss you – not only today but especially today. [...] You know that you get better-looking with every birthday so you never have to worry about having too many.'

The boxes in San Diego bear witness, among much else, to Freeman's enduring love for Hockney and her sustained admiration for his work. They are filled with exhibition invitations, catalogues, and programmes for performances that he had designed. 'You are a great theater magician', she wrote to him in a fax of 1997, 'and your Tristan is one of the most beautiful experiences of my life.'[67]

One of the programmes that she kept was for Giacomo Puccini's *Turandot*, staged at the San Francisco Opera in 1993. Reproduced on the cover is an image of Hockney's stage design, a blue-lit fantasy land.[68] Silhouetted trees, dangling their tendrils like giant fingers, act as an inbuilt proscenium arch around the scene of a palace garden in Peking: a flight of steps, lanterns, a chorus of acolytes. Like many of the artist's pictures, the design offers a portal into a luminous otherworld. But it also withholds that world – keeps it at a remove, behind the frame of the trees, forever untouchable. The device of the trees is comparable to the borders that Hockney included in several of his paintings of the mid-1960s – those sections of raw canvas that act as barriers between the domain of the viewer and the fictive space of the picture.

That blue vignette on the cover of the *Turandot* programme is a reminder of what Freeman had once represented for Hockney – and he for her. The wonderment of the young artist, in Freeman's eyes, had been that he came from a place unlike her own – one that she could never fully enter. They recognized and cherished the specialness or even the strangeness in one another.

In *Beverly Hills Housewife*, the artist and his muse stand slightly apart. One is visible, the other an implicit presence. Each inhabits a distinct space, and yet each perceives the presence of the other across a short divide. The degree of distance between them may have accounted, paradoxically, for their closeness over the years. It is possible, even, that Freeman divined in the beguiling young artist a reflection of her own difference, her own determined otherness, that day when he came to visit in 1966. In *Beverly Hills Housewife*, she remains – however fractionally – a stranger to the world around her, and to herself.

28 Programme for *Turandot*,
San Francisco Opera, 1993

29 David Hockney, *Betty Freeman Los Angeles*, 1983. Photographic collage, 25.4 × 30.5 cm (10 × 12 in.)

30 David Hockney, *Betty Freeman, April 21 1991*, from *112 L.A. Visitors*, 1990–1
Colour laser-printed still video portrait, *c.* 57.2 × 76.2 cm (22½ × 30 in.)

31 David Hockney, *Betty Freeman, February 1, 1994*, 1994.
Pencil on paper, 55.9 × 76.2 cm (22 × 30 in.)

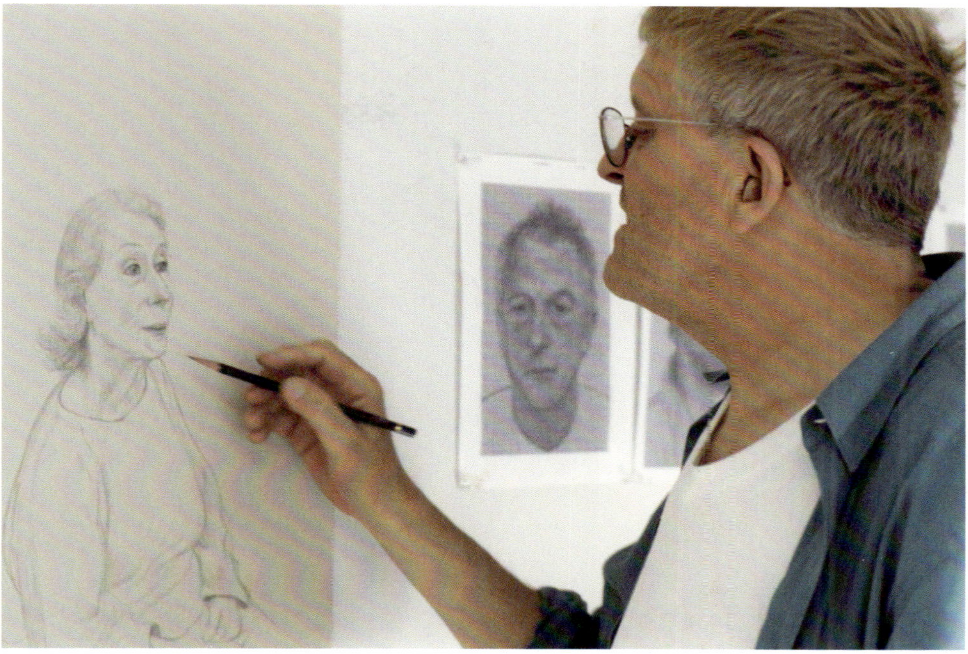

32 David Hockney drawing Betty Freeman, 20 July 1999.

33 Slim Aarons, *Poolside Pairs*, 1970

34 Publicity image for Season 13 of *Real Housewives of Beverly Hills*, 2023

35 Betty Freeman, 14 March 2004, with a German television crew for the filming of a documentary

Afterword

The Original Real Housewife

Beverly Hills Housewife was sold at auction, alongside numerous other works from Betty Freeman's collection, in May 2009. It achieved a record $7.9 million, a testament to its canonicity not only within Hockney's corpus, but in the history of twentieth-century art. The painting marked the onset of his high realist style, while reflecting a more private phase of becoming – a phase in which love, friendship, and the infinite possibility of Los Angeles might all be found embodied in a simple garden scene.

In the judgment of John Kasmin, Hockney's original dealer, the painting is not really a portrait at all.

> It's a picture of a collector in a setting. It's called *Beverly Hills Housewife*, it's not called *Betty*. It's a picture of something we didn't have in England, this lady who's a collector and her husband who collected animal heads. She looks slightly goofy, standing in front of a Turnbull. That's how she was, all the time. Daydreaming.[1]

In some respects, Freeman was the product of a bygone era. She was always on a plane of her own, slightly apart from the world. Sequestered by her wealth and her home, she was defined by both detachment and engagement, breezily indifferent to what did not concern her, and yet deeply committed to the people and the things she believed in. In another way, she was ahead of her time – pursuing her own course, breaking through the constraints of middle-class expectation to become a singular kind of impresario. Her upbringing had afforded her an independence of spirit that never left her, even during her first marriage. Hockney's painting, and her tongue-in-cheek naming of it, amounted to a re-examination of what it meant to be a rich housewife – a consideration of the way in which such an identity might emancipate rather than constrain.

Beverly Hills Housewife is a snapshot of a moment: a woman glimpsed in a given time and place. Even so, the vision of Freeman coming into the garden has longer historical reverberations, invoking portrayals of other female patrons in far-flung times and places: Titian's *Portrait of Isabella d'Este* (1536), John Singer Sargent's *Portrait of Isabella Stewart Gardner* (1888), fashion photographer Horst P. Horst's 1966 portrait of Gloria Vanderbilt for *Vogue*, or – at a more satirical level – Cindy Sherman playing the part of a formidable Manhattan lady in her 'Society Portraits' (2008). Hockney's picture not only ranges outwards, to the time and place of its creation, but backwards. Whether consciously or subliminally, it summons a larger iconography.

In the decades since it was painted, the force of the picture has also carried forward. The formula of a female subject envisioned in a sumptuous domestic space has retained a powerful hold in art and wider culture – and has helped to enshrine a particular 'LA look'. In 1970, the photographer Slim Aarons staged a photo shoot at the Richard Neutra-designed Kaufmann Desert House in Palm Springs, entitled 'Poolside Story'. The colour shots feature the fashion model Helen Dzo Dzo Kaptur and socialite Nelda Linsk in lounge suits (white lace and buttercup yellow), seated or standing beside the glassy swimming pool. One picture from the group, *Poolside Pairs*, shows the women reclining on loungers in the foreground, the pool and verdant lawns and house receding like a stage set against the distant backdrop of the San Jacinto Mountains (fig. 33). (At the far end of the pool, another pair of women stand, seemingly mid-conversation.)

In 2000, the fashion photographer Steven Meisel photographed two models, Georgina Grenville and Amber Valletta, for a Versace campaign entitled 'Four Days in LA'. In a striking move for the time, the pictures portrayed fashionable 'older' women rather than teenage models. Grenville and Valletta, made up to resemble near-identical society ladies on the cusp of middle age, stare out at the camera from a parade of luxurious interiors. Their feathered blow-dried hairstyles and kaleidoscopic array of outfits (fur coats, miniskirts, ball gowns) express a homage to – an arch pastiche of – the wealthy housewives of Los Angeles in the 1960s. Their expressions are deadeningly blank, radiating a sort

Gloria Vanderbilt at home, wearing a white lace dress designed by Mainbocher, 1966, photograph by Horst P. Horst for *Vogue*

of Diazepam-induced tranquillity. Whether or not Meisel was making a conscious allusion to Hockney's painting, the influence of *Beverly Hills Housewife* permeates the shoot – incidental but also inescapable.

Suffused by nostalgia for the era of *Beverly Hills Housewife* and Slim Aarons' 'Poolside Story', Meisel's photo shoot might be seen as the 'missing link' between Hockney's painting and the most significant pop-cultural treatment of Beverly Hills in recent years. For many people, the very phrase *Beverly Hills Housewife* is synonymous with the hit reality TV show *The Real Housewives of Beverly Hills*. First aired in 2010, the series charts the lives of a group of affluent women living in Los Angeles – their marriages, their extravagant homes, their shifting friendships and fallings-out (fig. 34).

Hockney's painting was not a conscious point of reference for the creators of the *Real Housewives* franchise, which began in 2006 with *The Real Housewives of Orange County*, and has since multiplied into eleven concurrent television shows across the United States (ranging from *The Real Housewives of New York* to *The Real Housewives of Salt Lake City*). The intention was to produce a real-life version of soap operas such as *Peyton Place* (1964–9) and *Desperate Housewives* (2004–12). And yet, like Hockney's painting, those shows had revolved around the figure of the upmarket housewife, variously reinforcing and subverting the archetype.

When Freeman posed for Hockney, she was playing with her image – aligning herself with a type: 'There's only one thing you can call this painting, since I'm dusting: *Beverly Hills Housewife*.' In posing for the painting, in giving it that title, she cast herself – allowed Hockney to cast her – as a character. Not a lie, but not the whole truth either. Far from being a passive muse or a compliant subject, she was an active participant, the self-conscious protagonist of the scene. This can be seen from the Polaroids that Hockney took in her garden, those small black-and-white snapshots on which the painting was based. Freeman's faint smile gives the game away. She sees herself being seen (or as John Berger put it, she 'turns herself into an object – and most particularly an object of vision'), but there is a cool irony about it all, a latent humour that punctures the pretence of an 'ordinary' middle-class woman at home.[4]

Both in the Polaroids and the final painting, Freeman steps into her image – becomes herself. She is in command, and yet she appears to hesitate fractionally, as if holding back the very moment of her grand entrance. The power of Hockney's portrayal rests on this ambiguity. There is a double sense of Freeman seizing the moment and standing apart from herself, eyeing herself with a certain droll scepticism.

As a painting about a woman's relationship with her own image, *Beverly Hills Housewife* is intriguingly close in spirit to the series *The Real Housewives of Beverly Hills*. Across its sixteen seasons and multiple cast changes, the show has presented women undergoing the same equivocations as Freeman appears to in the portrait – playing particular versions of themselves, but equally breaking out of the roles that are expected of them. According to Andy Cohen, executive producer of the *Real Housewives* franchise: 'There's no show that has given a platform for women over fifty in this way, in terms of expressing their sexuality and who they are and starting over in life and figuring things out. And I think that's brilliant.'[3]

What would a contemporary successor to Betty Freeman look like? Crystal Kung Minkoff is a television personality and entrepreneur in her forties, best known for appearing in three seasons of *The Real Housewives of Beverly Hills*. She joined the show in 2021, the first Asian-American to be featured, and soon distinguished herself as a cool-headed – somewhat understated – counterpoint to other cast members. Over coffee at the Beverly Hills Hotel, Minkoff admits to having worried that she would not cut it as a Beverly Hills housewife, despite having many of the credentials (she is rich, beautiful, and married to a Hollywood director).

> I was nervous that I was going to feel fraudulent. Or not measure up. I didn't grow up wealthy – I grew up in the Valley: middle class, comfortable, a very standard track. But ultimately, I actually think I'm *most* like the classic Beverly Hills housewife. Alex Baskin, who created the show said, 'You're the epitome – you're not loud, you're not messy.'[4]

Her very composure and restraint made her more a 'real' Beverly Hills housewife than any of her counterparts. Despite this, she was far from immune to the tensions and squabbles that make the show such compulsive viewing. One of her most notorious scenes involved a row with Sutton Stracke, a divorcee from South Carolina, in the garden of one of the other wives – the culmination of a long-building hostility. 'You're upset because you're jealous, period', Minkoff tells Stracke, as the quarrel comes to a head. 'Jealous of *what*?' Sutton cries back in fury. 'Your ugly leather pants?' Minkoff does not reply – simply smiles, in a glancing shot that frames her against the sunlit garden. The scene has become one of the classic, serially replayed confrontations of the entire series.

A criticism that is regularly made of *The Real Housewives* is that much of it is bewilderingly – or just mind-numbingly – banal. Daily lives, not least those of rich women in Beverly Hills, contain much that is trivial or unspectacular. But in this respect too, there is a curious parallel between the TV show and Hockney's picture, in which nothing very much is happening – or rather, in which the woman herself is the event. The scene in *Beverly Hills Housewife* seems poised between the mundane and the momentous; Freeman might be stepping into her garden without a thought, or we might be witnessing an epiphany.

What renders *The Real Housewives* compelling is the capacity for situations to tip, in the same way, from the banal to the profound and back again. Seemingly ludicrous scenarios belie real, often painful currents of feeling. At its most soap-operatic, *The Real Housewives* has the capacity to deliver an emotional blow. It comes in a mere look or phrase. Petty arguments act as outlets for deeper, hard-to-define feelings. (There is a parallel, here, with Freeman's long and increasingly acrimonious correspondence with Clyfford Still: lurking inside her letters was an ambivalence about what she wanted to do with her life – who she wanted to be.) In *The Real Housewives*, the images the women project mask as much as they reveal.

Like Freeman (or more specifically, the Freeman of Hockney's portrait), those women are symbols as well as individuals. They embody a time and a place and a way of life. It is telling, perhaps, that the way that Freeman stands in the painting – erect and totemic, almost like an object – prefigures the opening shot of each episode of *The Real Housewives*, in which the cast members are seen standing in a line, decked out in ball gowns and holding giant diamonds that serve as emblems of Beverly Hills. 'You have to represent Beverly Hills – what people think about Beverly Hills', Minkoff explains.

And just as Freeman's house frames her in the painting, transforming her into an emblem of her lifestyle, the houses of the 'real housewives' play a defining role. 'LA is a "home" kind of city, and your home is a character.' Minkoff says. Certain objects, too, become totems. If Freeman found herself forever associated with the pink dress of the painting, then the leather pants that Minkoff wore on that notorious episode of the show have become an indelible part of her image. 'At the time, that whole fight was so painful', she admits. 'I keep the pants bundled up in drawer. I can't look at them. My husband is like, "You should put them in a glass box" – like they're a memento. He doesn't have the emotional connection.'

By calling herself a Beverly Hills housewife, Freeman was also saying, in a roundabout way, that she was not one – or not the cliché that the phrase implies. In posing for the painting in the first place, she simultaneously revealed herself (made herself an exhibit, even) and held up a veil. That double action is reflected in her very stance – in her appearing to turn towards the viewer and equally to turn away, as though retreating into a private dimension.

'Her being halfway – that's how I always felt on the show', remarks Minkoff, observing Freeman's attitude in the painting. 'Obviously the painting was prior to the show. But then, to live in Beverly Hills or Bel Air, a fancy part of LA – where you have a beautiful mid-century house and clothes and staff – you don't live that kind of life without in some sense exposing yourself. You can say all day long, "I'm a very private person", but you're exposing yourself for *something*.' Even if you are merely projecting an image of yourself back to yourself, she suggests, there is a subtle exhibitionism at work – an act of display that presupposes an audience. 'Why do I have to have a mohair couch? I could live with IKEA. It's for public consumption.'

For Minkoff, being on a reality TV show instilled a continual, heightened awareness of what viewers would be seeing. 'I was always on the outside, looking in. I remember the producers would put their hands on my back and tell me: "Go into the scene, you're on the show now." It was a created version of Beverly Hills.'

There was always a friction between the image that the show demanded and who she felt herself to be. But that friction was only a more overt, more dramatic version of what exists in the lives of many people – the tension between how others perceive us and how we perceive ourselves. Los Angeles, more than many places, is a city that magnifies the contrast. It is a place of spectacle and celebrity, but also a place where an individual can retreat into their own home – their own world. In Beverly Hills, as Minkoff says: 'You have the element of Hollywood plus reality. It's so much bigger – there's so much to live up to.' As a viewer, she suggests, 'You picture the person. You fall in love with this idea – but it's not what you think it is.' Hockney knew this, and expressed it nowhere more powerfully than in *Beverly Hills Housewife*, where the subject is by turns a figure on stage – an actor in her own drama – and a woman glimpsed in private, as though caught unawares.

Hanging in Freeman's house for forty years, *Beverly Hills Housewife* represented a specific instant in her life – for all that it came to stand for her *entire* self. Now that Minkoff is no longer on *The Real Housewives*, she looks at the show as an equivalent portrait of her life at a given time – a particular version of her, frozen in aspic. For a while, when she watched the episodes back, that portrait seemed excessively glamorized.

> While I was filming it and while it was airing, I really struggled with accepting it. That was the challenge: *that's not me*. The disconnect was annoying. But then my husband told me: 'Yes, you're very "Valley", you hold onto your Chinese roots, but *you're also fancy*. You don't need to reject that part of you.'

The character that the show presented was not so much false as fragmentary: one facet of a larger whole. 'The audience doesn't want to watch someone that *really* looks like herself', she concedes. 'They want to find glimpses. They want the best "Beverly Hills" version of me. It's a part of who we are – and that collectively becomes who we are as a group, and that collectively becomes what Beverly Hills is.'

Being a Beverly Hills housewife is about precisely this, now as much as in 1966. Standing apart from oneself, apprehending the gap that exists between image and reality – and accepting that life is lived in that gap. Minkoff's experience carries echoes, incidental yet precise, of the life of Betty Freeman. Freeman, too, was not a born Beverly Hills woman. She came from elsewhere. Hockney's portrait captured her at a transitional moment, on the cusp of middle age, much as *The Real Housewives* captured Minkoff in her early forties. In both cases, the portrayal was necessarily partial – not wholly real – and yet it became an indelible part of each woman's life, no longer simply a mirror image but an active force. 'I do feel like the show has made me take a new path in my life, that I wouldn't otherwise have taken', says Minkoff. 'But it's taken me a while to feel comfortable with that.' It is tempting to imagine that Hockney's *Beverly Hills Housewife* gave Betty Freeman a similar impetus. It reflected her back to herself, showing her not simply who she was, but who she might become.

Notes

CHAPTER ONE

Flight to Los Angeles

1 Joan Didion, *Play It As It Lays* (New York: Farrar, Straus, and Giroux, 1970), p. 24.

2 Stephanie Barron, interview with the author, 11 April 2025.

3 Bob Mizer, quoted in Daniel Wenger, 'The obsessive photographer behind America's first gay magazine', *New Yorker*, 22 September 2016: <https://www.newyorker.com/culture/photo-booth/the-obsessive-photographer-behind-americas-first-gay-magazine>, accessed 13 October 2025.

4 David Hockney, quoted in Marc Fumaroli, *David Hockney: Dessins et Gravures*, exh. cat. (Paris: Galerie Claude Bernard, 1975), n.p. See Edmund White, 'The lineaments of desire', in Sarah Howgate and Barbara Stern Shapiro (eds), *David Hockney Portraits*, exh. cat. (London: National Portrait Gallery, 2006), pp. 48–61, p. 53.

5 White, 2006, p. 53.

6 John Rechy, *City of Night* (New York: Grove Press, 1963), p. 120.

7 Carey McWilliams, *Southern California: An Island on the Land* (Santa Barbara, CA, and Salt Lake City, UT: Peregrine Smith, 1973 [1946]), p. 239.

8 David Hockney, *David Hockney by David Hockney: My Early Years*, ed. Nikos Stangos (London: Thames & Hudson, 1976), p. 93. Curiously, though, the 'fantasy' of the boy in the apron collided with the real-life model of Hockney's friend, Mo McDermott, painted while showering at Hockney's London flat.

9 Of Goodman, Hockney said: 'He was very American, with crew-cut hair. I didn't really think of him as a lover. We were just sexy friends.' David Hockney, interviewed in August 2010. Quoted in Christopher Simon Sykes, *Hockney: The Biography, Volume I, 1937–1975 – A Rake's Progress* (London: Century, 2011), p. 113.

10 Andy Warhol and Pat Hackett, *POPism: The Warhol '60s* (New York: Harcourt Brace Jovanovich, 1980), p. 24.

11 On the trip to the set of *Naked City*, see Peter Webb, *Portrait of David Hockney* (London: Paladin, 1990), p. 83, and Sykes, 2011, pp. 128–9. Warhol's account in *POPism* suggests a different television show: 'I'd met Dennis a couple of months earlier through Henry, on the same day that I'd introduced Henry to the young English painter David Hockney. Dennis brought one of my Mona Lisa paintings on the spot and then he and Henry and David and I went up to the sound stage on West 125th Street where Dennis was doing an episode on the TV show "The Defenders".' Warhol and Hackett, 1980, p. 36.

12 Wayne Koestenbaum, *Andy Warhol* (New York: Viking, 2001), pp. 101–2.

13 Andy Warhol, diary entry for Thursday, 26 March 1981. Reproduced in Andy Warhol, *The Andy Warhol Diaries*, ed. Pat Hackett (New York: Warner Books, 1989), p. 366.

14 David Hockney, quoted in Mark Glazebrook, 'David Hockney: An interview', *David Hockney: Paintings, Prints and Drawings, 1960–1970*, exh. cat. (London: Whitechapel Art Gallery, 1970), pp. 8–15, p. 11.

15 Oliver Andrews, oral history conducted

by George M. Goodwin for UCLA Center for Oral History Research, 11 November 1976. <https://oralhistory.library.ucla.edu/catalog/21198-zz0008z90j>, accessed 4 February 2026.

16 Webb, 1990, pp. 83–5; Sykes, 2011, pp. 143–5.

17 Oliver Andrews, oral history conducted by George M. Goodwin for UCLA Center for Oral History Research, 11 November 1976.

18 Rechy, 1963, p. 141.

19 Hockney, 1976, p. 99.

20 John Kasmin, interview with the author, 29 July 2024.

21 John Kasmin, interview with Peter Webb, 14 January 1990. Quoted in Webb, 1990, p. 45.

22 John Kasmin, interview with the author, 29 July 2024.

23 John Kasmin, interview with the author, 29 July 2024.

24 David Hockney, interviewed in September 2010. Quoted in Sykes, 2011, p. 150.

25 Hockney, 1976, p. 68. Quoted in Webb, 1990, p. 85.

26 Alan Hollinghurst, *Isherwood is at Santa Monica* (Oxford: Sycamore Press, 1975).

27 David Hockney, quoted in *Hockney* (dir. Randall Wright, 2014).

28 Hockney, 1976, p. 98.

29 Susan King, 'Adventures with Bridges', *Los Angeles Times*, 6 July 2011: <https://www.latimes.com/archives/la-xpm-2011-jul-06-la-et-jack-larson-20110706-story.html>, accessed 13 October 2025.

30 Christopher Isherwood, letter to Don Bachardy, 26 October 1961; Don Bachardy, letter to Christopher Isherwood, 2 June 1967. See Katherine Bucknell (ed.), *The Animals: Love Letters between Christopher Isherwood and Don Bachardy* (London: Vintage Books, 2014), p. 88, p. 276.

31 Christopher Isherwood, *A Single Man* (London: Vintage Books, 2010 [1964]), p. 98.

32 John Kasmin, interview with the author, 29 July 2024

33 Marco Livingstone has pointed out that the woman's pose and situation loosely resemble that of the figure beneath a loggia in Giotto's scene of the *Annunciation to St Anne* in the Scrovegni Chapel. See Marco Livingstone, *David Hockney*, fourth edition (London: Thames & Hudson, 2017), p. 67.

34 David Hockney, interviewed in September 2010. Quoted in Sykes, 2011, p. 155.

35 Derek Boshier, interviewed in September 2010. Quoted in Sykes, 2011, p. 156.

36 Stuart Preston, 'A square world; Albers' geometrics English Pop; Liberman and Marca-Relli', *New York Times*, 4 October 1964, p. 21.

37 Hockney, 1976, p. 101.

38 Ibid., p. 103.

39 Webb, 1990, p. 102.

40 Quoted in Livingstone, 2017, p. 84.

41 Jack Larson, interviewed in September 2010. Quoted in Sykes, 2011, p. 181.

42 Hockney, 1976, p. 151.

43 Betty Freeman, video interview with Charlie Scheips and Gregory Evans, 2005, courtesy of the David Hockney Foundation. *Ubu Roi* ran at the Royal Court from 21 July to 27 August 1966, directed by Ian Cuthbertson and starring Max Wall as Père Ubu. *Drawings for Ubu Roi and Cavafy Etchings* opened at the Kasmin Gallery, London, on 21 July 1966. Freeman misremembered the play and exhibition

as having been earlier, around 1963, but stated that she saw both before meeting Hockney. She had also seen a production of *Ubu Roi* in the spring of 1965 at UCLA, directed by Marc Estrin. Betty Freeman, letter to Sam Francis, 10 May 1964, Sam Francis Papers, Getty Research Institute, Los Angeles.

44 Peter Schlesinger, interview with the author, 20 July 2023.

45 John Kasmin, interviewed in August 2010. Quoted in Sykes, 2011, p. 152. In 2005, Freeman recalled her first meeting with Hockney as having taken place in 1964, which would have been during the artist's first stay in Los Angeles: Betty Freeman, video interview with Charlie Scheips and Gregory Evans, 2005, courtesy of the David Hockney Foundation. It is possible, however, that they did not meet until 1966. According to Peter Webb, 'Hockney had been introduced to Betty Freeman by Jack Larson, and she remembers that he asked if he could paint her on her terrace *very soon after they met*'. Webb, 1990, p. 104 (my italics). By this account, the visit to Hillcrest Road in 1966 followed quickly upon their first meeting.

46 Betty Freeman, video interview with Charlie Scheips and Gregory Evans, 2005, courtesy of the David Hockney Foundation.

47 Stephanie Barron, interview with the author, 11 April 2025.

48 Peter Schlesinger, interview with the author, 20 July 2023.

49 Betty Freeman, video interview with Charlie Scheips and Gregory Evans, 2005, courtesy of the David Hockney Foundation. It is worth noting that Schlesinger has no memory of swimming in the pool that day. Email to the author, 3 October 2023.

50 Betty Freeman, video interview with Charlie Scheips and Gregory Evans, 2005, courtesy of the David Hockney Foundation.

CHAPTER TWO

Mrs Stanley Freeman

1 Norma Desmond, played by Gloria Swanson, in *Sunset Boulevard* (dir. Billy Wilder, 1950).

2 Robert Isadore Wishnick (1892–1980); Freda Monica Wishnick (née Frankel) (1893–1978). See Alfred E. Clark, 'Robert Wishnick, 87; founder and executive of chemical company; helped civic groups', *New York Times*, 15 February 1980, Section B, p. 18. Freeman had two younger brothers: John, born in 1922, and William, born in 1924.

3 Irving Cutler, *The Jews of Chicago* (Chicago: University of Illinois Press, 1996), pp. 56–65.

4 See William Wishnick, *The Witco Story* (New York: Newcomen Society in North America, 1976), pp. 10–14; and William Wishnick, interview by James J. Bohning at Witco Corporation, New York, 13 May 1991 (Philadelphia: Chemical Heritage Foundation, Oral History Transcript # 0093, p. 1–5). The latter source states that Robert was nine, rather than seven, when he broke his arm.

5 Ibid., p. 1.

6 *The Badger*, Vol. 30, 1916 (yearbook published by the junior class of the University of Wisconsin), p. 114.

7 Barbara Jepson, 'A cultivated ear: Betty Freeman's living room is the West Coast's center for new music', *Connoisseur*, February 1987, pp. 107–11, p. 108.

8 Betty Freeman, quoted in David Kopplin, 'Patron and photographer Betty Freeman: Where would modern music be without her?', *Performing Arts*, October 1998, pp. 18–21, p. 21.
9 Ralph P. Locke, 'The "Grand Composers" of the present day: Betty Freeman discusses how she chooses and supports them', 28 September 1991, reproduced in Ralph P. Locke and Cyrilla Barr (eds), *Cultivating Music in America: Women Patrons and Activists since 1860* (Berkeley, CA: University of California Press, 1997), pp. 59–64, p. 62.
10 Frank Lloyd Wright, *An Autobiography*, second edition (London: Faber & Faber, 1945), p. 99.
11 Hannah Green, 'Mister Nabokov', *New Yorker*, 6 February 1977: <https://www.newyorker.com/magazine/1977/02/14/mister-nabokov>, accessed 14 October 2025.
12 Victor Zuckerkandl, *Sound and Symbol: Music and the External World*, trans. Willard R. Trask (Princeton: Princeton University Press, 1956), p. 235.
13 Betty Freeman, quoted in Locke, 1991, p. 62.
14 'Betty Freeman's address to the California Arts Council, California Institute of Technology, Pasadena', 27 January 1983, revised April 2003. Typescript in Los Angeles Philharmonic Archives.
15 Shelley Freeman, interview with the author, 14 January 2026.
16 Betty Freeman, quoted in Locke, 1991, p. 62.
17 Peter Goulds, interview with the author, 23 May 2024.
18 Shirley Blum Neilson, interview with Hunter Drohojowska-Philp, 22 May 2008. Quoted in Hunter Drohojowska-Philp, *Rebels in Paradise: The Los Angeles Art Scene and the 1960s* (New York: Henry Holt and Company, 2011), p. 59.
19 Walter Hopps, interviewed 4 October 1975. Quoted in Betty Turnbull, *The Last Time I Saw Ferus* (Newport Beach, CA: Newport Harbor Art Museum, 1976), n.p.
20 Ed Ruscha, 'Introduction: This crazy but genius Walter Hopps', in Walter Hopps, *The Dream Colony: A Life in Art*, ed. Deborah Treisman, from interviews with Anne Doran (New York: Bloomsbury, 2017), pp. xv–xix, p. xviii.
21 Betty M. Asher, oral history conducted by Thomas H. Garver for the Archives of American Art, Smithsonian Institution, Washington DC, 30 June and 7 July 1980.
22 Irving Blum, oral history conducted by Paul Cummings for the Archives of American Art, Smithsonian Institution, Washington DC, 31 May to 31 June 1977.
23 Betty Freeman, letter to Clyfford Still, 5 November 1959. Clyfford Still Museum Archives.
24 Jepson, 1987, p. 110. Hopps claimed to have 'hooked Betty Freeman up with Clyfford Still' while he was running the Ferus Gallery between 1957 and 1962. Hopps, 2017, p. 105.
25 Betty Freeman, letter to Clyfford Still, 30 November 1960. Clyfford Still Museum Archives. Freeman donated *1955-H* to the Los Angeles County Museum of Art in 1991, after having loaned it there for two decades. It was formerly on loan to the Pasadena Art Museum from July 1968.
26 Betty Freeman, letter to Clyfford Still, 2 November 1960. Clyfford Still Museum Archives. Percy Bysshe Shelley, 'Ode to the West Wind', 1819,

in *The Selected Poetry and Prose of Shelley*, ed. Bruce Woodcock (Ware: Wordsworth Editions, 1994), pp. 401–3.

27 Betty Freeman, letter to Clyfford Still, 9 January 1961. Clyfford Still Museum Archives.

28 Virginia Dwan in conversation with Stephanie Barron and James Meyer, Los Angeles County Museum of Art, 14 March 2017: <https://www.lacma.org/art/exhibition/los-angeles-new-york-dwan-gallery-1959-1971>, accessed 11 September 2025.

29 Clyfford Still, letter to Betty Freeman, 16 January 1961. Clyfford Still Museum Archives.

30 On habits of collecting in Los Angeles in the twentieth century – and the prominent role of George Gershwin, Charles Laughton, Billy Wilder, and others – see Hopps, 2017, p. 105.

31 Betty Freeman, letter to Sam Francis, 22 November 1964. Sam Francis Papers, Getty Research Institute, Los Angeles.

32 Jepson, 1987, p. 110.

33 See photographs of Freeman with Still at <www.clyffordstillonline.org/archive>, accessed 22 August 2025. Colour photographs from the same archive indicate that one of the works they were viewing was *PH-225* (1956), on loan to LACMA from 1961 to 1963. The painting is now in the collection of the Modern Art Museum of Fort Worth.

34 Betty Freeman, notes from Clyfford and Patricia Still's visit to Los Angeles, 24 May 1961. Betty Freeman Papers, 1951–1969, Archives of American Art, Smithsonian Institution, Washington DC.

35 Clyfford Still, telegram to Betty Freeman, 18 January 1961. Clyfford Still Museum Archives.

36 Clyfford Still, draft letter to Betty Freeman, 29 June 1965. Clyfford Still Museum Archives.

37 Clyfford Still, letter to Betty Freeman, 19 August 1963. Clyfford Still Museum Archives.

38 Patricia Still ('Mrs Clyfford Still'), letter to Betty Freeman, 14 June 1961. Clyfford Still Museum Archives.

39 Notes and interviews relating to both of Freeman's books are held in the Betty Freeman Papers, 1951–1969, Archives of American Art, Smithsonian Institution, Washington DC.

40 Clyfford Still, letter to Betty Freeman, 1 September 1961. Clyfford Still Museum Archives.

41 Betty Freeman, letter to Clyfford Still, 9 October 1963. Clyfford Still Museum Archives.

42 See Betty Freeman, interview with Albrecht Dümling, Berlin, 3 November 1990, n.p. Still's letter to Freeman of 21 February 1964 corroborates her memory: 'Some knowledge of the limitations of Sartre, Croce, Kierkegaard, Manes, Loyols, Descartes, the Epistles of St Paul, as well as of our more contemporary Saints, Einstein, Marx, and Freud, could well be a starting line.' Clyfford Still Museum Archives.

43 Clyfford Still, letter to Betty Freeman, 18 March 1964. Clyfford Still Museum Archives.

44 Clyfford Still, letter to Betty Freeman, 17 September 1968. Clyfford Still Museum Archives.

45 Betty Freeman, letter to Clyfford Still, 21 September 1968. Betty Freeman Papers, 1951–1969, Archives of American Art, Smithsonian Institution, Washington DC.

46 Betty Freeman, interview with Albrecht Dümling, Berlin, 3 November 1990, n.p.
47 Betty Freeman, 'Sam Francis: Ideas and Paintings' (unpub. manuscript, 1969), p. 5.
48 Betty Freeman, letter to Sam Francis, 22 November 1964. Sam Francis Papers, Getty Research Institute, Los Angeles. The auction of the collection of Mr and Mrs Walter Ross took place at Parke-Bernet Galleries, Madison Avenue, 21 October 1964.
49 Opening statement, New Arts Society project meeting, 16 February 1967. Sam Francis Papers, Getty Research Institute, Los Angeles. See Gabrielle Selz, *Light on Fire: The Art and Life of Sam Francis* (Oakland, CA: University of California Press, 2021), pp. 212–14.
50 Freeman, 1969, pp. 43–4.
51 Betty Freeman, letter to Sam Francis, 31 May 1964. Sam Francis Papers, Getty Research Institute, Los Angeles.
52 Betty Freeman, quoted in Locke, 1991, p. 62.
53 Betty Freeman, letter to Clyfford Still, 19 March 1961. Clyfford Still Museum Archives.
54 Betty Freeman, quoted in Locke, 1991, p. 62.
55 In a lecture of 1983 (revised in 2003), she claimed to have continued playing into the 1970s, closer to the era of the musicales, the regular salon evenings of new music that she held at her home. Betty Freeman, 'Encounters', (unpub. typescript, May 2003), n.p. Los Angeles Philharmonic Archives.
56 Jepson, 1987, p. 110.
57 Shelley Freeman, interview with the author, 5 August 2024.
58 W. B. Yeats, 'To a Friend Whose Work Has Come to Nothing', 1913, in *The Collected Poems of W. B. Yeats*, second edition, ed. Richard J. Finneran (New York: Scribner 1996), p. 109.
59 Betty Freeman, letter to Clyfford Still, 25 July 1962. Freeman refers in the letter to the current exhibition of the collection of Mr and Mrs Ben Heller, which took place at the California Palace of the Legion of Honor, San Francisco, 30 April to 3 June 1962.
60 Robbie and Fanny Freeman, interview with the author, 9 July 2024.

CHAPTER THREE

Turning

1 Seneca, *Letters from a Stoic*, trans. Robin Campbell (London: Penguin, 1969), p. 49.
2 Andrew Causey, 'Mapping and representing', in Paul Melia (ed.), *David Hockney* (Manchester and New York: Manchester University Press, 1995), pp. 89–110, p. 100.
3 Betty Freeman, letter to Clyfford Still, 23 August 1961. Clyfford Still Museum Archives.
4 White, 2006, p. 48.
5 Preston, 1964.
6 Betty Freeman, video interview with Charlie Scheips and Gregory Evans, 2005, courtesy of the David Hockney Foundation.
7 Ibid.
8 *David Hockney: New Paintings and Drawings*, Landau-Alan Gallery, New York, 14 March to 8 April 1967.
9 Betty Freeman, interview with Jake Johnson, 17 September 2008. Quoted in Jake Johnson, 'Two studies of Harry Partch: Conversations with Danlee Mitchell and Betty Freeman', *Echo: A Music Centered Journal*, Vol. 12, No. 1, Fall 2014, n.p.

10 Betty Freeman, video interview with Charlie Scheips and Gregory Evans, 2005, courtesy of the David Hockney Foundation.

11 Peter Goulds, interview with the author, 23 May 2024.

12 W. H. Auden, 'Underneath an Abject Willow', March 1936, reproduced in W. H. Auden, *Collected Poems*, ed. Edward Mendelson (New York: Vintage, 1991), p. 140.

13 Julia Bloomfield, interview with the author, 6 September 2024.

14 Jaś Elsner, 'Reflections on the "Greek Revolution" in art: from changes in viewing to the transformation of subjectivity', in Simon Goldhill and Robin Osborne (eds), *Rethinking Revolutions Through Ancient Greece* (Cambridge: Cambridge University Press, 2006), pp. 68–95, pp. 78–81.

15 John Kasmin, interview with the author, 29 July 2024.

16 Henry Geldzahler, 'Introduction', in Hockney, 1976, pp. 9–26, p. 16.

17 'Its use of shadows isn't that strong. I like that. I mean, the shadows on the figures are just quite light, aren't they?' David Hockney, quoted in Susanna Avery-Quash, '"It's stuck with me all my life": David Hockney on Piero della Francesca's *Baptism of Christ*', *Art Newspaper*, 9 July 2024: <https://www.theartnewspaper.com/2024/07/09/its-stuck-with-me-all-my-life-david-hockney-on-piero-della-francescas-baptism-of-christ>, accessed 15 October 2025. Hockney expressed his debt to Piero and other canonical artists in the exhibition *The Artist's Eye* at the National Gallery, London, 1 July to 31 August 1981 – part of the 'Artist's Eye' series, begun in 1979, in which living artists responded to, and curated displays using, the gallery's collection. In 2024, the gallery brought together *Looking at Pictures on a Screen* (1977) and *My Parents* (1977), displaying them either side of Piero's *Baptism*, for the exhibition *Hockney and Piero: A Longer Look*, 8 August to 27 October 2024.

18 See Nanette Aldred, 'Figure paintings and double portraits', in Melia (ed.), 1995, pp. 68–88, p. 75, n. 3.

19 See Livingstone, 2017, p. 67.

20 For the three ages of man theory, see Adriano Marinazzo, 'L'enigma della "Flagellazione" di Piero della Francesca', *Minuti Menarini*, No. 3[illegible], March 2006, pp. 11–13.

21 Kenneth Clark, *Piero della Francesca* (London: Phaidon Press, 1951), p. 1[illegible] Clark's theory was that they allude to the 1459 Council of Mantua – Pope Pius's attempt to form an army to fight the Ottoman Turks, following the Fall of Constantinople in 1453. On the other hand, Ernst Gombrich contended that the foreground scene represented the repentance of Judas. See E. H. Gombrich, 'The repentance of Judas in Piero della Francesca's "Flagellation of Christ"', *Journal of the Warburg and Courtauld Institutes*, Vol. 22, No. 1/2, January–June 1959, p. 172.

22 Clark, 1951, p. 20.

23 Berenson goes on to overstate the painting's lack of psychological resonance, proposing that the figures are 'Not flesh and blood' but rather 'more akin to the exquisite architecture of the columns, entablatures, ceilings, pavements, all cut like precious stones.' Bernard Berenson, *Piero della Francesca, or The Ineloquent in Art* (New York: Macmillan, 1954), pp. 4–5.

24 Franck Mercier, 'Salvation in perspective: A new interpretation of Piero della Francesca's *Flagellation of Christ*', *Annales: Histoire, Sciences Sociales*, Vol. 72, No. 3, September 2017, pp. 489–523, p. 513.
25 Warhol and Hackett, 1980, p. 25.
26 See Laurence Leamer, *Capote's Women: A True Story of Love, Betrayal, and a Swan Song for an Era* (New York: G. P. Putnam's Sons, 2021).
27 Truman Capote, quoted in Leamer, 2021, p. 11.
28 Truman Capote, quoted in Gerald Clark, *Capote: A Biography* (New York: Carroll & Graf, 2001 [1988]), p. 283.
29 Betty Freeman, 'Sam Francis: Ideas and Paintings' (unpub. manuscript, 1969), p. 20.
30 'David Hockney in Paris: A conversation with Melvyn Bragg', *Listener*, Vol. 93, 22 May 1975, pp. 672–3. He seems to have overlooked, in this statement, the paintings that he made of Bradford in the late 1950s.
31 David Hockney, quoted in Nikos Stangos, *Pictures by David Hockney* (New York: Harry N. Abrams, Inc., 1976).
32 The invitation to produce the prints came from Kenneth Tyler, a printmaker who ran a workshop, Gemini, on Melrose Avenue. Sykes, 2011, p. 163.
33 Hockney, 1976, p. 102.
34 Livingstone, 2017, p. 55.
35 Hockney, 1976, p. 104.
36 Edmund White is wrong to claim that *Life Painting of Myself* (1962) is 'one of his first recognizable portraits'. White, 2006, p. 52.
37 Erwin Panofksy, *Early Netherlandish Painting*, Vol. 1 (Cambridge, MA: Harvard University Press, 1953), p. 194. Quoted in White, 2006, p. 59.
38 Hockney, 1976, p. 102.
39 Paul Melia, 'Showers, pools and power', in Melia (ed.), 1995, pp. 49–67, p. 61.

CHAPTER FOUR
Californianess

1 Alison Lurie, *The Nowhere City* (New York: Henry Holt and Company, 1965), p. 135.
2 Hockney, 1976, p. 104. See also Sykes, 2011, pp. 189–90.
3 Peter Schlesinger, interview with the author, 20 July 2023.
4 Don Bachardy, interview with Hunter Drohojowska-Philp, 29 June 2007. Quoted in Drohojowska-Philp, 2011, p. 128.
5 Peter Schlesinger, interview with the author, 20 July 2023
6 Lanchester moved into this house after Laughton's death in 1962, becoming a neighbour of Isherwood and Bachardy's.
7 David Fine, *Imagining Los Angeles: A City in Fiction* (Reno, NV: University of Nevada Press, 2004), p. 125.
8 Betty Freeman, interview with Albrecht Dümling, Berlin, 3 November 1990, n.p.
9 F. Scott Fitzgerald, 'Magnetism', *Saturday Evening Post*, 3 March 1928, pp. 5–7, 74, 76, 78, p. 5.
10 Mike Davis, *Ecology of Fear: Los Angeles and the Imagination of Disaster* (New York: Metropolitan Books, 1998), p. 202.
11 Christopher Isherwood, interview with Derek Hart in 'Christopher Isherwood: A Born Foreigner' (dir. Fred Burnley), first broadcast on BBC One as an *Omnibus* programme, 2 November 1969.
12 Betty Freeman, untitled speech to the Modern and Contemporary Art Council of LACMA, 7 March 1990. Typescript in Los Angeles Philharmonic Archives.

13 Irving Blum, quoted in Melia, 1995, p. 50.

14 Thomas Adès, interview with the author, 10 March 2025.

15 Betty Freeman, quoted in Locke, 1991, p. 63.

16 Eve Babitz, *Slow Days, Fast Company: The World, the Flesh, and L.A.* (New York: New York Review Books, 2016 [1977]), p. 7.

17 William Wilson, preface to Reyner Banham, 'A London–L.A. love affair: Somebody over there likes us', *Los Angeles Times, West*, 6 June 1971, pp. 9–14, p. 9.

18 Banham, 1971, p. 12.

19 'David Hockney in Paris: A conversation with Melvyn Bragg', *Listener*, Vol. 93, 22 May 1975, p. 673. Quoted in Webb, 2011, p. 90. Artists as diverse as Billy Al Bengston, Vija Celmins, Llyn Foulkes, and Dennis Hopper referred in their work to the surf culture, freeways, storefronts, and signage of the city. See Cécile Whiting, *Pop LA: Art and the City in the 1960s* (Berkeley, CA: University of California Press, 2006), pp. 8–10.

20 Ed Ruscha, *Nine Swimming Pools and a Broken Glass* (Los Angeles: Blair Litho, 1968) is an artist's book containing ten photographs – one of a series of books that Ruscha began in 1962 with *Twentysix Gasoline Stations*.

21 He adds: 'It's full of illusions and it allows its people to indulge in all of these illusions.' Ed Ruscha, quoted in Robert Landau and John Pashtag, 'A conversation with Edward Ruscha', in Ed Ruscha, *Leave Any Information at the Signal: Writings, Interviews, Bits, Pages*, ed. Alexandra Schwartz (Cambridge, MA: MIT Press, 2002), pp. 243–5, p. 244.

22 Reyner Banham, *Los Angeles: The Architecture of Four Ecologies* (Berkeley, CA: University of California Press, 2009 [1971]), p. 1[illegible]5. Elsewhere in the book, he wrote: '[T]he freeways seem to have fixed Los Angeles in canonical and monumental form, much as the great streets of Sixtus V fixed Baroque Rome'. Ibid., p. 17.

23 Peter Plagens, 'The ecology of evil', *Artforum*, Vol. 7, No. 4, December 1972, pp. 39–41.

24 Didion, 1970, p. 162.

25 David Hockney, quoted in Glazebrook, 1970, p. 11.

26 Banham, 1971, p. 12. Banham identified Richard Hamilton, who had visited Los Angeles in 1963 to speak at the Duchamp retrospective at the Pasadena Art Museum, as an important precursor.

27 Banham, 1971, p. 14.

28 Shelley Freeman, interview with the author, 5 August 2024.

29 Mitchell Owens, 'Art/Architecture: Classical furniture upstaged by flower power', *New York Times*, 15 August 1999, Section 2, p. 33.

30 Anon., 'A house furnished with flowers', *Life*, 27 July 1962, pp. 48–55, p. 54. Eszter Haraszty and Bruce David Colen, *Needlepainting: A Garden of Stitches* (New York: Liveright, 1974), p. xiv.

31 Betty Freeman, letter to Clyfford Still, 9 January 1961. Clyfford Still Museum Archives.

32 Haraszty and Colen, 1974, p. xiv.

33 Betty Freeman, letter to Clyfford Still, 5 November 1959. Clyfford Still Museum Archives.

34 Julia Bloomfield, interview with the author, 6 September 2024.

35 The divan was a 1930s piece that Geldzahler had rejected for the collection of the Metropolitan Museum of Art, and bought privately. See David Shapiro, 'David Hockney paints a portrait,' *Art News*, Vol. 68, No. 3, May 1969, pp. 28–31.

36 Meyer Schapiro, '"Muscipula Diaboli": The symbolism of the Mérode Altarpiece', in Meyer Schapiro, *Late Antique, Early Christian and Mediaeval Art: Selected Papers* (New York: George Braziller, 1979), pp. 1–11, p. 10.

37 Michel Foucault, *The Order of Things: An Archaeology of the Human Sciences*, trans. Alan Sheridan (New York: Vintage Books, 1971), p. 14.

38 Foucault, 1971, p. 7.

39 Foucault, 1971, p. 10.

40 Gene Baro, 'A changed Englishman, William Turnbull', *Art in America*, Vol. 54, No. 2, March/April 1966, pp. 102–3, p. 102.

41 Hockney, 1976, p. 98.

42 Carole Rathfon, 'William Turnbull', *Artforum*, Vol. 4, No. 9, May 1966, p. 15: <https://www.artforum.com/events/william-turnbull-2-236163/>, accessed 30 October 2025. See Jules Langsner (ed.), *Sculpture and Painting*, exh. cat. (Newport Beach, CA: Fine Arts Patrons of Newport Harbor, 1966).

43 William Turnbull, quoted in *Beyond Time* (dir. Alex Turnbull, 2011).

44 Ibid.

45 William Turnbull, quoted in Langsner (ed.), 1966, p. 6.

46 William Turnbull, 'Images without temples', in Theo Crosby and John Bodley (eds), *Living Arts*, Vol. 1 (London: Institute of Contemporary Arts, 1963), pp. 14–27, p. 15.

47 Francis Bacon, quoted in Michael Peppiatt, 'Portrait of the artist at work', in Kosme de Barañano (ed.), *Francis Bacon: Lo Sagrado y lo Profano / The Sacred and the Profane*, exh. cat. (Valencia: Institut Valencià d'Art Modern, 2003), pp. 149–69, p. 160.

48 James Turrell, quoted in Michael Govan, 'Inner light: The radical reality of James Turrell', in Michael Govan and Christine Y. Kim (eds), *James Turrell: A Retrospective*, exh. cat. (Los Angeles: Los Angeles County Museum of Art; New York: Prestel, 2013), pp. 13–35, p. 13.

49 Anthony Haden-Guest, 'A bright and lucid land: David Hockney's Los Angeles', *New West Magazine*, 26 April 1976, pp. 80–1, p. 81.

50 Thomas K. Wolfe, 'There goes (VAROOM! VAROOM!) that Kandy Kolored (THPHHHHHHH!) tangerine-flake streamline baby (RAHGHHHH!) around the bend (BRUMMMMMMMMMMMMMMMM ...)', *Esquire*, November 1963, pp. 114–18, p. 114.

51 Billy Al Bengston, quoted in Ben Bamsey, 'Billy Al Bengston: Where is the art world going?', *Artworks*, 18 September 2020: <https://artworksmag.com/billy-al-bengston/>, accessed 2 November 2025.

52 Nicholas Wilder, oral history conducted by Ruth Bowman for the Archives of American Art, Smithsonian Institution, Washington DC, 18 July 1988.

53 Larry Bell, interview with Hunter Drohojowska-Philp, 'Artists Talk: LA Legends, with Larry Bell, Billy Al Bengston, Ed Moses, and Ed Ruscha', Claremont Graduate University, 18 January 2017.

54 Jules Langsner, *Four Abstract Classicists*, exh. cat. (Los Angeles: Los Angeles County Museum of Art, 1959), p. 10. See also, Peter Plagens, *Sunshine Muse:*

Art on the West Coast, 1945–1970 (Berkeley, CA: University of California Press, 1999 [1974]), p. 120: 'The patented "look" was elegance and simplicity, and the mythical material was plastic'.

55 James Turrell, quoted in Christine Y. Kim, 'James Turrell: A life in art', in Govan and Kim, 2013, pp. 37–48, p. 43.

56 Peter Alexander (1939–2020), *Small Cloud Box*, 1966, cast polyester resin. Los Angeles County Museum of Art.

57 Peter Alexander, quoted in Carolina A. Miranda, 'California light and space artist Peter Alexander dies at 81', *Los Angeles Times*, 28 May 2020: <https://www.latimes.com/entertainment-arts/story/2020-05-28/peter-alexander-obit-light-and-space-artist-dies-at-81>, accessed 5 October 2025.

58 David Hockney, oral history conducted by Lawrence Weschler for the Archives of American Art, Smithsonian Institution, Washington DC, 3 September 1984.

59 Jules Langsner, 'Local avant-garde', *Art News*, Vol. 55, No. 7, November 1956, p. 63. Quoted in Sarah Schrank, *Art and the City: Civic Imagination and Cultural Authority in Los Angeles* (Philadelphia: University of Pennsylvania Press, 2009), p. 117.

60 Art Seidenbaum, 'A look at bomb craters after cultural explosion', *Los Angeles Times*, 12 April 1968. Quoted in William Hackman, *Out of Sight: The Los Angeles Art Scene of the Sixties* (New York: Other Press, 2015), p. 155.

61 Julia Bloomfield, interview with the author, 6 September 2024.

62 According to some accounts, Syndell's death was a suicide, and he was run down by Hopps's Stanford classmate James Newman. See Drohojowska-Philp, 2011, p. 21.

63 Ed Moses interviewed by Hunter Drohojowska-Philp, 'Artists Talk: LA Legends, with Larry Bell, Billy Al Bengston, Ed Moses, and Ed Ruscha', Claremont Graduate University, 18 January 2017.

64 Irving Blum, oral history conducted by Paul Cummings for the Archives of American Art, Smithsonian Institution, Washington DC, 31 May to 31 June 1977. See also Calvin Tompkins, 'A touch for the now', *New Yorker*, 29 July 1991, pp. 33–57, p. 44.

65 Irving Blum, oral history conducted by Paul Cummings for the Archives of American Art, Smithsonian Institution, Washington DC, 31 May to 31 June 1977.

66 Edward Kienholz, oral history conducted by Lawrence Weschler for UCLA Center for Oral History Research, 1977.

67 Larry Bell, interview with Hunter Drohojowska-Philp, 15 September 2007. Quoted in Drohojowska-Philp, 2011, p. 85.

68 Frank Gehry, quoted in Barbara Isenberg, *Conversations with Frank Gehry* (New York: Alfred A. Knopf, 2009), p. 57.

69 David Hockney, oral history conducted by Lawrence Weschler for the Archives of American Art, Smithsonian Institution, Washington DC, 3 September 1984.

70 Judy Chicago, *Through the Flower: My Struggle as a Woman Artist* (Garden City, NY: Doubleday & Company Inc., 1975), p. 35.

71 Peter Schlesinger, interview with the author, 20 July 2023.

72 James Dean, quoted in Jason Colavito, *Jimmy: The Secret Life of James Dean* (Lanham, MD: Applause Books, 2021), p. 151.

73 Edward Kienholz, quoted in Anon.,

'Art: Beanery built for art', *Life*, Vol. 60, No. 2, 14 January 1966, pp. 78–80, p. 83.

74 Patrick Procktor, quoted in Grace Glueck, 'New place to go: Ersatz Beanery; lines form at local copy of pub on coast', *New York Times*, 25 November 1965, p. 45.

75 Peter Goulds, interview with the author, 23 May 2024.

76 See Laura Paulson and Allison Unruh, 'The collection of Betty Freeman', in *Betty Freeman: Legacy of a Patron*, auction catalogue (New York: Christie's, 2009), pp. 13–15, p. 14.

77 Lauren O'Neill Butler, 'Interviews: Virginia Dwan', *Artforum*, 11 February 2014: <https://www.artforum.com/columns/virginia-dwan-on-her-life-and-work-219331/>, accessed 4 November 2025.

78 Betty Freeman, letter to Sam Francis, 10 May 1964. Sam Francis Papers, Getty Research Institute, Los Angeles.

79 Betty Freeman, letter to Sam Francis, 1 February 1968 [?]. Sam Francis Papers, Getty Research Institute, Los Angeles.

80 Nicholas Wilder, oral history conducted by Ruth Bowman for the Archives of American Art, Smithsonian Institution, Washington DC, 18 July 1988.

81 William Wilson, 'The fast life and artful times of Nicholas Wilder: A walk down memory lane with L.A.'s wunderkind art dealer of the '60s and '70s', *Los Angeles Times*, 27 November 1988, *Calendar*, pp. 4–5, pp. 100–1.

82 Nicholas Wilder, quoted in Wilson, 1988.

83 David Hockney, interviewed in September 2010. Quoted in Sykes, 2011, p. 179.

84 Ibid.

85 Peter Schlesinger, interviewed in April 2010. Quoted in Sykes, 2011, p. 189.

86 Nicholas Wilder, oral history conducted by Ruth Bowman for the Archives of American Art, Smithsonian Institution, Washington DC, 18 July 1988.

87 Hockney had been invited by Sid Felsen of Gemini Press to produce a new series of lithographs of Los Angeles.

88 Evans had befriended Hockney in Paris in the early 1970s. By the time of the photograph, the two were lovers and close companions.

89 Kenneth Clark, in 'The Fallacies of Hope', episode 12 of *Civilization* (1969, BBC Television). See also Sykes, 2011, pp. 190–1.

90 Charles Nordhoff, *California for Health, Pleasure and Residence* (New York: Harper and Brothers, 1873), p. 137. See McWilliams, 1973, pp. 96–7. See also Melia, 1995, p. 64.

91 Norman M. Klein, *The History of Forgetting: Los Angeles and the Erasure of Memory* (London and New York: Verso, 2008 [1990]), p. 29.

92 Mike Davis, *City of Quartz: Excavating the Future in Los Angeles* (New York: Vintage Books, 1992), p. 20. See also, Klein, 2008, pp. 20–38.

93 This is not to say that there was not a racialized undercurrent to some of the early marketing of Los Angeles – such attitudes were endemic in American society at the time.

94 Davis conceptualized the divergent responses as a 'master dialectic of sunshine and *noir*'. Davis, 1992, p. 23. Klein takes as disapproving a stance against noir as he does against boosterism, calling the genre 'essentially a mythos about white male panic – the white knight in a cesspool of urban decay'. Klein, 2008, p. 79.

95 Nordhoff, 1873, p. 137.

96 These competing, antithetical visions of the city were reflected in the title of the exhibition *Sunshine & Noir:*

Art in L.A. 1960–1997, Louisiana Museum of Modern Art, Copenhagen, in 1997 (travelling to Kunstmuseum Wolfsburg and Castello di Rivoli, Turin) – itself an echo of Davis's thesis. See also Stephanie Barron, 'Introduction: The making of Made in California', in Stephanie Barron, Sheri Bernstein, and Ilene Susan Fort (eds), *Made in California: Art, Image, and Identity*, exh. cat. (Los Angeles: Los Angeles County Museum of Art, 2001), pp. 27–48, p. 27, p. 35. See also Sheri Bernstein, 'Selling California, 1900–1920', ibid., pp. 65–102.

97 David Hockney, quoted in Robert Wennersten, 'Hockney in LA,' *London Magazine*, August/September 1973, p. 35. See Whiting, 2006, p. 114.

98 Bragg, 1975, p. 673. Quoted in Webb, 2011, p. 90.

99 Betty Freeman, video interview with Charlie Scheips and Gregory Evans, 2005, courtesy of the David Hockney Foundation.

100 Howard N. Fox, 'Tremors in Paradise, 1960–1980', in Barron, Bernstein, and Fort (eds), 2000, pp. 193–234, p. 198.

101 Reyner Banham, 'Beverly Hills, too, is a ghetto', *Listener*, 5 September 1968, pp. 296–8.

102 Joan Didion, 'Los Angeles notebook', *Slouching Towards Bethlehem* (New York: Delta, 1968), pp. 217–21, p. 220.

103 Davis, 1992, p. 22.

104 Melia, 1995, p. 66.

105 Ibid., p. 67

106 See Aldred, 1995, p. 76, who perceives a movement away from a 'polemical tone' (in Hockney's earlier London paintings) towards 'a more naturalistic depiction of modern life'.

107 Anonymous Los Angeles artist, quoted in Maurice Tuchman, 'A decade of Edward Kienholz', *Artforum*, Vol. 4, No. 8, April 1966, pp. 41–5, p. 42.

108 Edward Kienholz, oral history conducted by Lawrence Weschler for UCLA Center for Oral History Research, 1977. Quoted in Carrie Rickey, 'Unpopular culture (travels in Kienholzland)', *Artforum*, Vol. 21, No. 10, Summer 1983, pp. 42–8, p. 44.

109 Edward Kienholz, interviewed by William Hackman, September 1985. Quoted in Hackman, 2015, p. 38.

110 Edward Kienholz, quoted in Rickey, 1983, p. 44.

111 Kienholz's mid-career retrospective, organized by Maurice Tuchman, was held at LACMA, 30 March to 15 May 1966. *Back Seat Dodge '38* had made its first appearance at the Dwan Gallery in September 1964; Virginia Dwan had advised Kienholz to truncate the car to fit it through the door. The exhibition was shut down by the police before an intervention by city councilman Tom Bradley: Dwan in conversation with Stephanie Barron and James Meyer, LACMA, 14 March 2017: <https://www.lacma.org/art/exhibition/los-angeles-new-york-dwan-gallery-1959-1971>, accessed 11 September 2025.

112 Lyn Kienholz, quoted in Edward Wyatt, 'In sunny southern California, a sculpture finds its place in the shadows', *New York Times*, 2 October 2007, p. 32.

113 Marcel Duchamp, quoted in Walter Hopps, 'A note from the Underworld', in Hopps (ed.), *Kienholz: A Retrospective*, exh. cat. (New York: Whitney Museum of American Art, 1996), pp. 24–37, p. 33.

114 Letter from Warren M. Dorn, supervisor, Fifth District, to Edward W. Carter, president, Board of Governors, Los Angeles County Museum of Art, 17 March 1966. Jan Butterfield

scrapbook, Archives of American Art, Smithsonian Institution, West Coast Regional Center, Huntington Library, San Marino, CA. At the time, Dorn had his sights set on running for governor of California against the incumbent Pat Brown. See Schrank, 2009, pp. 129–31; Michael Fallon, *Creating the Future: Art and Los Angeles in the 1970s* (Berkeley, CA: Counterpoint, 2014), p. 166; and Francis Frascina, *Art, Politics and Dissent: Aspects of the Art Left in Sixties America* (Manchester: Manchester University Press, 1990), pp. 89–94.

115 Warren M. Dorn, quoted in Lawrence Weschler, 'Cars and carcasses: The last word on Ed Kienholz', *New Yorker*, 6 May 1996: <https://www.newyorker.com/magazine/1996/05/13/cars-and-carcasses>, accessed 7 November 2025.

116 Betty Freeman, letter to Sam Francis, 22 November 1964. Sam Francis Papers, Getty Research Institute, Los Angeles.

117 Tuchman, 1966, p. 43.

118 Edward Kienholz, oral history conducted by Lawrence Weschler for UCLA Center for Oral History Research, 1977.

119 *Five Car Stud* was first exhibited at *documenta 5*, curated by Harald Szeemann, Kassel, Germany, 30 June to 8 October 1972.

120 Edward Kienholz, quoted in Rickey, 1983, p. 46.

CHAPTER FIVE

The Dreamer

1 David Hockney, *A Grand Procession of Dignitaries in the Semi-Egyptian Style* (1961), oil on canvas. Private collection. Made at the RCA, the work was one of four paintings exhibited in the *Young Contemporaries* show of 1962 (RBA Galleries, London, opening 7 February). Compare the statuesque pharaonic figures in *Man in a Museum (or You're in the Wrong Movie)* (1962) and *The Actor* (1963), in each of which the Egyptian idol is twinned with a modern element: a museum visitor or a floral-patterned sofa.

2 John Kasmin, quoted in Simon Garfield, 'Time, Gentlemen', *Independent*, 25 July 1992: <https://www.independent.co.uk/arts-entertainment/arts-time-gentlemen-kasmin-is-a-small-man-with-a-big-name-as-a-dealer-in-contemporary-art-now-after-30-years-he-is-saying-goodbye-to-cork-street-simon-garfield-was-just-asking-him-about-it-when-an-old-friend-1535723.html>, accessed 12 May 2026.

3 C. P. Cavafy, 'Waiting for the Barbarians' (1898), *Collected Poems*, trans. Edmund Keeley and Philip Sherrard (London: Chatto & Windus, 1975), p. 14.

4 David Hockney, undated quotation: <https://www.thedavidhockneyfoundation.org/chronology/1961>, accessed 12 August 2025.

5 Hockney, 1976, p. 89.

6 A tempera-on-paper facsimile of the wall painting, made in 1921 by Charles K. Wilkinson, is in the Metropolitan Museum of Art, New York.

7 Geldzahler, 1976, pp. 9–23, p. 18.

8 Clyfford Still, letter to Betty Freeman, 1 September 1961. Clyfford Still Museum Archives.

9 Harry Partch, quoted in *The Dreamer That Remains* (dir. Stephen Pouliot, 1972).

10 Robbie and Fanny Freeman, interview with the author, 9 July 2024.

11 Shelley Freeman, interview with the author, 5 August 2024.

12 Betty Freeman, letter to Clyfford Still, 25 January 1961. Clyfford Still Museum Archives.

13 Johnson, 2014, n.p.

14 Betty Freeman, letter to Sam Francis, 1 February 1968. Sam Francis Papers, Getty Research Institute, Los Angeles.

15 Shelley Freeman, email to the author, 23 July 2025.

16 Pamela Kort, interview with the author, 9 January 2025.

17 White, 2006, p. 55.

18 Pamela Kort, interview with the author, 9 January 2025.

19 Shelley Freeman, interview with the author, 14 January 2026.

20 Betty Freeman, postcard to Sam Francis, 7 April, year unknown (1980s). Sam Francis Papers, Getty Research Institute, Los Angeles.

21 Robert Wilson, interview with the author, 7 July 2025.

22 Thomson had begun his musical portraits in 1928, in response to the abstract 'word portraits' of his friend Gertrude Stein. See Anthony Tommasini, *Virgil Thomson's Musical Portraits* (New York: Pendragon Press, 1986), p. 9, pp. 179–83.

23 Betty Freeman, 'Corfu Summer with Virgil Thomson, 1981' (unpub. typescript, August 1982), n.p. Los Angeles Philharmonic Archives.

24 Ibid.

25 Virgil Thomson, quoted in Betty Freeman, unpublished memoir of Virgil Thomson. Quoted in Anthony Tommasini, *Virgil Thomson: Composer on the Aisle* (New York and London: W. W. Norton & Company, 1997), p. 519.

26 Franco Assetto, quoted in ibid., p. 521.

27 Ibid., p. 522.

28 Aristotle, *The Nicomathean Ethics of Aristotle*, trans. by Drummond Percy Chase (London: J. M. Dent & Sons, 1911), p. 221 (my italics).

29 On Partch's 'vagabond life' in the 1960s, see Anthony Tommasini, 'A Medici of modern arts in tune with composers', *New York Times*, 19 March 1998, Section E, pp. 1, 10; and Kopplin, 1998, p. 18. Partch's homelessness in 1965 was intermittent; he had experienced longer phases in the past. In 1952, while living in the Bay Area, he was 'sleeping in his old Studebaker': Bob Gilmore, *Harry Partch: A Biography* (New Haven: Yale University Press, 1998), p. 217.

30 Harry Partch, quoted in *The Dreamer That Remains* (dir. Stephen Pouliot, 1972).

31 In the event, his presentation took place on 8 February 1965. Partch showed colour slides of his musical instruments, along with two films – *Music Studio* (1958) and *Rotate the Body in all its Planes* (1961) – and he played fragments of several of his compositions with the help of two assistants. Gilmore, 1998, p. 328.

32 Partch was living in a house in Del Mar that belonged to Helen Raitt, whose husband Russell worked in the Scripps Institute of Oceanography at the University of California, San Diego. The property was offered on the assumption that Partch 'would soon have a position as a composer-in-residence' at the university. See Gilmore, 1998, p. 323.

33 Also on the board were Leonard Stein (director of the Schoenberg Institute) and Lawrence Morton, each of whom was closely involved in the 'Monday Evening Concerts' in Los Angeles. As Freeman drolly recorded in 2003: 'Leonard Stein chose the composers and I subsidized whomever he chose.' Freeman, 2003.

34 Ray Duncan, 'Pasadena, the old order changeth', *Los Angeles*, December 1963, p. 39. Quoted in Hackman, 2015, p. 111.

35 Christy Fox, 'Society and art worlds converge', *Los Angeles Times*, 9 October 1963. Quoted in ibid., p. 110.

36 Betty Freeman, quoted in Locke, 1991, p. 63.

37 Freeman, 2003, n.p.

38 Betty Freeman, quoted in Locke, 1991, p. 63.

39 Harry Partch, *Genesis of a Music* (Madison, WI: University of Wisconsin Press, 1974 [1949]), p. 406.

40 Betty Freeman, quoted in Gilmore, 1998, p. 322.

41 Harry Partch, 'W. B. Yeats, 1941', reproduced in Harry Partch, *Bitter Music: Collected Journals, Essay, Introductions, and Librettos*, ed. Thomas McGeary (Chicago: University of Illinois Press, 1991), pp. 165–8, p. 165.

42 Ibid., p. 167.

43 Freeman, 2003, n.p.

44 Performer-composers Steve Reich and Philip Glass were driving cabs in the same city; Freeman was introduced to both of them around 1972 by John Rockwell, music critic for the *Los Angeles Times*. See Kopplin, 1998. See also Locke, 1991, pp. 61–2.

45 'Los Angeles Legend: Betty Freeman', *meetthecomposer.org*, 2005. Printout at Los Angeles Philharmonic Archives.

46 Cathy Curtis, 'Photographer is able to bring artists' complexities into focus', *Los Angeles Times*, 16 January 1989, Part VI, p. 12.

47 Harry Partch, letter to Betty Freeman, 22 April 1965. Quoted in Gilmore, 1998, pp. 329–30.

48 Harry Partch, letter to Lauriston Marshall, 12 January 1965. Quoted in Gilmore, 1998, p. 324.

49 Walter Hopps, 'Walter Hopps interview', interview conducted by Joanne L. Ratner, ed. William R. Hackman, in Lars Nittve and Helle Crenzien (eds), *Sunshine & Noir: Art in L.A. 1960–1967*, exh. cat. (Louisiana: Museum of Modern Art, 1997), pp. 137–43, p. 142.

50 Gilmore, 1998, p. 363.

51 Ibid., p. 376.

52 Harry Partch, liner notes for *Delusion of the Fury* (CBS Records, 1971).

53 The work premiered on 9 January 1969 at the UCLA Playhouse, directed by John Crawford. See Kopplin, 1998, p. 19.

54 Danlee Mitchell, quoted in Johnson, 2014, n.p.

55 Harry Partch, letter to John Cage, 14 October 1967. Quoted in Gilmore, 1998, p. 324, n. 17.

56 John Adams, quoted in Tommasini, 1998, p. 10.

57 Betty Freeman, interview with Albrecht Dümling, Berlin, 3 November 1990, n.p.

58 Harry Partch, 'A Quarter-Saw Section of Motivations and Intonations', 1967, reproduced in Partch, 1991, pp. 196–9, p. 196.

59 Betty Freeman, untitled address at the New Music Summer School of the International Music Institute, Darmstadt, 1 August 1988. Typescript in Los Angeles Philharmonic Archives.

60 Gilmore, 1998, p. 376. On Freeman's upset, see Danlee Mitchell in Johnson, 2014.

61 See, for example, Horace's dedication in Book I of the Odes (23 BC): 'Maecenas, descended from olden kings, / my rampart and sweet admiration.' Horace, *The Complete Odes and Epodes*, trans. W. G. Shepherd (London: Penguin, 1983), p. 13.

62 Betty Freeman, interview with Jake Johnson, 27 September 2008. Quoted in Johnson, 2014, n.p.

CHAPTER SIX

A Modern Medici

1 Ed Ruscha, email to the author, 16 December 2025.

2 Betty Freeman, 'On patronage', speech for 'Panel for ARTTABLE, Santa Monica', 5 March 1992. Typescript in Los Angeles Philharmonic Archives.

3 Betty Freeman, quoted in Locke, 1991, p. 61.

4 Betty Freeman, quoted in Curtis, 1989, p. 12.

5 His 1963 book *An Anthology of Chance Operations* brought together writings and musical scores by a number of artist contemporaries. La Monte Young and Jackson MacLow (eds), *An Anthology of Chance Operations: Concept Art, Anti Art, Indeterminacy, Plans of Action, Diagrams, Music, Dance Constructions, Improvisation, Meaningless Work, Natural Disasters, Compositions, Mathematics, Essays, Poetry* (New York: George Maciunas and Jackson MacLow, 1963).

6 Billy Name, quoted in Legs McNeil and Gillian McCain, *Please Kill Me: The Uncensored Oral History of Punk* (New York: Grove Press, 1996), p. 4.

7 In 1964, Young began his still-unfinished improvisatory magnum opus *The Well-Tuned Piano*, founded on a system of 'just intonation'. Young has performed the work on numerous occasions. The first recording in 1964 lasted less than an hour. A version of 1981 was more than five hours long. Another in 1987 lasted six and a half hours.

8 Betty Freeman, 'Life and Times, Segment 3: A patron of composers', radio interview with Vicki Curry, 14 July 2006. Quoted in Jake Johnson, 'The Music Room: Betty Freeman's musical soirées', *Twentieth-Century Music*, Vol. 14, No. 3, 2017, pp. 391–409, p. 397.

9 Freeman, 2003, n.p.

10 David Hockney, interviewed in December 2010. Quoted in Sykes, 2011, p. 186

11 Peter Schlesinger, interview with the author, 20 July 2023.

12 She pointed out in 1990 that Young's performances took place in the Museum of Modern Art, New York; that John Cage's premieres typically took place in museums rather than concert halls; and that the Museum of Contemporary Art, Los Angeles, would be hosting a retrospective for Cage ('Rolywholyover A Circus', 12 September to 28 November 1993). Betty Freeman, interview with Albrecht Dümling, Berlin, 3 November 1990, n.p.

13 Betty Freeman, letter to Sam Francis, 1 February 1968. Sam Francis Papers, Getty Research Institute, Los Angeles.

14 Betty Freeman, interview with Albrecht Dümling, Berlin, 3 November 1990, n.p.

15 Nelson Goodman, *Languages of Art. An Approach to a Theory of Symbols* (Indianapolis: Bobbs-Merrill Company, 1968), p. 33. Compare Stein's own narration of the incident through the imagined voice of her partner, Alice B. Toklas, in *The Autobiography of Alice B. Toklas* (New York: The Literary Guild, 1933), p. 14: 'After a little while I murmured to Picasso that I liked his portrait of Gertrude Stein. Yes, he said, everybody says that she does not look like it but that does not make any difference, she will, he said.'

16 Gertrude Stein, *Picasso* (London: B. T. Batsford Ltd., 1938), p. 8.

17 Ernest Hemingway, *A Moveable Feast* (New York: Bantam, 1965 [1964]), p. 25, p. 14.

18 Betty Freeman Papers, 1965–2008, University of California, San Diego.

19 Betty Freeman, preface to unpublished manuscript entitled 'Salotto: Composers and Their Music'. Los Angeles Philharmonic Archives. Quoted in Johnson, 2017, p. 398.

20 'Evenings on the Roof' was a monthly series of concerts of chamber music in Los Angeles, famed for its championing of modernist compositions. In its first incarnation, from 1939 to 1954, it was hosted by Peter and Frances Mullen Yates at their hillside home on Micheltorena Street. In 1954, the music scholar Lawrence Morton took over as artistic director, and 'Evenings on the Roof' became 'Monday Evening Concerts'. See Dorothy Lamb Crawford, *Evenings On and Off the Roof: Pioneering Concerts in Los Angeles, 1939–1971* (Berkeley, CA: University of California Press, 1995). In 1965, the concerts moved to the Los Angeles County Museum of Art – see Locke, 1991, p. 63. A concurrent point of reference for Freeman, although perhaps only a vague one, was the work of Paul Fromm (1906–87), a German-Jewish émigré and wine importer, who in 1952 had created the Fromm Music Foundation, dedicated to the promotion of new music.

21 Alan Rich, 'Introduction' to the unpublished book 'Music Room, 1981–1994', tapes edited by Alan Rich, photographs by Betty Freeman, 2004, pp. 1–2. Los Angeles Philharmonic Archives.

22 Allan Kozinn, 'Alan Rich, Los Angeles music critic, dies at 85', *New York Times*, 26 April 2010, Section B, p. 13.

23 On the musicales and their cultural impact, see Jake Johnson, *The Music Room: A Story of Art, Friendship, and Gathering in Betty Freeman's Beverly Hills Home* (Oxford: Oxford University Press, 2025). Freeman and Rich worked together on a book about the Music Room events from 1981 to 1994 (covering the final years at Judith Rosen's house), but it was never published: Freeman seems to have been disheartened by peer reviews of the manuscript.

24 Alan Rich, quoted in Elaine Woo, 'Philanthropist gave thousands to composers', *Los Angeles Times*, 7 January 2009: <https://www.latimes.com/archives/la-xpm-2009-jan-07-me-freeman7-story.html>, accessed 25 January 2026.

25 Peter Sellars, interview with the author, 2 April 2025.

26 Betty Freeman, quoted in Eugene Patron, '"At Home" in Beverly Hills', *Daily Telegraph*, 5 November 1988, page unknown.

27 Melinda Wortz, 'An evening at Betty Freeman's (11/12/86 – or why I would rather hear music there than anywhere)', inscribed in Music Room ledger ('Il Salotto Musicale'), Los Angeles Philharmonic Archives.

28 Roger Nichols, 'Pierre Boulez: obituary', *Guardian*, 6 January 2016: <https://www.theguardian.com/music/2016/jan/06/pierre-boulez>, accessed 8 February 2026.

29 Julia Bloomfield, interview with the author, 6 September 2024.

30 John Kasmin, interview with the author, 29 July 2024.

31 Thomas Thompson, 'The importance of being Fleischmann', *New York Times Magazine*, 11 April 1976, p. 194.

32 Betty Freeman, interview with Albrecht Dümling, Berlin, 3 November 1990, n.p. In one of his numerous missives to

Freeman over the years, Fleischmann wrote: 'I know of no one who does as much for the music of our time as you do: famous composers and unknown ones, successful ones and struggling ones – and they have all been enabled to create because of you.' Ernest Fleischmann, letter to Betty Freeman, 27 January 1993. Los Angeles Philharmonic Archives.

33 Peter Sellars, interview with the author, 2 April 2025.

34 Rich, 2004, pp. 1–2.

35 The musicale was held on 13 February 1983. Recording in Los Angeles Philharmonic Archives.

36 Peter Sellars, interview with the author, 2 April 2025.

37 Jepson, 1987, p. 108.

38 Betty Freeman, quoted in Jepson, 1987, p. 111.

39 Rich, 2004, p. 2.

40 The other performers that night were Joan La Barbara and Morton Subotnick. Photographs from the occasion show guests including Joan Quinn, Robert Fitzpatrick, Dorrance Stalvey, Rhonda Kess, Carl Stone, Maurice Tuchman, Eva Sottes, Miki Benoff, Suzanne and Tibor Zada, Philip Glass, and Eli Gold.

41 'David Hockney, Joan La Barbara, and Morton Subotnick', Betty Freeman Music Room recording, 4 April 1982. Los Angeles Philharmonic Archives. *Le Sacre du Printemps / Le Rossignol / Oedipus Rex*, directed by John Dexter, premiered at the Metropolitan Opera, New York, on 3 December 1981.

42 Betty Freeman, quoted in Anthony Tommasini, 'The music world's modern-day Medici', *Boston Sunday Globe*, 11 November 1990, pp. 91, 95, p. 95.

43 Pamela Kort, interview with the author, 9 January 2025.

44 Stephanie Barron, interview with the author, 11 April 2025.

45 Dan Flavin (1933–96), *monument 4 for those who have been killed in ambush (to P.K. who reminded me about death)*, 1966, fluorescent light and metal fixtures. The title refers to those killed in the Vietnam War; 'P.K.' refers to Paul Katz, a friend of the artist's and a photographer at the Guggenheim Museum.

46 Thomas Adès, interview with the author, 10 March 2025.

47 Betty Freeman, letter to Dan Flavin, 9 January 1985. Betty Freeman Papers, 1965–2008, University of California, San Diego.

48 Claes Oldenburg (1929–2022), *Typewriter Eraser*, 1976, painted aluminium, stainless steel, ferroconcrete, and bronze. Quoted in Barbara Haskell, *Claes Oldenburg: Object Into Monument*, exh. cat. (Pasadena, CA: Pasadena Art Museum; Los Angeles: Ward Ritchie Press, 1971), p. 122. Oldenburg refers specifically here to an earlier sculpture of 1970 in which the disk of the eraser is depicted as squashed by its fall.

49 Thomas Adès, interview with the author, 10 March 2025.

50 Betty Freeman, quoted in Jepson, 1987, p. 111.

51 'Marta Feuchtwanger and Rhonda Kess', Betty Freeman Music Room recording, 13 February 1983. Los Angeles Philharmonic Archives.

52 See Ralph P. Locke and Cyrilla Barr, 'Introduction: Music patronage as a "female-centered cultural process"', in Locke and Barr (eds), 1997, pp. 1–15.

53 On the longer context of 'salon culture' in the nineteenth and twentieth centuries, and Freeman's place within it, see Johnson, 2017, pp. 391–409.

54 Peter Sellars, interview with the author, 2 April 2025.

55 Tommasini, 1997, p. 519.

56 Pamela Kort, interview with the author, January 2025.

57 Stephanie Barron, interview with the author, 11 April 2025.

58 Betty Freeman, interview with Frank K. Oteri, Newmusicbox.org, 10 July 2000. Printout at Los Angeles Philharmonic Archives.

59 Betty Freeman, interview with Albrecht Dümling, Berlin, 3 November 1990, n.p.

60 Patron, 1988.

61 Tommasini, 1998, p. 1.

62 Ibid., p. 10.

63 Russell Platt, 'Betty Freeman: Her Majesty', *New Yorker*, 7 January 2009: <https://www.newyorker.com/culture/goings-on/betty-freeman-her-majesty>, accessed 5 October 2025.

64 Betty Freeman, 'A creative approach to new music', transcript of a speech given by Freeman at Kyoto City University of Art, 1990, reproduced in *Upbeat: A Publication for Los Angeles Philharmonic Subscribers*, October 1990, pp. 14–15, p. 15.

65 Thomas Adès, interview with the author, 10 March 2025.

66 Patron, 1988.

67 Betty Freeman, quoted in Locke, 1991, p. 63.

68 Tommasini, 1998, p. 10.

69 Peter Sellars, interview with the author, 2 April 2025.

70 Thomas Adès, interview with the author, 10 March 2025.

71 Pamela Kort, interview with the author, 9 January 2025.

72 Robert Wilson, interview with the author, 7 July 2025.

73 Betty Freeman, quoted in Kopplin, 1998, pp. 19–20.

CHAPTER SEVEN

Betty

1 David Hockney, quoted in Garfield, 1992.

2 Betty Freeman, letter to Clyfford Still, 16 July 1968. Clyfford Still Museum Archives.

3 Peter Sellars, interview with the author, 2 April 2025.

4 Betty Freeman, quoted in Tommasini, 1998, p. 10.

5 Zuckerkandl, 1956, p. 235.

6 Pamela Kort, interview with the author, 9 January 2025.

7 Harry Partch, letter to Betty Freeman, 4 December 1966. Quoted in Gilmore, 1998, p. 356.

8 Pamela Kort, interview with the author, 9 January 2025.

9 Julia Bloomfield, interview with the author, 6 September 2024.

10 Thomas Adès, interview with the author, 10 March 2025.

11 Peter Schlesinger, interview with the author, 20 July 2023.

12 Betty Freeman, letter to Harry Partch, 29 March 1972. Quoted in Gilmore, 1998, p. 370.

13 Betty Freeman, handwritten note (date 23 November 1985) on the typescript copy of Partch's first draft of the text for *The Dreamer That Remains*. Quoted in Gilmore, 1998, p. 372.

14 Harry Partch, letter to Betty Freeman, 18 November 1972. Quoted in Gilmore, 1998, p. 378.

15 Thomas Adès, interview with the author, 10 March 2025.

16 Robert Wilson, interview with the author, 7 July 2025.

17 Betty Freeman, video interview with Charlie Scheips and Gregory Evans, 2005, courtesy of the David Hockney Foundation.

18 Robbie and Fanny Freeman, interview with the author, 9 July 2024.

19 Julia Bloomfield, interview with the author, 6 September 2024.

20 Joan Didion, 'The Women's Movement', *The White Album* (New York: Farrar, Straus and Giroux, 1979), pp. 109–18, p. 113.

CHAPTER EIGHT

Other Selves

1 David Hockney, quoted in *Hockney* (dir. Randall Wright, 2014).

2 Julia Bloomfield, interview with the author, 6 September 2024.

3 Thomas Adès, interview with the author, 10 March 2025.

4 Betty Freeman, video interview with Charlie Scheips and Gregory Evans, 2005, courtesy of the David Hockney Foundation.

5 Trip Gabriel, 'At home with: David Hockney – Acquainted with the light', *New York Times*, 21 January 1993, Section C, pp. 1, 6, p. 1.

6 Aristotle, *Nicomachean Ethics*, Book IX, p. 267.

7 Vladimir Nabokov, *Ada, or Ardor: A Family Chronicle* (London: Penguin, 1969), p. 428.

8 Geoff Dyer, *The Ongoing Moment* (London: Little, Brown, 2005).

9 Roland Barthes, *Camera Lucida: Reflections on Photography*, trans. Richard Howard (London: Flamingo, 1984), p. 27.

10 Celia Birtwell, quoted in *Hockney* (dir. Randall Wright, 2014).

11 David Hockney, interviewed in August 2010. Quoted in Sykes, 2011, p. 140.

12 Mo McDermott, a wistful, gentle character who made sculptures of cut-out trees and the like, was a constant of Hockney's London life at this time, as reflected by his recurring presence in the 1973 film *A Bigger Splash*.

13 Celia Birtwell, interviewed in November 2010. Quoted in Sykes, 2011, p. 217.

14 David Hockney, interviewed in December 2010. Quoted in Sykes, 2011, p. 219.

15 David Hockney, *Celia*, 1970, pencil and crayon on paper. Private collection.

16 'C'est une perte de temps considérable.' Jean-Auguste-Dominique Ingres, letter to Jean-François Gilibert, 27 February 1826, reproduced in Boyer d'Agen (ed.), *Ingres d'après une correspondance inédite* (Paris: H. d'Aragon, 1909), pp. 132–3. The accuracy and finesse of Ingres's drawings reflect the 'troubadour style' of the day – a style associated with the faithful recreation of historical episodes, including medieval legends.

17 David Hockney, letter to Henry Geldzahler, 15 December 1971. Quoted in Sykes, 2011, pp. 266–7.

18 The work was exhibited in *David Hockney: A Drawing Retrospective*, Royal Academy of Arts, London, 9 November 1995 to 28 January 1996.

19 The work belonged latterly to Robert Wilson, who kept it on the wall of his New York home until his death in July 2025.

20 Displayed in *Encounters: New Art from Old*, National Gallery, London, 14 June to 17 September 2000. Hockney began to draw using a camera lucida in March 1999. The device, which he had long suspected to have been used by Ingres and countless other artists from the Renaissance onwards, consists of 'nothing more than a small prism (mounted at the end of a metal arm) through which the subject is refracted and reconstituted as a virtual image

on a sheet of paper'. Marco Livingstone, 'David Hockney', in Richard Morphet (ed.), *Encounters: New Art from Old*, exh. cat. (London: National Gallery, 2000), pp. 152–63, p. 158.

21 David Hockney, 'Did Ingres use a camera?', *Independent*, 24 June 1999, p. 25.

22 Jean-Auguste-Dominique Ingres (1780–1867), *Study for the Portrait of Louis-François Bertin (1766–1841)*, undated, black chalk and graphite, 34.9 × 34.5 cm. Metropolitan Museum of Art, New York. Bertin's appearance in the final painting of 1832 has been described by art historian Geraldine Pelles as 'at once intense, suspicious and aggressive'. Geraldine Pelles, *Art, Artists and Society: Origins of a Modern Dilemma; Painting in England and France, 1750–1850* (Englewood Cliffs, NJ: Prentice Hall, 1963), p. 82.

23 Locke's reflection on the ancient Greek thinker Socrates, part of his theory of personal identity, continues: 'And to punish Socrates waking for what sleeping Socrates thought, and waking Socrates was never conscious of, would be no more of right than to punish one twin for what his brother-twin did, where of he knew nothing …' John Locke, *An Essay Concerning Human Understanding*, thirtieth edition (London: William Tegg & Co., 1849 [1689]), p. 227.

24 David Hockney, quoted in Lawrence Weschler, 'True to life', in David Hockney, *Cameraworks* (London: Thames & Hudson, 1984), pp. 6–41, p. 9.

25 *Pearblossom Hwy. 11–18th April 1986 #1*, 1986, chromogenic print. This was followed by a second, larger version – both are in the collection of the Getty Museum, Los Angeles. The image was originally conceived for an article in *Vanity Fair* that reenacted the journey taken in Vladimir Nabokov's novel *Lolita* (1955) by the antihero, Humbert Humbert; the magazine ultimately decided not to use Hockney's work.

26 David Hockney, *Christopher Isherwood Talking to Bob Holman*, 1983, chromogenic print collage on paperboard, edition of twenty.

27 Betty Freeman, quoted in Kopplin, 1998, p. 20.

28 Robert Wilson, interview with the author, 7 July 2025.

29 Betty Freeman, 'Introduction', in Betty Freeman and Robert Wilson, 'A Faxing Friendship: Robert Wilson + Betty Freeman' (unpub. manuscript, 2008), n.p.

30 Betty Freeman, 'Introduction', in *Music People & Others: Photographs by Betty Freeman* (Berlin: American Music Week, 1990) n.p.

31 Freeman and Wilson, 2008, n.p.

32 Freeman, 1969, p. 24.

33 Pamela Kort, interview with the author, 9 January 2025.

34 Stephanie Barron, interview with the author, 11 April 2025.

35 Shelley Freeman, interview with the author, 14 January 2026.

36 Shelley Freeman, interview with the author, 5 August 2024.

37 Thomas Adès, interview with the author, 10 March 2025.

38 Betty Freeman, 1932, n.p.

39 Betty Freeman, interview with Albrecht Dümling, Berlin, 3 November 1990, n.p.

40 Ibid.

41 Robbie and Fanny Freeman, interview with the author, 9 July 2024.

42 Julia Bloomfield, interview with the author, 6 September 2024.

43 Robert Wilson, interview with the author, 7 July 2025.

44 Betty Freeman, interview with Albrecht Dümling, Berlin, 3 November 1990, n.p.

45 Betty Freeman, 'A creative approach to new music', 1990, pp. 14–15.

46 Peter Sellars, interview with the author, 2 April 2025.

47 Thomas Adès, interview with the author, 10 March 2025.

48 Betty Freeman, interview with Albrecht Dümling, Berlin, 3 November 1990, n.p.

49 Harry Partch, quoted in *The Dreamer That Remains* (dir. Stephen Pouliot, 1972).

50 Jepson, 1987, pp. 110–11.

51 Danlee Mitchell, quoted in Johnson, 2014, n.p.

52 Her debut show took place at the Otis/Parsons Gallery, Los Angeles, in 1985, and for the next twenty years, she exhibited internationally in galleries and musical institutions.

53 Betty Freeman, interview with Albrecht Dümling, Berlin, 3 November 1990, n.p.

54 Peter Sellars, interview with the author, 2 April 2025.

55 Betty Freeman, 'Introduction', in *Music People & Others: Photographs by Betty Freeman*, 1990, n.p. An earlier iteration of the exhibition in 1987–8 had travelled between the Galleria Decalage, Milan; the Los Angeles Philharmonic; and the Next Wave Festival of the Brooklyn Academy of Music.

56 Peter Sellars, interview with the author, 2 April 2025.

57 Shelley Freeman, interview with the author, 14 January 2026.

58 Betty Freeman, 'Birth of the Music Room', preface to the unpublished book 'Music Room, 1981–1994', tapes edited by Alan Rich, photographs by Betty Freeman, 2004, p. 3. Los Angeles Philharmonic Archives.

59 Robbie and Fanny Freeman, interview with the author, 9 July 2024.

60 Julia Bloomfield, interview with the author, 6 September 2024.

61 Betty Freeman, video interview with Charlie Scheips and Gregory Evans, 2005, courtesy of the David Hockney Foundation.

62 Betty Freeman, fax to David Hockney, 14 February 2006. Betty Freeman Papers, 1965–2008, University of California, San Diego.

63 Tom Christie, 'Paintings with friends', *LA Weekly*, 16–22 June 2006, p. 17.

64 Julia Bloomfield, interview with the author, 6 September 2024.

65 Thomas Adès, interview with the author, 10 March 2025.

66 Robert Wilson, interview with the author, 7 July 2025.

67 Betty Freeman, fax to David Hockney, 28 January 1997. Betty Freeman Papers, 1965–2008, University of California, San Diego.

68 The production took place at the War Memorial Opera House, conducted by Donald Runnicles and Ian Robertson, and directed by Peter McClintock, with production design by Hockney and costumes by Ian Falconer.

AFTERWORD

The Original Real Housewife

1 John Kasmin, interview with the author, 29 July 2024.

2 John Berger, *Ways of Seeing* (London: Penguin, 1972), p. 47.

3 Naomi Fry, 'Andy Cohen still loves the Housewives', *New Yorker*, 19 February 2023: <https://www.newyorker.com/culture/the-new-yorker-interview/andy-cohen-still-loves-the-housewives>, accessed 15 January 2026.

4 This and all subsequent quotes are from our conversation of 10 November 2025.

Bibliography

Adam, Peter, *David Hockney and his Friends*, Bath, 1997

Aldred, Nanette, 'Figure paintings and double portraits', in Paul Melia (ed.), *David Hockney*, Manchester and New York, 1995, pp. 68–88

Andrews, Oliver, oral history conducted by George M. Goodwin for UCLA Center for Oral History Research, 11 November 1976

Anon., 'Art: Beanery built for art', *Life*, Vol. 60, No. 2, 14 January 1966, pp. 78–80

Asher, Betty M., oral history conducted by Thomas H. Garver for the Archives of American Art, Smithsonian Institution, Washington DC, 30 June and 7 July 1980

Avery-Quash, Susanna, '"It's stuck with me all my life": David Hockney on Piero della Francesca's *Baptism of Christ*', *Art Newspaper*, 9 July 2024: <https://www.theartnewspaper.com/2024/07/09/its-stuck-with-me-all-my-life-david-hockney-on-piero-della-francescas-baptism-of-christ>, accessed 15 October 2025

Babitz, Eve, *Slow Days, Fast Company: The World, the Flesh, and LA*, New York, 2016 [1977]

Bamsey, Ben, 'Billy Al Bengston: Where is the art world going?', *Artworks*, 18 September 2020: <https://artworksmag.com/billy-al-bengston/>, accessed 2 November 2025

Banham, Reyner, 'Beverly Hills, too, is a ghetto', *Listener*, 5 September 1968, pp. 296–8

—, 'A London–LA love affair: Somebody over there likes us', *Los Angeles Times, West*, 6 June 1971, pp. 9–14

—, *Los Angeles: The Architecture of Four Ecologies,* Berkeley, CA, 2009 [1971]

Baro, Gene, 'A changed Englishman, William Turnbull', *Art in America*, Vol. 54, No. 2, March/April 1966, pp. 102–3

Barringer, Tim, and Edith Devaney, *82 Portraits and 1 Still-life*, exh. cat., London, 2016

Barron, Stephanie, 'Introduction: The making of made in California', in Stephanie Barron, Sheri Bernstein, and Ilene Susan Fort (eds), *Made in California: Art, Image, and Identity*, exh. cat., Los Angeles, 2000, pp. 27–48

Barthes, Roland, *Camera Lucida: Reflections on Photography*, trans. Richard Howard, London, 1982

Benefield, Richard, et al, *David Hockney: A Bigger Picture*, exh. cat., San Francisco and New York, 2013

Berenson, Bernard, *Piero della Francesca: The Ineloquent in Art*, New York, 1954

Berger, John, *Ways of Seeing*, London, 1972

Berman, Russell A., 'British expatriates and German exiles in 1930s–1940s Los Angeles', in Kevin R. McNamara (ed.), *The Cambridge Companion to the Literature of Los Angeles,* Cambridge, 2010, pp. 49–58

Bernstein, Sheri, 'Selling California, 1900–1920', in Stephanie Barron, Sheri Bernstein, and Ilene Susan Fort (eds), *Made in California: Art, Image, and Identity*, exh. cat., Los Angeles, 2000, pp. 65–102

Blum, Irving, oral history conducted by Paul Cummings for the Archives of American Art, Smithsonian Institution, Washington DC, 31 May to 31 June 1977

Bragg, Melvyn, 'David Hockney in Paris: A conversation with Melvyn Bragg', *Listener*, Vol. 93, 22 May 1975, pp. 672–3

Brook, Vincent, *Land of Smoke and Mirrors: A Cultural History of Los Angeles*, New Brunswick, NJ and London, 2013

Bucknell, Katherine (ed.), *The Animals: Love Letters between Christopher Isherwood and Don Bachardy*, London, 2014

Carr, Henry, *Los Angeles, City of Dreams*, New York, 1935

Causey, Andrew, 'Mapping and representing', in Paul Melia (ed.), *David Hockney*, Manchester and New York, 1995, pp. 89–110

Chicago, Judy, *Through the Flower: My Struggle as a Woman Artist*, Garden City, NY, 1975

Christie, Tom, 'Paintings with friends', *LA Weekly*, 16–22 June 2006, p. 17

Clark, Alfred E., 'Robert Wishnick, 87; founder and executive of chemical company; helped civic groups', *New York Times*, 15 February 1980, Section B, p. 18

Clark, Gerald, *Capote: A Biography*, New York, 2001 [1988]

Clark, Kenneth, *Piero della Francesca*, London, 1951

Colavito, Jason, *Jimmy: The Secret Life of James Dean*, Lanham, MD, 2024

Crawford, Dorothy Lamb, *Evenings On and Off the Roof: Pioneering Concerts in Los Angeles, 1939–1971*, Berkeley, CA, 1995

Curtis, Cathy, 'Photographer is able to bring artists' complexities into focus', *Los Angeles Times*, 16 January 1989, Part VI, p. 12

Cusset, Catherine, *David Hockney: A Life*, trans. Teresa Lavender Fagan, London, 2020

Cutler, Irving, *The Jews of Chicago*, Chicago, 1996

Davis, Mike, *City of Quartz: Excavating the Future in Los Angeles*, New York, 1992

—, *Ecology of Fear: Los Angeles and the Imagination of Disaster*, New York, 1998

Didion, Joan, 'Los Angeles notebook', *Slouching Towards Bethlehem*, New York, 1968, pp. 217–21

—, *Play It As It Lays*, New York, 1970

—, 'The Women's Movement', *The White Album*, New York, 1979, pp. 109–18

Drohojowska-Philp, Hunter, *Rebels in Paradise: The Los Angeles Art Scene and the 1960s*, New York, 2011

Dyer, Geoff, *The Ongoing Moment*, London, 2005

Eagleton, Terry, *Across the Pond: An Englishman's View of America*, New York and London, 2013

Elsner, Jaś, 'Reflections on the "Greek Revolution" in art: From changes in viewing to the transformation of subjectivity', in Simon Goldhill and Robin Osborne (eds), *Rethinking Revolutions Through Ancient Greece*, Cambridge, 2006, pp. 68–95

Fallon, Michael, *Creating the Future: Art and Los Angeles in the 1970s*, Berkeley, CA, 2014

Fine, David, *Imagining Los Angeles: A City in Fiction*, Reno, NV, 2004

Fitzgerald, F. Scott, 'Magnetism', *Saturday Evening Post*, 3 March 1928, pp. 5–7, 74, 76, 78

Foucault, Michel, *The Order of Things: An Archaeology of the Human Sciences*, trans. Alan Sheridan, New York, 1971, p. 14

Fox, Howard N., 'Tremors in Paradise, 1960–1980', in Stephanie Barron, Sheri Bernstein, and Ilene Susan Fort (eds), *Made in California: Art, Image, and Identity*, exh. cat., Los Angeles, 2000, pp. 193–234

Frascina, Francis, *Art, Politics and Dissent: Aspects of the Art Left in Sixties America*, Manchester, 1990

Freeman, Betty, 'Sam Francis: Ideas and Paintings' (unpub. manuscript, 1969). Getty Research Institute, Los Angeles

—, 'Corfu Summer with Virgil Thomson, 1981' (unpub. typescript, August 1982). Los Angeles Philharmonic Archives

—, untitled address at New Music Summer School, International Music Institute, Darmstadt, 1 August 1988. Typescript in Los Angeles Philharmonic Archives

—, untitled speech to the Modern and Contemporary Art Council of LACMA, 7 March 1990. Typescript in Los Angeles Philharmonic Archives

—, 'A creative approach to new music', speech given at Kyoto City University of Art, 1990, reproduced in *Upbeat: A Publication for Los Angeles Philharmonic Subscribers*, October 1990, pp. 14–15

—, 'Introduction', in *Music People & Others: Photographs by Betty Freeman,* Berlin: American Music Week, 1990), n.p

—, 'On patronage', speech for 'Panel for ARTTABLE, Santa Monica', 5 March 1992. Typescript in Los Angeles Philharmonic Archives

—, 'Betty Freeman's address to the California Arts Council, California Institute of Technology, Pasadena', 27 January 1983, revised April 2003. Typescript in Los Angeles Philharmonic Archives

—, 'Encounters' (unpub. typescript, May 2003 [1983]). Los Angeles Philharmonic Archives

—, 'Birth of the Music Room', preface to the unpublished book 'Music Room, 1981–1994', tapes edited by Alan Rich, photographs by Betty Freeman, 2004, p. 3. Los Angeles Philharmonic Archives

Friedman, Martin (ed.), *Hockney Paints the Stage*, exh. cat., Minneapolis, New York, and London, 1983

Fry, Naomi, 'Andy Cohen still loves the Housewives', *New Yorker*, 19 February 2023: <https://www.newyorker.com/culture/the-new-yorker-interview/andy-cohen-still-loves-the-housewives>, accessed 15 January 2026

Fumaroli, Marc, *David Hockney, Dessins et Gravures*, exh. cat., Paris, 1975, n.p.

Gabriel, Trip, 'At home with: David Hockney – Acquainted with the light', *New York Times*, 21 January 1993, Section C, pp. 1, 6

Garfield, Simon, 'Time gentlemen', *Independent*, 25 July 1992: <https://www.independent.co.uk/arts-entertainment/arts-timegentlemen-kasmin-is-a-small-manwith-a-big-name-as-a-dealer-incontemporary-art-now-after-30-yearshe-is-saying-goodbye-to-cork-streetsimon-garfield-was-just-asking-himabout-it-when-an-old-friend-1535723.html>, accessed 12 May 2026

Gayford, Martin, *A Bigger Message: Conversations with David Hockney*, third edition, London, 2025

Geldzahler, Henry, 'Introduction', in David Hockney, *David Hockney by David Hockney: My Early Years*, ed. Nikos Stangos, London, 1976, pp. 9–26

Gilmore, Bob, *Harry Partch: A Biography,* New Haven, 1998

Glazebrook, Mark, 'David Hockney: An interview', *David Hockney: Paintings, Prints and Drawings, 1960–1970*, exh. cat., London, 1970, pp. 8–15

Glueck, Grace, 'New place to go: Ersatz Beanery; lines form at local copy of pub on coast', *New York Times*, 25 November 1965, p. 45

Gombrich, E. H., 'The repentance of Judas in Piero della Francesca's "Flagellation of Christ"', *Journal of the Warburg and Courtauld Institutes*, Vol. 22, No. 1/2, January–June 1959, p. 172

Goodman, Nelson, *Languages of Art: An Approach to a Theory of Symbols*, Indianapolis, 1968

Govan, Michael, 'Inner light: The radical reality of James Turrell', in Michael Govan and Christine Y. Kim (eds), *James Turrell: A Retrospective*, exh. cat., Los Angeles and New York, 2013, pp. 13–35

Green, Hannah, 'Mister Nabokov', *New Yorker*, 6 February 1977: <https://www.newyorker.com/magazine/1977/02/14/mister-nabokov>, accessed 14 October 2025

Griffiths, Paul, 'Paying homage to a musical muse, Betty Freeman', *New York Times*, 24 May 2001, Section E, p. 6

Hackman, William, *Out of Sight: The Los Angeles Art Scene of the Sixties*, New York, 2015

Haden-Guest, Anthony, 'A bright and lucid land: David Hockney's Los Angeles', *New West Magazine*, 26 April 1976, pp. 80–1

Haraszty, Eszter, and Bruce David Colen, *Needlepainting: A Garden of Stitches*, New York, 1974

Haskell, Barbara, *Claes Oldenburg: Object into Monument*, exh. cat., Pasadena, CA and Los Angeles, 1971

Hockney, David, *David Hockney by David Hockney: My Early Years*, ed. Nikos Stangos, London, 1976

—, oral history conducted by Lawrence Weschler for the Archives of American Art, Smithsonian Institution, Washington DC, 3 September 1984

—, 'Vogue par David Hockney', French *Vogue*, No. 662, December 1985–January 1986, cover, pp. 219–59

—, *That's the Way I See It*, ed. Nikos Stangos, London, 1993

—, 'Did Ingres use a camera?', *Independent*, 24 June 1999, p. 25

—, *Secret Knowledge: Rediscovering the Lost Techniques of the Old Masters*, London, 2001

Hockney, David, and Hans Werner Holzwarth, *David Hockney: A Bigger Book*, Cologne, 2016

Hollinghurst, Alan, *Isherwood is at Santa Monica*, Oxford, 1975

Hopps, Walter, 'A note from the Underworld' in Walter Hopps (ed.), *Kienholz: A Retrospective*, New York, 1996, pp. 24–37

—, 'Walter Hopps interview', conducted by Joanne L. Ratner, ed. William R. Hackman, in Lars Nittve and Helle Crenzien (eds), *Sunshine & Noir: Art in LA 1960–1967*, exh. cat., Louisiana, 1997, pp. 137–43

—, *The Dream Colony: A Life in Art*, ed. Deborah Treisman, from interviews with Anne Doran, New York, 2017

Isenberg, Barbara, *Conversations with Frank Gehry*, New York: Alfred A. Knopf, 2009

Isherwood, Christopher, *A Single Man*, London, 2010 [1964]

Jepson, Barbara, 'A cultivated ear: Betty Freeman's living room is the West Coast's center for new music', *Connoisseur*, February 1987, pp. 107–1[illegible]

Johnson, Jake, 'Two studies of Harry Partch: Conversations with Danlee Mitchell and Betty Freeman', *Echo: A Music Centered Journal*, Vol. 12, No. 1, Fall 2014, n.p

—, 'The Music Room: Betty Freeman's musical soirées', *Twentieth-Century Music*, Vol. 14, No. 3, 2017, pp. 391–409

—, *The Music Room: A Story of Art, Friendship, and Gathering in Betty Freeman's Beverly Hills Home*, Oxford, 2025

Kienholz, Edward, oral history conducted by Lawrence Weschler for UCLA Center for Oral History Research, 1977

Kim, Christine Y., 'James Turrell: A life in art', in Michael Govan and Christine Y. Kim (eds), *James Turrell: A Retrospective*, exh. cat., Los Angeles and New York, 2013, pp. 37–48

King, Susan, 'Adventures with Bridges', *Los Angeles Times*, 6 July 2011: <https://

www.latimes.com/archives/la-xpm-2011-jul-06-la-et-jack-larson-20110706-story.html>, accessed 13 October 2025

Klein, Norman M., *The History of Forgetting: Los Angeles and the Erasure of Memory*, London and New York, 2008 [1990]

Koestenbaum, Wayne, *Andy Warhol*, New York, 2001

Kopplin, David, 'Patron and photographer Betty Freeman: Where would modern music be without her?', *Performing Arts*, October 1998, pp. 18–21

Kozinn, Allan, 'Alan Rich, Los Angeles music critic, dies at 85', *New York Times*, 26 April 2010, Section B, p. 13

Landau, Robert, and John Pashtag, 'A conversation with Edward Ruscha', in Ed Ruscha, *Leave Any Information at the Signal: Writings, Interviews, Bits, Pages*, ed. Alexandra Schwartz, Cambridge, MA, 2002, pp. 243–45

Langsner, Jules, (ed.), *Four Abstract Classicists*, exh. cat., Los Angeles, 1959

— (ed.), *Sculpture and Painting*, exh. cat., Newport Beach, CA, 1966

Leamer, Laurence, *Capote's Women: A True Story of Love, Betrayal, and a Swan Song for an Era,* New York, 2021

Livingstone, Marco, 'David Hockney', in Richard Morphet (ed.), *Encounters: New Art from Old*, exh. cat., London, 2000, pp. 152–63

—, *David Hockney*, fourth edition, London: Thames & Hudson, 2017

Livingstone, Marco, Edith Devaney, et al, *David Hockney: A Bigger Picture*, exh. cat., London, 2012

Locke, Ralph P., 'The "Grand Composers" of the present day: Betty Freeman discusses how she chooses and supports them', 28 September 1991, reproduced in Ralph P. Locke and Cyrilla Barr (eds), *Cultivating Music in America: Women Patrons and Activists since 1860,* Berkeley, CA, 1997, pp. 59–64

Locke, Ralph P., and Cyrilla Barr, 'Introduction: Music patronage as a 'female-centered cultural process', in Ralph P. Locke and Cyrilla Barr (eds), *Cultivating Music in America: Women Patrons and Activists since 1860,* Berkeley, CA, 1997, pp. 1–15

Lurie, Alison, *The Nowhere City*, New York, 1965

McLung, William Alexander, *Landscapes of Desire: Anglo Mythologies of Los Angeles,* Berkeley, CA, 2002

McNeil, Legs, and Gillian McCain, *Please Kill Me. The Uncensored Oral History of Punk*, New York, 1996

McWilliams, Carey, *Southern California: An Island on the Land*, Santa Barbara, CA, and Salt Lake City, UT, 1973 [1946]

—, *California: The Great Exception*, Berkeley, CA and Los Angeles, 1999 [1949]

Marinazzo, Adriano, 'L'enigma della "Flagellazione" di Piero della Francesca', *Minuti Menarini*, No. 324, March 2006, pp. 11–13

Melia, Paul 'Showers, pools and power', in Paul Melia (ed.), *David Hockney*, Manchester and New York, 1995, pp. 49–67

Mercier, Franck, 'Salvation in perspective: A new interpretation of Piero della Francesca's *Flagellation of Christ*', *Annales Histoire, Sciences Sociales*, Vol. 72, No. 3, 2017, pp. 489–523

Miranda, Carolina A., 'California light and space artist Peter Alexander dies at 81', *Los Angeles Times*, 28 May 2020: <https://www.latimes.com/entertainment-arts/story/2020-05-28/peter-alexander-obit-light-and-space-artist-dies-at-81>, accessed 5 October 2025

Nabokov, Vladimir, *Ada, or Ardor: A Family Chronicle*, London, 1969

Nittve, Lars, and Helle Crenzien (eds), *Sunshine & Noir: Art in LA 1960–1967*, exh. cat., Louisiana, 1997

Nordhoff, Charles, *California for Health, Pleasure and Residence,* New York, 1873

Obrist, Hans Ulrich, *The Hockney Interviews,* London, 2025

O'Neill Butler, Lauren, 'Interviews: Virginia Dwan', *Artforum*, 11 February 2014: <https://www.artforum.com/columns/virginia-dwan-on-her-life-and-work-219331/>, accessed 4 November 2025

Owens, Mitchell, 'ART/ARCHITECTURE; Classical furniture upstaged by flower power', *New York Times*, 15 August 1999, Section 2, p. 33

Panofksy, Erwin, *Early Netherlandish Painting, Vol. 1*, Cambridge, MA, 1953

Partch, Harry, *Genesis of a Music*, Madison, WI, 1974 [1949]

—, 'W. B. Yeats, 1941', reproduced in Harry Partch, *Bitter Music: Collected Journals, Essay, Introductions, and Librettos*, ed. Thomas McGeary, Chicago, 1991, pp. 165–8

Patron, Eugene, '"At Home" in Beverly Hills', *Daily Telegraph*, 5 November 1988, page unknown

Paulson, Laura, and Allison Unruh, 'The collection of Betty Freeman', in *Betty Freeman: Legacy of a Patron*, auction catalogue, New York, 2009, pp. 13–15

Peppiatt, Michael, 'Portrait of the artist at work', in Kosme de Barañano (ed.), *Francis Bacon: Lo Sagrado y lo Profano / The Sacred and the Profane*, exh. cat., Valencià, 2003, pp. 149–169

Plagens, Peter, 'The ecology of evil', *Artforum*, Vol. 7, No. 4, December 1972, pp. 39–41

—, *Sunshine Muse: Art on the West Coast, 1945–1970*, Berkeley, CA, 1999 [1974]

Platt, Russell, 'Betty Freeman: Her Majesty', *New Yorker*, 7 January 2009: <https://www.newyorker.com/culture/goings-on/betty-freeman-her-majesty>, accessed 5 October 2025

Preston, Stuart, 'A square world; Albers' geometrics; English Pop; Liberman and Marca-Relli', *New York Times*, 4 October 1964, p. 21

Rathfon, Carole, 'William Turnbull', *Artforum*, Vol. 4, No. 9, May 1966, p. 15: <https://www.artforum.com/events/william-turnbull-2-236163/>, accessed 30 October 2025

Rechy, John, *City of Night*, New York, 1963

Rich, Alan, 'Introduction' to the unpublished book 'Music Room, 1981–1994', tapes edited by Alan Rich, photographs by Betty Freeman, 2004, pp. 1–2. Los Angeles Philharmonic Archives

Rickey, Carrie, 'Unpopular culture (travels in Kienholzland)', *Artforum*, Vol. 21, No. 10, Summer 1983, pp. 42–8

Rosenthal, Norman, et al, *David Hockney*, exh. cat., London and Paris, 2025

Ruscha, Ed, *Nine Swimming Pools and a Broken Glass*, Los Angeles, 1968

—, 'Introduction: This crazy but genius Walter Hopps', in Walter Hopps, *The Dream Colony: A Life in Art*, ed. Deborah Treisman from interviews with Anne Doran, New York, 2017, pp. xv–xix

Sayag, Alain, and David Hockney, *David Hockney: Photographs*, exh. cat., London and New York, 1982

Schama, Simon, 'California dreamer', *New Yorker*, 18 March 1996, pp. 98–101

Schapiro, Meyer, '"Muscipula Diaboli": The symbolism of the Mérode Altarpiece', in Meyer Schapiro, *Late Antique, Early Christian and Mediaeval Art: Selected Papers*, New York, 1979, pp. 1–11

Schimmel, Paul (ed.), *Helter Skelter: LA Art in the 1990s*, exh. cat., Los Angeles, 1992
Schrank, Sarah, *Art and the City: Civic Imagination and Cultural Authority in Los Angeles*, Philadelphia, 2009
Selz, Gabrielle, *Light on Fire: The Art and Life of Sam Francis*, Oakland, CA, 2021
Shapiro, David, 'David Hockney paints a portrait,' *Art News*, Vol. 68, No. 3, May 1969, pp. 28–31
Spender, Stephen, and Pierre Restany, *David Hockney: Tableaux et Dessins / Paintings & Drawings*, exh. cat., Paris, 1974
Stangos, Nikos, *Pictures by David Hockney*, New York, 1976
Starr, Kevin, *Americans and the California Dream: 1850–1915*, New York, 1973
—, 'The Case Study House Program and the impending future: Some regional considerations', in Elizabeth A. T. Smith (ed.), *Blueprints for Modern Living: History and Legacy of the Case Study Houses*, exh. cat., Los Angeles and Cambridge, MA, 1989, pp. 131–43
Stein, Gertrude, *The Autobiography of Alice B. Toklas*, New York, 1933
—, *Picasso*, London, 1938
Stephens, Chris, 'Between contingency and longevity: William Turnbull and the long tradition of sculpture', in Jon Wood (ed.), *William Turnbull: International Modern Artist*, London, 2022, pp. 213–221
Stephens, Chris, and Andrew Wilson (eds), *David Hockney*, exh. cat., London, 2017
Sykes, Christopher Simon, *Hockney: The Biography, Volume I, 1937–1975 – A Rake's Progress*, London, 2011
—, *Hockney: The Biography, Volume II, 1975–2012 – A Pilgrim's Progress*, London, 2014
Thompson, Thomas, 'The importance of being Fleischmann,' *New York Times Magazine*, 11 April 1976, p. 194
Tommasini, Anthony, *Virgil Thomson's Musical Portraits*, New York, 1986
—, 'The music world's modern-day Medici', *Boston Sunday Globe*, 11 November 1990, pp. 91, 95
—, *Virgil Thomson: Composer on the Aisle*, New York and London, 1997
—, 'A Medici of modern arts in tune with composers', *New York Times*, 19 March 1998, Section E, pp. 1, 10
Tompkins, Calvin, 'A touch for the now', *New Yorker*, 29 July 1991, pp. 33–57
Tuchman, Maurice, 'A decade of Edward Kienholz', *Artforum*, Vol. 4, No. 8, April 1966, pp. 41–5, p. 42
— (ed.), *David Hockney: A Retrospective*, exh. cat., Los Angeles, 1988
Turnbull, Betty, *The Last Time I Saw Ferus*, Newport Beach, CA, 1976
Turnbull, William, 'Images without temples', in Theo Crosby and John Bodley (eds), *Living Arts, Vol. 1*, London, 1963, pp. 14–27
Ulin, David L., *Writing Los Angeles: A Literary Anthology*, New York, 2002
Warhol, Andy, and Pat Hackett, *POPism: The Warhol '60s*, New York, 1980
—, *The Andy Warhol Diaries*, ed. Pat Hackett, New York, 1989
Webb, Peter, *Portrait of David Hockney*, London, 1990
Wenger, Daniel, 'The obsessive photographer behind America's first gay magazine', *New Yorker*, 22 September 2016: <https://www.newyorker.com/culture/photo-booth/the-obsessive-photographer-behind-americas-first-gay-magazine>, accessed 13 October 2025
Wennersten, Robert, 'Hockney in LA,' *London Magazine*, August/September 1973, p. 35
Weschler, Lawrence, 'True to life', in *David Hockney, Cameraworks*, London, 1984, pp. 6–41

—, 'Cars and carcasses: The last word on Ed Kienholz', *New Yorker*, 6 May 1996: <https://www.newyorker.com/magazine/1996/05/13/cars-and-carcasses>, accessed 7 November 2025

—, *True to Life: Twenty-Five Years of Conversations with David Hockney*, Berkeley, CA, Los Angeles and London, 2008

White, Edmund, 'The lineaments of desire', in Sarah Howgate and Barbara Stern Shapiro (eds), *David Hockney Portraits*, exh. cat., London, 2006, pp. 48–61

Whiting, Cécile, *Pop LA: Art and the City in the 1960s*, Berkeley, CA, 2006

Wilder, Nicholas, oral history conducted by Ruth Bowman for the Archives of American Art, Smithsonian Institution, Washington DC, 18 July 1988

Wilson, William, preface to Reyner Banham, 'A London–LA love affair: Somebody over there likes us', *West*, *Los Angeles Times*, 6 June 1971, p. 9

—, 'The fast life and artful times of Nicholas Wilder: A walk down memory lane with LA's wunderkind art dealer of the '60s and '70s', *Los Angeles Times*, 27 November 1988, *Calendar*, pp. 4–5, pp. 100–1

Wishnick, William, *The Witco Story*, New York, 1976

—, interview by James J. Bohning at Witco Corporation, New York, 13 May 1991, Chemical Heritage Foundation, Philadelphia, Oral History Transcript #0093

Wolfe, Thomas K., 'There goes (VAROOM! VAROOM!) that Kandy Kolored (THPHHHHHH!) tangerine-flake streamline baby (RAHGHHHH!) around the bend (BRUMMMMMMMMMMMMMMMM ...)', *Esquire*, November 1963, pp. 114–18

Woo, Elaine, 'Philanthropist gave thousands to composers', *Los Angeles Times*, 7 January 2009: <https://www.latimes.com/archives/la-xpm-2009-jan-07-me-freeman7-story.html>, accessed 25 January 2026

Wright, Frank Lloyd, *An Autobiography*, second edition, London, 1945

Wyatt, Edward, 'In sunny southern California, a sculpture finds its place in the shadows', *New York Times*, 2 October 2007, p. 32

Young, La Monte, and Jackson MacLow (eds), *An Anthology of Chance Operations: Concept Art, Anti Art, Indeterminacy, Plans of Action, Diagrams, Music, Dance Constructions, Improvisation, Meaningless Work, Natural Disasters, Compositions, Mathematics, Essays, Poetry*, New York, 1963

Zuckerkandl, Victor, *Sound and Symbol: Music and the External World*, trans. Willard R. Trask, Princeton, 1956

FILM AND VIDEO

'Betty Freeman', untitled interview with Charlie Scheips and Gregory Evans, 2005, courtesy of the David Hockney Foundation

A Bigger Splash (dir. Jack Hazan), 1973

'Artists Talk: LA Legends, with Larry Bell Billy Al Bengston, Ed Moses, and Ed Ruscha', Claremont Graduate University, 18 January 2017

'Christopher Isherwood: A Born Foreigner' (dir. Fred Burnley), interview with Christopher Isherwood by Derek Hart first broadcast on BBC One as an *Omnibus* programme, 2 November 1969

The Dreamer That Remains (dir. Stephen Pouliot), 1972

Hockney (dir. Randall Wright), 2014

Portrait of David Hockney (dir. David Pearce), 1972

'Virginia Dwan in conversation with Stephanie Barron and James Meyer', Los Angeles County Museum of Art, 14 March 2017

Acknowledgments

Thank you to Marc Valli, who understood from the outset my fascination with Betty Freeman and commissioned this book, and to Andrew Brown, who edited it with sensitivity and flair. I am grateful to my agent, Isobel Dixon, and to everyone at Blake Friedmann Literary Agency. Huge thanks are also due to the stellar teams at Thames & Hudson on both sides of the Atlantic, in particular to Harry Burton, Harriet Clarke, Hannah Corbett, and Curtis Garner. The David Hockney Foundation has supported and facilitated this project – thank you especially to Tracy Bartley, Shannan Kelly, and Brittnee Zuckerman.

Above all, I am grateful to those who shared their memories of Betty Freeman and her all-but-lost era: Thomas Adès, Stephanie Barron, Julia Bloomfield, Shelley Butler, Albrecht Dümling, Robbie and Fanny Freeman, John Kasmin, Pamela Kort, John Rockwell, Ed Ruscha, Peter Schlesinger, Peter Sellars, and the late Robert Wilson. With special thanks, also, to Crystal Kung Minkoff – the other Beverly Hills housewife of the book.

Picture Credits

Sources of colour illustrations by figure number

1, 2 Photo Richard Schmidt. © David Hockney

3, 4 Photo John Kasmin. Courtesy the artist and Lyndsey Ingram

5 Photo Sotheby's. © David Hockney

6 Photo Richard Schmidt. The David Hockney Foundation. © David Hockney

7 Convent of San Marco, Florence

8 Galleria Nazionale delle Marche, Urbino

9, 10 Photo Richard Schmidt. © David Hockney

11 © Ed Ruscha. Courtesy the artist and Gagosian

12 Photo Christie's Images Limited

13 National Gallery, London

14 Museo del Prado, Madrid

15 Photo Richard Schmidt. © David Hockney

16 Photo Prudence Cuming Associates. © The Estate of William Turnbull. Courtesy Offer Waterman, London

17 Photo Restricted gift of Mr. and Mrs. Frederic G. Pick/Bridgeman Images. © David Hockney

18 Photo © Museum Associates/LACMA. © The Edward and Nancy Kienholz Foundation, Inc. Courtesy of L.A. Louver, Venice, CA.

19 Photo Richard Schmidt. © David Hockney

20 Photo John Kasmin. Courtesy the artist and Lyndsey Ingram

21 The Metropolitan Museum of Art, New York. Rogers Fund, 1930

22 Photo Christie's Images Limited. © Claes Oldenburg. Courtesy Paula Cooper Gallery

23 Photo Christie's Images Limited. © 2025 Stephen Flavin/Artists Rights Society, New York and DACS, London. Courtesy Ronald Davis Studio

24, 25 *A Bigger Splash*, Dir. Jack Hazan, 1973

26, 27 Betty Freeman Papers. MSS 227. Special Collections & Archives, UC San Diego Library

28 Photo David Hockney. Betty Freeman Papers. MSS 227. Special Collections & Archives, UC San Diego Library

29, 30 Courtesy the Los Angeles Philharmonic Archives. © David Hockney

31 Photo Richard Schmidt. © David Hockney

32 Betty Freeman Papers. MSS 227. Special Collections & Archives, UC San Diego Library. Drawing © David Hockney

33 Slim Aarons/Getty Images

34 Emily Shur/Bravo via Getty Images

35 Photo Laszlo Molnav. Betty Freeman Papers. MSS 227. Special Collections & Archives, UC San Diego Library

Sources of black-and-white illustrations by page number

2 Betty Freeman Papers. MSS 227. Special Collections & Archives, UC San Diego Library

6, 9 Photo Richard Schmidt. © David Hockney

12 *Young Physique*, October 1961

15 Photo The Museum of Modern Art, New York/Scala, Florence

29, 30, 33 © David Hockney

38 Courtesy Shelley Butler

45 Courtesy Clyfford Still Museum Archives

50, 51 Betty Freeman Papers. MSS 227. Special Collections & Archives, UC San Diego Library

57 © J. Paul Getty Trust. Getty Research Institute, Los Angeles (2004.R.10)

69 Photo John Kasmin. Courtesy the artist and Lyndsey Ingram

83 © Ed Ruscha. Courtesy the artist and Gagosian

109, 129 Betty Freeman Papers. MSS 227. Special Collections & Archives, UC San Diego Library

130 Photo Steve Goldstein. Courtesy the Los Angeles Philharmonic Archives

136 © Betty Freeman. All rights reserved 2026/Bridgeman Images

145 Photo Leigh Wiener. Betty Freeman Papers. MSS 227. Special Collections & Archives, UC San Diego Library

148, 151 Photo Betty Freeman. Courtesy the Los Angeles Philharmonic Archives

159, 168, 170, 172, 175 Betty Freeman Papers. MSS 227. Special Collections & Archives, UC San Diego Library

187 Courtesy the Los Angeles Philharmonic Archives

190 Photo Philipp Lachenmann

195 Horst P. Horst/Condé Nast via Getty Images

Index

Page references in *italics* indicate black-and-white illustrations; **bold** indicates fig. numbers in plate sections.